# APPLIED HUMAN FACTORS IN MEDICAL DEVICE DESIGN

# APPLIED HUMAN FACTORS IN MEDICAL DEVICE DESIGN

*Edited by*

Mary Beth Privitera

ELSEVIER

ACADEMIC PRESS

An imprint of Elsevier

Academic Press is an imprint of Elsevier
125 London Wall, London EC2Y 5AS, United Kingdom
525 B Street, Suite 1650, San Diego, CA 92101, United States
50 Hampshire Street, 5th Floor, Cambridge, MA 02139, United States
The Boulevard, Langford Lane, Kidlington, Oxford OX5 1GB, United Kingdom



**Library of Congress Cataloging-in-Publication Data**

A catalog record for this book is available from the Library of Congress

**British Library Cataloguing-in-Publication Data**

A catalogue record for this book is available from the British Library

ISBN: 978-0-12-816163-0

For information on all Academic Press publications visit our website at https://www.elsevier.com/books-and-journals

*Publisher:* Mara Conner
*Acquisition Editor:* Fiona Geraghty
*Editorial Project Manager:* Ali Afzal-Khan
*Production Project Manager:* Nirmala Arumugam
*Cover Designer:* Alan Studholme

Typeset by TNQ Technologies



*Thanks to HS Design for permission to use product images from their portfolio. All rights reserved.*

# Contents

# III

# Human factors in design

## 7. Applied human factors in design

Mary Beth Privitera, M. Robert Garfield, and Daryle Gardner-Bonneau



## 8. Combination devices

Jeffrey Morang and Mary Beth Privitera



## 9. Applying design principles to instructional materials

Renée Bailey



# IV

# Formative design evaluation & reporting

## 10. Heuristic analysis, cognitive walkthroughs & expert reviews

Mary Beth Privitera

### 11. Simulated use formatives

Deborah Billings Broky, Tressa J. Daniels, and Melissa R. Lemke



# V

## Safety related risk

### 12. Use-focused risk analysis

Sophia Kalita and Melissa R. Lemke



### 13. Root cause analysis

Tim Reeves, Christina Mendat, and Eric Shaver



### 14. Known use error analysis & post market surveillance

Mary Beth Privitera



# VI

## Usability validation & reporting

### 15. Human factors validation (summative usability) testing including residual risk analysis

Mary Beth Privitera, Merrick Kossack, and Kate Cox



### 16. Human factors validation testing of combination products

Tim Reeves, Christina Mendat, and Liz Mauer

# Contributors

This book was developed in collaboration with AAMI faculty. For more information about AAMI training, please visit www.aami.org.


*AAMI Human Factors Faculty Leaders:*

**Renée Bailey** Agilis Consulting Group, Orlando, FL, United States

**Tressa J. Daniels** UXD & Human Factors, BD Medical, San Diego, CA, United States

**Daryle Gardner-Bonneau** Bonneau and Associates, Portage, MI, United States

**Merrick Kossack** Emergo by UL - Human Factors Research & Design, Chicago, IL, United States

**Melissa R. Lemke** Agilis Consulting Group, LLC., Cave Creek, AZ, United States

**Mary Beth Privitera** HS Design, Gladstone, NJ, United States

**Tim Reeves** Human Factors MD, LLC., Charlotte, NC, United States

*Supported by:*

**Tor Alden** HS Design, Gladstone, NJ, United States

**Jennifer N. Avari Silva** Washington University in St Louis, SentiAR, Inc., St Louis, MO, United States

**Russell J. Branaghan** Arizona State University, Mesa, AZ, United States

**Deborah Billings Broky** Agilis Consulting Group, LLC., Cave Creek, AZ, United States

**L. Bryant Foster** Research Collective, Tempe, AZ, United States

**Kate Cox** HS Design, Gladstone, NJ, United States

**Ian Culverhouse** Rebus Medical, Bristol, United Kingdom

**Ashley French** Agilis Consulting Group, LLC., Cave Creek, AZ, United States

**Emily A. Hildebrand** Research Collective, Tempe, AZ, United States

**Jessie Huisinga** Agilis Consulting Group, LLC., Cave Creek, AZ, United States

**Sophia Kalita** Agilis Consulting Group, LLC., Cave Creek, AZ, United States

**Liz Mauer** Human Factors MD, LLC., Charlotte, NC, United States

**Christina Mendat** Human Factors MD, LLC., Charlotte, NC, United States

**Jeffrey Morang** Sanofi, Cambridge, MA, United States

**M. Robert Garfield** Abbot, St. Paul, MN, United States

**Eric Shaver** Human Factors MD, LLC., Charlotte, NC, United States

**Jonathan R. Silva** Washington University in St Louis, SentiAR, Inc., St Louis, MO, United States

**Leah K. Taylor** Agilis Consulting Group, LLC., Cave Creek, AZ, United States

# Author biographies

This book was supported by AAMI Faculty Members listed below.

**Mary Beth Privitera, M.Design, PhD, FIDSA**
**Principal, HFE/Research, HS Design, Inc.**
**Co-Chair & Faculty, Association for the Advancement of Medical Instrumentation Human Engineering Committee**
**Professor, Director, Medical Device Innovation & Entrepreneurship Program**
**Department of Biomedical Engineering, University of Cincinnati**

Dr. Mary Beth Privitera, M.Design, FIDSA, is internationally known as an expert in medical product design, specifically in the area of applied human factors. She is a principal at HS-Design, responsible for human factors and research. Additionally, she serves as faculty and co-chair of the Association for the Advancement of Medical Instrumentation's Human Engineering Committee.

Privitera also holds an appointment as Professor at the University of Cincinnati's Department of Biomedical Engineering and works collaboratively among the Colleges of Medicine, Engineering and Design. She is currently the Co-Founder and Director of the Medical Device Innovation and Entrepreneurship Program. Her previous academic appointments include industrial design and the Department of Emergency Medicine.

She has worked on devices intended for use across the practice of medicine including advanced technologies for: augmented reality for electrophysiologists, nasogastric tube placement, peripheral artery disease, intracranial aneurysm treatments, autoregulatory index monitoring for neuro ICU patients, combination drug delivery devices utilizing biometric technologies. She has observed care in nearly every area of the hospital throughout the United States and conducted studies human factors internationally. Her current research focuses on applied ergonomics, design and clinical applications involving additive manufacturing processes. She has authored several peer reviewed articles and a book titled "*Contextual Inquiry for Medical Device Design*," promoting best practices for phase zero medical device development. She has been a member of the AAMI Human Engineering Committee since 2004, an author of AAMI/ANSI HE 75 and has led the AAMI faculty team in the development of this book.

**Melissa R. Lemke**
**Managing Director of Human Factors Engineering/Agilis Consulting Group, LLC**

Melissa R. Lemke is an industry recognized expert and thought leader in medical device and regulatory human factors engineering, and she leads the human factors consulting services of Agilis Consulting Group, LLC as the Managing Director of Human Factors Engineering. She is a biomedical engineer with formal training in human factors who understands how to design strategies and lead cross functional teams through successful human factors processes, scientific methods, and regulatory submissions. Since 2003, Melissa has managed, designed, and executed human factors analyses and user interface optimization for clients developing medical devices and combination products for a variety of intended users both inside and outside the U.S.

Throughout her career, Melissa has contributed to numerous approved pre-market and post-market human factors usability and labeling submissions involving medical products such as surgical devices, infusion pumps, endoscopes, in vitro diagnostics, reprocessing, implantable devices, mobile and web-based technologies, over the counter products, home use devices as well as a variety of drug delivery and combination products. She has designed, executed, and managed human factors evaluations with diverse intended users such as physicians, nurses, pharmacists, surgical technicians, biomedical technicians, field engineers as well as lay users, caregivers, and adolescents. Melissa has specialized training and expertize conducting studies and evaluating designs with users who have a variety of physical, sensory, and cognitive limitations.

Melissa received her Master of Science in Biomedical Engineering from Marquette University as well as specializations in rehabilitation biosystems and biomechanics during her graduate and undergraduate studies. During her thesis research, she evaluated the accessibility and usability of numerous medical products for people with disabilities and developed a mobile usability lab. She also managed and participated in various research projects as part of the Rehabilitation Engineering Research Center on Accessible Medical Instrumentation (RERC-AMI), which culminated in research findings that were leveraged in the Americans with Disabilities Act Accessibility Guidelines (ADAAG).

Melissa is an active member of the Human Factors Engineering and Home Care and EMS Environments Committees of the Association for the Advancement of Medical Instrumentation (AAMI) as well as the Human Factors and Ergonomics Society (HFES). She is a contributing author to the original HE75 (2009): *Human Factors Engineering - Design of Medical Devices*, including lead co-author of the accessibility considerations section. Most recently, she led the committee updates of the revised HE75 for the sections on Accessibility and Home Healthcare. Melissa is an AAMI University faculty member who co-teaches *Human Factors for Medical Devices* as well as *Applying Human Factors to Improve Instructional Materials as Part of the User Interface*. She has published and presented numerous works related to human factors applied to medical device design as well as accessible design throughout her career. She began her profession in biomedical and human factors engineering after her brother sustained a spinal cord injury in 1999. Melissa remains passionate about helping clients develop and bring to market medical products that are safe, effective, and usable for people of all abilities through the application of sound human factors engineering during the design process.

**Tressa Daniels**

Tressa Daniels is the Associate Director of User Experience Design and Human Factors Engineering at Becton Dickinson in San Diego. Tressa has worked in the field of Human Factors Engineering for 21 years. Her expertize is in user interaction design, ethnographic research and executing human factors analyses of consumer and medical products, including infusion pumps, oral medication dispensing systems, migraine machines and DNA Sequencers. Formerly, Tressa worked at Illumina, CareFusion, HP, Intel and Xerox and was a Human Factors instructor at Woodbury University in Burbank, CA and is also a member of AAMI Faculty in Human Factors in Medical Device Design. She is a member of the Human Factors and Ergonomics Society (HFES) as well as the Association for the Advancement of Medical Instrumentation (AAMI). She serves on AAMI's Human Factors Engineering Standards Committee as well as hosts regular Webinars for AAMI. Tressa holds a Bachelors Degree in Psychology as well as a Masters in Human Factors Engineering and Applied Experimental Psychology.

**Renée Bailey**

**Director of Instructional Design & Creative Solutions/Agilis Consulting Group, LLC**

As Agilis' Director of Instructional Design & Creative Solutions, Renée Bailey brings more than 20 years of experience designing and developing instructional materials and strategies for medical device manufacturers, pharmaceutical, and other high-profile companies. Renée joined the Agilis Consulting Group in 2014 after 5 years of consulting in the medical device industry.

In her role as a Certified Expert Practitioner in evidence-based instructional design, Renée actively participates within a team dedicated to efficiently guiding clients while navigating the changing landscape of product regulatory pathways. Renée stays well informed of the U.S. and international regulations related to labeling and human factors to ensure instructional materials meet appropriate guidelines. She leads Agilis' instructional design team in the development of instructional materials, in collaboration with Agilis clients, to successfully meet FDA and international requirements for human factors submissions.

Renée is an expert in applying a scientific, systematic and scalable methodology based in Human Performance Technology (HPT) to produce effective instructional materials for healthcare professionals, lay user patients and caregivers, and clinical educators. Her process is focused on achieving safe and effective interactions between end-users and medical devices. Renée's instructional products include instructions for use, quick guides, training materials and programs (instructor-led, self-directed, eLearning), instructional videos, and on-device and packaging labels. Renée regularly creates and optimizes instructional materials for diagnostic kit devices, pen injectors, auto-injectors, inhalers, on-body delivery devices, life-supporting devices, reprocessing procedures for medical instruments, over-the-counter medical products, surgical devices, and robotic systems.

Prior to joining Agilis, Renée was a Performance Consultant for various Fortune 100/500 companies and a diverse list of clients which includes Medtronic, Pfizer Pharmaceuticals, Endo Pharmaceuticals, Deloitte & Touche, and Roche-Genentech. She gained vast experience and expertize leading large scale/high profile projects and project teams throughout her career. Renée strategically collaborates with global cross-functional teams and stakeholders to improve and streamline their complex, multi-level processes for producing effective user instructions across device and product platforms.

Renée is an industry thought leader and an active conference speaker on topics related to human factors engineering and regulatory requirements, evidence-based instructional labeling and training, post-market surveillance, and processes related to medical product clearance, approval and global market success. She also serves as Faculty for AAMI's human factors course, *Applying Human Factors to Improve Instructional Materials as Part of the User Interface*.

**Tim Reeves, PhD CHFP**
**Senior Technical Director**
**Human Factors MD, LLC**

Tim is the Senior Technical Director of Human Factors MD and founded the company in early 2001. Tim provides technical oversight to the team. He has more than 25 years of commercial experience evaluating and designing usable, effective, and safe medical devices. He is a Certified Human Factors Professional, and as a member of AAMI's Human Factors Committee, is a contributor to HE:75. He has been an invited speaker at several professional meetings, including the AAMI/FDA sponsored Human Factors and Patient Safety for Medical Devices and the first IBC's conference on Human Factors and Combination Products. Tim is the Lead Instructor and Subject Matter Expert for AAMI's popular Human Factors for Medical Device Design course. He has a PhD in cognitive psychology and human factors from the University of Toronto.

**Daryle Gardner-Bonneau** recently retired as the principal of Bonneau and Associates, a human factors consultancy in Portage, MI. She received the PhD in human factors from The Ohio State University (OSU) in 1983, and has earned master's degrees in psychology, industrial and systems engineering, and brass pedagogy (music), also from OSU. She is a fellow of the Human Factors and Ergonomics Society and a member of the Association for the Advancement of Medical Instrumentation (AAMI). Daryle has been involved in standards work for over 20 years. In addition to serving on AAMI's Human Factors Engineering, Alarms, and Home Care Device committees, she has served for over 10 years as the U.S. Technical Advisory Group (TAG) Chair for ISO/TC159 – Ergonomics and two of its subcommittees, and is currently the convener of ISO/TC159/SC1/WG2 on Ergonomics Process Standards. Daryle has contributed chapters to several books, and is the co-editor/co-author of *Human factors and medical device design (2011) and Human factors and voice interactive systems* (second edition, 2008). Her 35-year career has included academic, industry and consulting positions, and human factors research and consulting in the health care, telecommunications and aviation domains. Since retiring, she is spending increasing amounts of time on her hobbies, including Civil War history and genealogy.

**Merrick Kossack** has been practicing human factors engineering for over 25 years, most of that in the medical device industry. In his current role as Research Director of Human Factors Engineering (HFE) for Emergo by UL's Human Factors Research & Design (HFR&D) team (formerly UL-Wiklund), he works with medical technology developers to understand their HFE needs, develop strategies to meet those needs, and deliver the necessary HFE services. He manages the HFR&D team services in Chicago and supports the efforts in the group's other offices in Concord, MA, Utrecht, The Netherlands, Cambridge, UK, and Tokyo, Japan. His past work has involved technologies ranging from medical robotics to diagnostic and therapeutic devices to combination products.

Immediately prior to joining the Emergo by UL team, Merrick served as Principal Human Factors Engineer at Intuitive Surgical where he led the human factors engineering efforts developing the da Vinci Xi Surgical System. His responsibilities ranged from integrating human factors engineering into the organization's established design and development processes, to conducting usability studies, to providing the overall human factors strategy for each project to satisfy regulatory needs. Merrick has been a contributor to the Association for Advancement of Medical Instrumentation (AAMI) Human Factors Engineering committee responsible for work on today's relevant human factors industry standards. Merrick also taught human factors in medical device development at the University of California-Santa Cruz and is a member of AAMI's faculty helping to teach their *Human Factors for Medical Devices* course.

Merrick received his M.S. degree in Human-Machine Systems Research from the Georgia Institute of Technology and his B.S. Degree in Industrial Engineering from the University of Illinois.

Other contributors include the following:

**Jeffrey C. Morang, MS Human Factors – Ergonomics Usability Leader/Senior Human Factors Engineer, Sanofi Member (2014 – 2017), Association for the Advancement of Medical Instrumentation Human Engineering Committee**

Jeffrey Morang, MS Human Factors – Ergonomics, is a well-known human factors medical device design expert. He is currently a Usability Leader at Sanofi responsible for the application of human factors for drug-device combination products.

Jeff received his MS in Human Factors – Ergonomics from San Jose State University and has over 15 years of experience practicing human factors engineering in the industries of aviation, defense and medical device design focusing on enhancing product design as well as fulfilling regulatory requirements. Within the medical device design domain he has worked on devices for point-of-care diagnostics, intensive care units, emergency departments, surgical-assisted robotic surgery, operating theater diagnostics, drug delivery combination devices, neurological treatments, hospital sterile processing, diabetes care, and military/combat medical support. Jeff has conducted human factors research and studies across the globe and continues to promote the practice through guest lectures at local universities and international conferences and symposiums.

**Ashley French Hall**
**Human Factors Consultant/Agilis Consulting Group, LLC**

Ashley French Hall is a Biomedical Engineer and Human Factors Consultant with Agilis Consulting Group, LLC. Ashley has expertize in human factors engineering, usability testing, and optimization of user interface designs for medical devices and combination products, including experience with a wide range of devices such as surgical, diagnostic, home use, mobile, and over-the-counter devices as well as combination products such as auto-injectors, multi-use pens, syringes, inhalers, and wearable infusion devices. She also has experience designing and conducting studies with diverse end users such as lay patients and caregivers, adolescents, nurses, physicians, phlebotomists, and pharmacists. Since 2014, Ashley has led and contributed to human factors projects such as early stage heuristic analyses, formative evaluations, validation studies, post-market analyses, and post-market supplemental validation studies. She has led numerous projects through regulatory submissions that culminated in successful human factors validation studies and FDA clearance. Ashley also is experienced with international human factors industry standards and has conducted usability testing in multiple European countries.

**Leah Taylor**
**Human Factors Consultant/Agilis Consulting Group**

Leah Taylor is a Biomedical Engineer and Human Factors Consultant with Agilis Consulting Group, LLC. Leah works with Agilis clients to design, conduct and manage medical device and combination product human factors projects in support of global regulatory submissions. She graduated from the University of Iowa with a M.S. in Biomedical Engineering. Her graduate studies included assessment of performance in the operating room and collaboration with cross-functional teams of surgeons, medical residents, nurses and different types of medical technicians. Her previous work at the Mayo Clinic conducting research funded by the Department of Defense to assist in rehabilitation of wounded warriors.

**Deborah Billings Broky, PhD**
**Senior Human Factors Consultant/Agilis Consulting Group, LLC**

Dr. Deborah Billings Broky is a Senior Human Factors Consultant with Agilis Consulting Group, LLC, where she is responsible for managing, planning, executing, moderating and reporting human factors testing with the goal to evaluate and validate medical devices and products in alignment with U.S. and international regulatory guidance and standards. Since 2014, Billings Broky has worked with Agilis to design and implement human factors methodologies during all phases of product development, recommend data-driven design modifications to optimize user interface designs, and support clients with successful U.S. and international human factors regulatory submissions including client support during regulatory meetings and reviews with FDA.

**Sophia V. Kalita**
**Human Factors Consultant/Agilis Consulting Group, LLC**

Sophia V. Kalita is a Biomedical Engineer and Human Factors Consultant with Agilis Consulting Group, LLC. Sophia is experienced in applying human factors principles to the design, evaluation and validation of medical devices and products. Prior to joining Agilis Consulting Group, Sophia worked for a global medical device manufacturer as an R&D engineer where her focus was product design and manufacturing improvement. She advanced to project management where she led cross-functional teams to design and develop valuable medical devices. During this time, Sophia's accomplishments included managing and ensuring the success of human factors activities.

**Jessie Huisinga, PhD**
**Senior Human Factors Consultant/Agilis Consulting Group**
**Research Associate Professor/Department of Physical Therapy and Rehabilitation Science, University of Kansas Medical Center**

Dr. Jessie Huisinga is a Senior Human Factors Consultant with Agilis Consulting Group, LLC and an expert in assessing human performance with extensive experience working with individuals with neurological impairments. She has a background in Biomedical Engineering and Biomechanics, with specialized training in Neurology in order to evaluate movement patterns and task performance in persons with performance limitations. Dr. Huisinga works with Agilis clients to develop and optimize medical device and combination product user interfaces by applying various human factors methods as well as accessible design considerations. Dr. Huisinga has experience designing and conducting human factors formative and validation studies to help clients achieve successful human factors regulatory submissions and bring safe and effective medical devices and combination products to the market. She has experience assessing a diverse spectrum of home and professional use medical devices and products as well as conducting in-home and actual use usability studies.

**Christina Mendat, PhD**
**Managing Director**
**Human Factors MD, LLC**

Christina is a partner and Managing Director of Human Factors MD, a human factors consultancy that works exclusively on medical devices, including combination products. Christina is an expert at translating research findings such as user needs, requirements, product strengths and weaknesses into compelling design directions and solutions. She brings over a decade of experience in human factors having spent more than 5000 h in surgical suites, medical device usability studies, and outpatient facilities. She has presented papers to the Human Factors and Ergonomics Society and the American Psychology Society and has been providing multiple workshops and lectures on navigating the latest standard, HE75 and human factors integration in quality management systems. Christina has a PhD in experimental psychology and ergonomics from North Carolina State University.

**Eric Shaver, PhD**
**Technical Director**
**Human Factors MD, LLC**

Eric is a Technical Director with Human Factors MD. He has over two decades of human factors experience and has spent a good portion of those years on the litigation side of human factors and safety, specifically in warnings and risk communication, in addition to product development. At Human Factors MD, he contributes technically to all human factors activities throughout the product development life cycle. This includes participating in ideation sessions, understanding user needs, developing design requirements, providing input on early prototypes, developing device labeling, leading formative and validation studies, and conducting expert reviews. Prior to joining Human Factors MD, he was in charge of human factors for a medical device company in the Seattle area. Eric holds a PhD in Ergonomics Psychology from North Carolina State University.

**Liz Mauer, MHCI**
**Technical Director**
**Human Factors MD, LLC**

Liz has over 15 years of diverse experience in human factors research, design, and testing. She has particular expertize in planning, conducting, and analyzing findings from user research studies and translating those findings into directions for useful, usable, and desirable products. She is also experienced in applying the requirements of IEC 62366, HE75, and various FDA Guidance documents to support the development of safe and effective medical products. Liz holds a Masters Degree in Human-Computer Interaction from Carnegie Mellon University.

**Dr. Ian Culverhouse**

Dr. Ian Culverhouse is Co-Founder of Rebus Medical Ltd., UK and an experienced human factors consultant having worked with companies such as Roche, AstraZeneca, Smith and Nephew, Eli Lilly and Bosch Healthcare. Ian has a wealth of experience in applying HF to the design of medical devices throughout the development process, supporting manufacturers maximize their return on integrating HF into their business. In relation to Contextual Inquiry, Ian has led multiple global CI studies, including in-home and clinical environments in the UK, Germany and across the USA. Ian's PhD investigated the application of early stage interactive prototyping techniques. Today he advocates the philosophy of inclusion of early stage user testing to maximize the opportunity for learning and influencing design decisions.

**M. Robert Garfield**
**Senior Human Factors Engineer**
**Abbott, St. Paul, Minnesota, USA**

M. Robert "Bobby" Garfield is an experienced human factors engineer and industrial designer who has spent his career working for industry leading med-tech manufacturers and consulting agencies. Garfield is adept in the application of human factors in medical product design and has worked on programs ranging from handheld autoinjectors to next-generation robotic surgery systems. He is an alumnus of the University of Cincinnati (MDes, BSID) and Fitchburg State University (MBA).

**Kate Cox**
**Senior Human Factors Engineer**
**HS Design**

Kate is a Senior Human Factors Engineer and has been with HSD since September 2012. During this time, she has quickly become a key player of the design research and Human Factors team, specializing in applying human factors to the design and development of innovative medical devices for start-up and Fortune 500 companies. She is experienced in all aspects of the product development process, beginning with initial contextual inquiry research to defining requirements, analyzing risk, conducting usability testing, and developing HFE reports. She was worked on numerous devices over the years, from operating room equipment and tools, to consumer medical products, and laboratory devices. Kate has also been an integral member of the QMS team, working to further lock down quality controls within HSD to comply with ISO 13485. Kate has a Bachelor's and Master's Degree in Biomedical Engineering from Stevens Institute of Technology.

**Tor Alden, IDSA, MS**
**Principal**
**HS Design, Inc.**

As Principal of HSD, Tor Alden, MS, IDSA, brings his 25-year experience in user centric design, user research, strategic thinking and innovative product development to all HSD programs. His passionate collaboration to solve complex problems with innovative medical and digital health companies has led to over 45 patents and multiple design awards.

Tor is an avid speaker and writer consistently contributing to the technology, education, and design industries. Actively involved in patient safety, Tor serves on the AAMI Human Factors committee providing human factors use and usability guidance. Alden has also served on multiple advisory boards, and has held both Chapter and Medical section chairs for IDSA including serving as an IDSA International Design Excellence Awards 2018/2019 juror. HS Design is an ISO 13484 certified product development firm specializing in medical, life science, pharmaceutical and consumer healthcare markets. The firm's 40-year expertize in product design centers on solving complex usability and system problems for medical devices, high-technology products, and new ventures. HSD's expertize includes user research, user interface design, industrial design, human factors, mechanical, electrical and software engineering leading to full prototypes of complex systems through pilot launch.

**Jennifer Silva, M.D.**

Jennifer Silva is Director of Pediatric Electrophysiology and Associate Professor of Pediatrics at Washington University School of Medicine/St. Louis Children's Hospital, and serves as the Faculty Fellow in Entrepreneurship for Washington University SOM. She serves on committees within the Heart Rhythm Society (Chair, Women in Electrophysiology; Member, Communications Committee) and Pediatric and Congenital Electrophysiology Society, and serves on the NIH-SBIR study section for Cardiovascular Innovation. The scope of her research has been on developing and identifying clinical applications of new and emerging technologies within cardiac electrophysiology.

**Jonathan R. Silva, PhD FAHA**
**Associate Professor of Biomedical Engineering**
**Washington University in St. Louis**

**Russell J. Branaghan, PhD.**

Russell J. Branaghan is Associate Professor of Human Systems Engineering in the Ira A. Fulton Schools of Engineering at Arizona State University. There, he is co-founder and Co-chair of the Master of Science in User Experience program and Director of the User Experience Laboratory (XLab). He teaches courses in Human Factors, Human-Computer Interaction, Healthcare Human Factors, Research Methods, Statistics, and Memory & Cognition. Russ also serves as President of Research Collective, a human factors and user experience laboratory and consulting group in Tempe, Arizona.

**Emily A. Hildebrand, PhD**

Emily is a cognitive scientist specializing in human factors, with 10 + years of healthcare-specific experience. She leads usability, product design, and user-experience-related projects for Fortune 100 and Fortune 500 clients across a variety of fields. She also has extensive experience in product failure analysis and expert witness litigation support for medical devices. She has performed research on medical device usability, reusable medical device reprocessing, and workflow processes at the VA and Mayo Clinic. Her research culminated in guidance recommendations to FDA and AAMI for improving the usability of medical device interfaces. Emily contributes to field as an active member of human factors and reprocessing related committees within AAMI and as a member of the HFES Healthcare and Product Design technical groups.

**L. Bryant Foster, M.S.**

Bryant's performed human factors research for dozens of medical devices including surgical instruments, point-of-care devices, diagnostics, combination products, home-use devices, OTC products, and more. He's an active member of the Human Factors Engineering committee within the Association for the Advancement of Medical Instrumentation (AAMI); teaches a Human Factors and Design Controls course for the Regulatory Affairs Professional Society (RAPS); and is as an active member in the Human Factors and Ergonomics Society (HFES) including the Healthcare and Product Design technical group, and he presents at the Annual Meetings and Healthcare Symposiums. Bryant's written articles about human factors, usability, and human-centered design for several periodicals.

**Paula Labat-Rochecouste**
**Director of Human Factors and User Research**
**Human Center Ltd. UK**

# Foreword by Hanniebey D.N. Wiyor

What do we know about human interaction with medical devices? Given what we know, how then should we design device interfaces, clinical tasks and procedures so that medical device use can result in desirable clinical outcomes without causing harm to the patient? Wouldn't it be appropriate to minimize use-related hazards and risks? After all, we will be a patient someday.

Ensuring patient safety by minimizing potential use-related risk requires the application of human factors principles. A poorly designed user interface, one that lacks considerations of usability, increases the risk to patients. As medical devices become more diverse and complex, and are used with increased frequency in potentially multiple environments by users with variable skill and training levels, the importance of safe use is of interest to all involved: patients, caregivers, providers, hospitals/pharmacies, manufacturers, as well as regulatory agencies.

The above is quite possible if human factors engineering processes are incorporated into device development. Human factors/usability engineering focuses on the interactions between people and devices: the device user interface. This includes all physical controls and display elements as well as packaging, labeling, and training. The goals of the human factors processes are to minimize use-related hazards and risk and then confirm the efforts were successful, ultimately demonstrating users can use the device safely and effectively.

This book, authored under the leadership of AAMI Human Engineering Faculty and Committee members, describes human factors processes with the intent to improve patient safety and support device users through "Applied Human Factors in Medical Device Design."

**Hanniebey D. Wiyor, Ph.D.**
*LT U.S. Public Health Service*
*Regulatory Officer, Human factors Engineer*
*U.S. Food & Drug Administration*

# Foreword by Molly Follette Story

As technology advances, modern medical devices can perform increasingly amazing tasks; but those devices require varying amounts and types of interaction with human beings, whose basic capabilities have not changed.

Medical device developers are responsible for ensuring that the intended users of a device can use the device well because *it does not matter how innovative, safe, effective, reliable and affordable the medical device is unless its users are able to use the device successfully for its intended purpose.*

Human factors engineering (known in many parts of the world as usability engineering) is the science of studying and optimizing the relationship between the built environment and human beings, who tend to be error-prone. We cannot change very much about humans, but we can support development of medical devices that are error-resistant, and that humans can use well to diagnose diseases, treat medical conditions, and achieve and maintain good health.

This book was developed to support AAMI's human factors educational courses, which is going on its 11th year and has been well attended throughout its history. The book covers the breadth of the topic and digs more deeply into some specific areas; it also describes and provides insights into key relevant current US and international standards and guidance.

It is my hope that this book will help its readers understand how to practice good human factors/usability engineering for medical devices, and that future medical devices will be much better than they are today: safer, more effective, more useable, preferable, and even emotionally gratifying.

**Molly Follette Story, Ph.D.**
*Co-chair, AAMI Human Factors Engineering Committee, Cambridge, MA, United States*

# PART I

# Introduction

CHAPTER

# 1

# Introduction & background


Mary Beth Privitera[a], Ashley French[b], Melissa R. Lemke[b]

[a]HS Design, Gladstone, NJ, United States; [b]Agilis Consulting Group, LLC., Cave Creek, AZ, United States


OUTLINE




*Applied Human Factors in Medical Device Design*
https://doi.org/10.1016/B978-0-12-816163-0.00001-3

"*To err is human.*" **Alexander Pope**

# 1. Introduction

If the devil is in the details, redemption lies in the oversight of those details—especially when it comes to successful product design. From cars to coffee pots, we rely on products to help us live our daily lives; and we trust that those products adhere to safety and usability standards. Designers and manufacturers focus on details like interface design, user interaction and testing because users of all levels demand products that work easily and reliably. In the medical device industry, those details take on life-or-death importance. Agency-approved devices must meet regulatory requirements that include the application of human factors, a process which also optimizes usability. The value of applied human factors extends beyond analyzing raw data and user-testing. Human factors adds critical layers of contextual, environmental, and user characteristics to solve distinctive problems with distinctive designs.

The terms human factors is the scientific discipline concerned with the interactions between humans and other elements of a system. The profession applies theoretical principals, data, and methods in order to optimize design for the purposes of well-being and overall system performance (Carayon et al., 2018). Ergonomics [ergon (work) nomos (the study of)] is the study of interaction between human beings and the tools, tasks, and environments of work and everyday living.

The start of the field is attributed to the growth of aviation innovation during WWII. As technology advanced and the aviation industry grew, there was a critical need to improve human performance and reduce errors. Human factors (HF) is an applied field which looks at balancing the user needs, human capabilities, and limitations during the product design process, including sensing, perception, cognition, and action (anthropometry and movement).

# 2. Background

HF is a discipline that focuses on the variables which affect the performance of individuals using equipment (Sawyer, 1996). As such, in 1996, Dick Sawyer, a representative from the US Office of Health and Industry Programs, along with a CDRH (FDA Center for Devices and Radiologic Health) workgroup published "Do it by Design. An Introduction to Human Factors in Medical Devices" as the initial primer with the intent to impact the design of medical devices toward safe and effective use. The primer was directed to encourage manufacturers to improve the safety of medical devices and equipment by reducing the likelihood of use error.

According the FDA Human Factors Guidance (2016), HF considerations in the development of medical devices involve three major components of the device-user system: (1) device users, (2) device use environments, (3) device user interfaces (FDA, 2016). Fig. 1.1, below, highlights each component and its respective characteristics which must be considered in the device design in order to achieve the correct device use outcome. Lack of consideration of each component could result in use error.

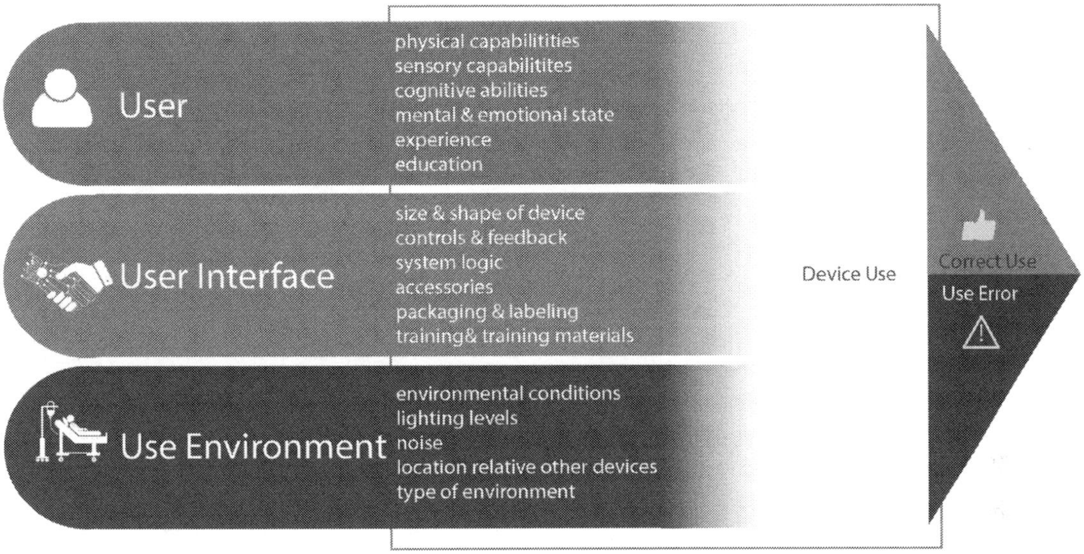


**FIG. 1.1** Major HF components of a device-user system interactions which result in device use outcomes. *Adapted from FDA/CDRH, 2016 Guidance.*

As medical devices become increasingly diverse in capabilities and environments in which they are used become busier, the need for specialized training and potential for use error increases (MRHA, 2017). This book discusses HF methodologies in detail to assure optimized and validated usability in medical device design.

## 2.1 Purpose of applying human factors in medical device design

A user's behavior is directly influenced by the user interface design. Misleading or illogical interfaces in the context of the user can induce errors by even the most proficient users. It has been demonstrated that design-induced errors in the use of medical devices can lead to patient injuries and deaths (Sawyer, 1996). Further, building on the HF component, the user as described in Fig. 1.1, Table 1.1 below describes the relationship of HF Attributes of Users to principles of design.

In designing a user interface based on the HF attributes and design principles found in Table 1.1, it is likely that the resulting product design will:

- Accommodate a range of users in variable environments, even under stressful situations.
- Be less prone to user error and possibly require less training.

### *2.1.1 History of human factors in medical device design*

The application of HF in healthcare can be traced back to the mid-1990's, when human factors engineering concepts of active (errors committed by users) and latent (errors as a result of organizational level of the design, e.g. inadequate procedures or incomplete training) failures (Cafazzo & St-Cyr, 2012). In this early work, the importance of considering team and

**TABLE 1.1** The relationship of human factors attributes to design principles.

| HF attributes of users | Examples of design principles |
|---|---|
| Perception: the ability to detect, identify and recognize sensory input. (Sawyer, 1996) | To minimize perception time, decision time and/or manipulation time |
| | To provide sensory feedback to users |
| Cognition: refers to higher-level mental phenomena, such as memory, information processing, use rules and strategies, hypothesis formation, and problem solving. (Sawyer, 1996) | To allow for error detection and recovery |
| | To match technology and users' mental models |
| | To ensure consistency of interface design with users' expectations |
| Physical HFE/Actions: vision, hearing, manual dexterity, strength and reach. (Sawyer, 1996) | To reduce or mitigate need for excessive physical exertion |
| | To optimize opportunities for physical movements, such as reaching controls |
| Expectancies: people are predisposed to react to new situations according to established habits. | To take advantage of existing conventions in the general population as well as those in the healthcare community. |
| Mental Modals: people form abstract concepts about how complex phenomena actually work | To accommodate for individual differences in how complex situations are mentally integrated. |

organizational factors in the design of safety critical systems as well as avoiding a culture of blaming the users were highlighted in order to embrace a culture of understanding the root causes of adverse events (Cafazzo & St-Cyr, 2012). Furthering these efforts, the Institute of Medicine's report "To Err is Human" (Kohn, Corrigan, & Donaldson, 2000) provided a catalyst for the use of human factors in the healthcare domain. This publication is widely referenced and encompassed considerable understanding of actions needed to solve safety problems within the US healthcare system.

As a result, HF practitioners embraced healthcare as a domain of study and developed many regulations, standards, and guidelines. Prominent sources include the American National Standards Institute (ANSI)/Association for the Advancement of Medical Instrumentation (AAMI) HE 75 standard on medical device design, FDA Human Factors Guidance, and IEC regulations regarding usability. As these standards and regulations have been generated in recent years, the discipline has made significant progress toward addressing patient safety problems in healthcare; however, it often remains misunderstood or unfamiliar.

The HF discipline ethos promotes the fundamental rejection that humans are primarily at fault when making errors in the use of a socio-technical system (Cafazzo & St-Cyr, 2012). The fundamental tenet is that when a use error occurs it is attributable to the design of the system; that some aspect of use performance, cognition or behavior was not fully considered during the design process in order to avoid the circumstance. Human factors addresses problems by modifying the design of the system to better aid people in accomplishing tasks (Russ et al., 2013).

### 2.1.2 Role of a human factors engineer in medical device development

In medical device development, a HF team member assists in the design of products, services, and virtual environments which optimize human well-being and overall system performance by applying human factors theory, principles and data regarding the relationship between humans and the respective technology. A HF team member works collaboratively on a multi-disciplinary team that includes other HF professionals, designers, engineers, clinical providers, and regulatory professionals. In the corporate environment, a HF professional may be part of the quality, regulatory, design, or human factors team. Regardless of team membership, HF experts typically interact and work closely with all product development team members, including: marketing, engineering, design, quality, regulatory, and clinical support.

Most HF professionals have a minimum Bachelor of Science in the following disciplines: human factors engineering, ergonomics, cognitive/experimental psychology, industrial design, or engineering psychology; and most employers expect further work experience in the area of medical devices with a demonstrated track record of success.

The responsibilities of a medical device human factors professional include the ability to conduct various types of studies (contextual inquiry, usability tests) and to assist in the analysis/specifics for new device design through analytical and generative techniques. All human factors professionals work in highly collaborative environments, often as part of cross-functional teams with the aim to deliver devices which are 'easy to use.'

They maintain professional attributes such as:

- A proven ability to focus on the user's needs and to translate those needs into design recommendations
- Expert knowledge of interaction design/user-centered design principles
- Expertise in planning all aspects of and executing on rapid user testing methods and quickly providing turn-around data-driven recommendations (including in-person and remote usability testing)
- Ability to organize, run, and report on customer site visits
- Ability to work independently and efficiently on complex projects
- Ability to deliver high-quality work on tight schedules

HF practitioners are often passionate in their determination to ensure the design of a safe product or system. They may insist the root cause of a use failure is a design flaw and that not actively addressing these flaws or weaknesses may lead to further mishaps. Conversely, they may promote enhanced training materials or new policies regarding device use; however, this may not be enough in order to mitigate an error situation and design changes may be required. This debate is an example of one that regularly occurs in the application of human factors in medical device design.

Human factors is a scientific discipline that requires years of training and most professionals hold relevant graduate degrees (Russ et al., 2013). It is problematic in practice to believe that HF consists of a limited set of principles that can be learned during brief training as the depth of understanding in application will be missing. This lack of depth will result in ineffective application of HF, often missing situational context and the ability to correctly identify design weaknesses.

Russ (2013) maintains that HF professionals are bound together by the common goal of improving design for human use, and represent different specialty areas and methodological skill sets. In reality, not all human factors engineers and scientists have the same skill set (Russ et al., 2013). A HF specialization is often acquired through a variety of coursework and the pursuit of post-graduate degrees with each university offering specialized focus areas. These specialties include: aging, augmented cognition, cognitive engineering or decision making, communication, human performance modeling, industrial ergonomics, macro ergonomics, perception and performance, product design, safety, training, and usability (Russ et al., 2013).

## 2.2 Promoting patient safety through applied ergonomics

The "most common root causes of sentinel events are human factors, leadership and communication," says Ronald Wyatt, M.D. medical director, Office of Quality and Patient Safety at the Joint Commission (2015). The goal of patient safety has led to a number of improvements by design, using a number of paradigms which have been focused on reducing injuries, errors and improving evidence-based practice (Karsh, Holden, Alper, & Or, 2006).

A focus of HF in healthcare and patient safety has been in the design of useable, safe medical devices and health IT (Carayon, 2018). In addition, HF has contributed to the understanding of human error and identifying the mechanisms of human error involved in patient safety. These could be the result of performance obstacles which may endanger patients by making it difficult for clinicians to perform tasks and procedures safely, or system resilience, or the ability of systems to anticipate and adapt to the potential surprise and failure of an important attribute.

HF contributes to patient safety via four objectives of system design. These are (Carayon, 2018):

- The identification and removal of system hazards from the design through maintenance phases. This is accomplished through designing a system according to HF principles which account for errors and hazards.
- If obstacles cannot be avoided due to intrinsic requirements, then strategies should be designed to mitigate the impact on performance by enhancing other systems elements.
- Enhancements of resilience and supporting adaptability and flexibility in human work, such as allowing problem or variance control at its source, will enable users to detect, adapt to and/or recover from errors, hazards, disruptions, and disturbances.
- When there is a change in the work system, the change must be balanced in effect on the entire work system and then the entire system needs to be optimized or balanced, i.e. HF system design cannot focus on one element of work in isolation.

In the Joint Commission's article "Human Factors Analysis in Patient Safety Systems" (2015), several strategies for addressing HF in a process or system are described. The goal of these strategies are to develop more reliability; to establish and sustain an adaptable health care organization that is attuned to the possibility of failure, is empowered and equipped to respond and learn; and the ability to contain or reduce hazardous conditions prior to patient harm. Table 1.2 below discusses the reliability of human factors strategies promoting patient safety (Joint Commission, 2015).

While Table 1.2 is organizationally based, FDA HF Guidance (2016) promotes the reduction of use-related hazards in medical devices in following ANSI/AAMI/ISO 14971 risk management options found in Table 1.3.

**TABLE 1.2** Human factors strategies promoting patient safety (Joint Commission, 2015).

| | |
|---|---|
| Most reliable | Forcing functions which prevent incorrect actions |
| | Computerized automation (which reduce cognitive load) |
| | Human-machine redundancy (visually checking then scanning) |
| Somewhat reliable | Checklists for high risk procedures |
| | Forced pause in a process to recheck details and steps e.g., Time Out procedures |
| | Reminders e.g., electronic medical record prompts |
| | Standardization of equipment and supplies across the organization |
| | Planned error recovery opportunities which build in double checks, redundant confirmations |
| Least reliable | Education and Training |
| | Rules, policies and procedures |

**TABLE 1.3** Strategies promoting safe devices (ISO/IEC, 2007).

| | | |
|---|---|---|
| Most preferred and effective strategy | Inherent safety by design | Examples: |
| | | Use connectors that cannot be connected to the wrong component |
| | | Remove features that can be mistakenly selected or eliminate an interaction when it could lead to use error |
| | | Improve detectability or readability of controls, labels & displays |
| | | Automate device functions that are prone to use error when performed manually |
| Alternative strategy when above is not possible | Protective measures in the device itself or in manufacturing process | Examples: |
| | | Incorporate safety mechanisms such as physical guards, software interlocks |
| | | Include warning screens to advise user of essential conditions |
| | | Use alerts for hazardous conditions |
| | | Use technologies that require less or no maintenance |
| Least effective strategy | Information for safety | Examples: |
| | | Provide written information in user manual that highlight and discuss the use-related hazard |
| | | Train users to avoid the hazard |

### *2.2.1 Impact on the future of clinical practice and patient experience*

"Our medical system is frankly not designed to optimize all the moving pieces of a patient's complex care. Many obstacles to better care are not about individual practitioners' decisions but are more about systemic barriers" (Yurkiewicz, 2018).

HF is found in the organizational management of healthcare. This includes building a culture of safety and knowledge of human factors in patient safety as a training for all those involved in the delivery of care. Issues include mental workload, distractions, physical environment, physical demands, device and product design, teamwork as well as process design (Desai, Medvedev, & Yanco, 2016). Desai et al. (2016) further posits that a positive safety culture is one where the organization is open, just, informed and where reporting and learning from the error is the norm.

Organizational HF has the potential to provide users more control over their work systems, to have support throughout the care compendium and advance opportunities to develop new skills. This patient safety model divides the system into five elements: people, tools and technology, tasks, environments, and outcomes. This aligns with applying HF in device design as it is a key part of the system. Device design and packaging often lack consideration for cognitive limitations and frequently can be the cause of patient safety incidents.

# 3. Applicable human factors agency guidance's and standards (Ashley French, Melissa R. Lemke)

International HF standards and national HF guidances are required in order to obtain a successful FDA submission for new medical devices and devices with modifications to their user interface. Different regulatory requirements exist in the United States as compared to other regions of the world (e.g., Europe) and a manufacturer's HF strategy will determine the approach used to meet these standards.

## 3.1 Determine which standards are applicable to U.S. submissions

Adhering to relevant human factors standards and guidance throughout device development can lead to improved device design, reduced use-related risk, and often reduced time to market with an optimized product design toward usability. Thus, it is important for the manufacturer to know from the beginning which of the existing human factors standards to follow, and which are applicable for each particular device.

Determining which standards are applicable to a U.S. medical device submission can be more manageable for manufacturers if the following 3 steps are followed:

1. Search for **general** U.S. human factors standards and FDA recognized guidance that apply to all medical devices and products;
2. Search for any **specific** U.S. human factors standards and FDA recognized guidance that apply to your particular device (e.g., device for home use, infusion pump, etc.);
3. Search for FDA recognized international standards that apply to your device.

It is up to the manufacturer to identify which standards and guidances are recognized by the FDA and applicable to a specific product, so it is important to know where to find these documents.

### 3.1.1 General standards that apply to all medical devices

Some of the general standards listed in Table 1.1 are applicable to all medical devices, like FDA CDRH's final guidance: *Applying Human Factors and Usability Engineering to Medical Devices* and ANSI/AAMI/IEC 62366-1:2015 - *Medical Devices Part 1: Application of usability engineering to medical devices.* Manufacturers should consider these general human factors guidance documents across all types of products throughout the device development process. Table 1.4 below includes examples of general human factors guidance that apply to all medical devices.

## 3.2 Searching for specific applicable standards

The list of standards and guidance documents in this chapter is not exhaustive. Regulatory agencies and standards development organizations also periodically update or create new human factors standards and guidance documents. Therefore, manufacturers should regularly conduct applicable searches (as recommended below) for standards and guidance relevant to their specific device types to ensure relevant information is considered as part of device development and FDA submissions.

Use the links below to search for U.S. and international human factors standards and guidance:

- AAMI: http://my.aami.org/store/Default.aspx
- FDA: https://www.fda.gov/RegulatoryInformation/Guidances/default.htm
- ISO: https://www.iso.org/standards-catalogue/browse-by-ics.html
- IEC: https://webstore.iec.ch/

**TABLE 1.4** Examples of general U.S. human factors medical device standards & guidance applicable to all medical devices and products.

| Title/version | Description |
|---|---|
| FDA CDRH's final guidance: *Applying Human Factors and Usability Engineering to Medical Devices* that was issued on February 3, 2016. | This guidance contains recommendations to assist industry in following appropriate human factors and usability engineering processes to maximize the likelihood that new medical devices will be validated as safe and effective for the intended users, uses, and use environments.<br>This guidance applies to all medical devices applying for FDA submission, specifically. |
| ANSI/AAMI HE75:2009/(R) 2013, *Human factors engineering – Design of medical devices.* | This FDA recognized industry standard is a comprehensive reference that includes general human factors principles, management of use error risk, design elements, and integrated human factors solutions.<br>This industry standard applies to all medical devices applying for FDA submission. |
| FDA CDRH final guidance: *Guidance on Medical Device Patient Labeling; Final Guidance for Industry and FDA Reviewers* published on April 19, 2001. | This guidance serves to assist manufacturers in development of patient labeling, and to assist reviewers in their review and evaluation of medical device patient labeling to help make critical information understandable to the end user.<br>This guidance applies to all devices with patient labeling applying for FDA submission with medical device patient labeling, specifically any information associated with a device targeted to a patient or lay caregiver. |

## 3.3 Human factors medical device standards for U.S. Submissions

### *3.3.1 U.S. human factors medical device guidance*

The following organizations have issued United States regulatory guidance for human factors for medical devices:

- The Food and Drug Administration (FDA), and its associated branches:
  - The Center for Devices and Radiological Health (CDRH)
  - The Center for Drug Evaluation and Research (CDER)
  - The Center for Biologics Evaluation and Research (CBER)
- Association for the Advancement of Medical Instrumentation (AAMI)
- American National Standards Institute (ANSI)

Table 1.4 provides examples of the most commonly used general human factors medical device standards published for U.S. device applications and describes the information included in each document. These general human factors standards and guidance documents are applicable across all medical devices and products.

Note that the standards and guidance listed in this chapter do not represent an exhaustive list of all standards available. In addition, standards and guidance documents are constantly being updated and replaced by newer (or final) versions. It is important that a manufacturer always perform a search of standards for each of the organizations (e.g., FDA, AAMI, IEC, etc.) to identify all human factors guidance relevant to the specific device type that is seeking FDA (See Section 3.4) FDA/AAMI recognized international standards applicable to U.S. submissions), as nuances of user-device interactions or use-related risks for specific device types can have a separate set of regulatory guidance principles. However, the resources outlined in Tables 1.4–1.6 provide the human factors references to build upon as a manufacturer starts to identify all guidance relevant to their specific medical device.

### *3.3.2 Specific U.S. standards that only apply to certain devices*

Other standards and guidelines are specific to a certain medical device type (e.g., medical devices intended for home use, or over-the-counter products), intended use of the device (e.g., a process such as instrument sterilization), or device component that performs a special function (e.g., software design). To determine the more specific standards that are applicable for a particular product, the manufacturer should identify details and user interface aspects of the new product, including the device type, intended use, and certain device components that may be associated with human factors guidance.

Table 1.5 below describes the most commonly used human factors for medical device standards and guidance in the U.S. that are applicable for only particular types of medical devices and products. It describes which type(s) of products the standard or guidance is applicable to, as well as the information included in each document.

**TABLE 1.5** Examples of specific U.S. human factors medical device guidance documents applicable only to particular devices or products.

| Relevant device type | Title/version | Description |
|---|---|---|
| Combination products | FDA CDER's draft guidance: *Human Factors Studies and Related Clinical Study Considerations in Combination Product Design and Development* that was published in February 2016. | This guidance provides information to industry and FDA staff on the underlying principles of human factors studies during the development of combination products.<br><br>This guidance applies specifically to combination product medical devices applying for FDA submission. |
| Combination products | FDA CDER guidance: *Safety Considerations for Product Design to Minimize Medication Errors* that was published in April 2016. | This guidance provides best practices on how to improve drug product and container closure design and also describes examples of product designs that have resulted in post-market medication errors.<br><br>This guidance specifically applies to prescription and non-prescription drugs and biologic products regulated by CDER. |
| Devices for home use | FDA CDRH guidance: *Design Considerations for Devices Intended for Home Use* that was published (and updated) on November 24, 2014. | This guidance is intended to assist manufacturers in designing and developing home use medical devices that comply with applicable standards of safety and effectiveness and other regulatory requirements.<br><br>This guidance applies specifically to home use medical devices applying for FDA submission. |
| Devices for use in non-clinical settings | AAMI TIR 49:2013, *Design of Training and Instructional Materials for Medical Devices used in Non-Clinical Environments.* | This guidance identifies special considerations for devices that are intended to be used in nonclinical settings only, although much of the guidance can be generally applied to design of training and instructional materials for clinical devices as well. In particular, training and instructional material technology and terminology should successfully guide the user when performing tasks.<br><br>This guidance applies to devices used in non-clinical settings only. |
| Reusable devices | FDA CDRH and CBER guidance: *Reprocessing Medical Devices in Health Care Settings: Validation Methods and Labeling* published on March 17, 2015. | This document provides guidance on development and validation of reprocessing instructions for medical devices.<br><br>This guidance applies to reusable devices, or single-use devices that are initially supplied as non-sterile and require the user to process the device prior to use. |
| Infusion pumps | FDA CDRH's guidance: *Infusion Pumps Total Product Life Cycle, Guidance for Industry and FDA Staff* that was published on December 2, 2014. | This guidance identifies device features and use-related considerations that manufacturers should address through the product life cycle.<br><br>This guidance applies to infusion pumps. |

## 3.4 FDA/AAMI recognized international human factors medical device standards that are applicable to U.S. products

The following organizations have issued international regulatory standards for human factors for medical devices, which are recognized by the FDA and/or AAMI:

- International Electrotechnical Commission (IEC)
- International Organization for Standardization (ISO)

Table 1.6 describes commonly used human factors medical device standards published for international applications and provides information on where the standards are applicable.

**TABLE 1.6** Examples of international human factors medical device standards that may apply to U.S. submissions.

| Title/version | Description |
|---|---|
| ANSI/AAMI/IEC 62366-1:2015 - *Medical Devices Part 1: Application of Usability Engineering to Medical Devices.* | This standard specifies a process for a manufacturer to analyze, specify, develop and evaluate the usability of a medical device as it relates to safety. This usability engineering (human factors engineering) process permits the manufacturer to assess and mitigate risks associated with correct use and use errors, i.e., normal use. It can be used to identify but does not assess or mitigate risks associated with abnormal use.<br><br>FDA recognized standard, applicable to all medical devices. |
| ANSI/AAMI/IEC/TR 62366-2:2016 - *Medical Devices Part 2: Guidance on the Application of Usability Engineering to Medical Devices.* | This document is a prescriptive technical report, and contains background information and provides guidance that addresses specific areas that experience suggests can be helpful for those implementing a usability engineering (human factors engineering) process both as defined in IEC 62366-1:2015 and as supporting goals other than safety.<br><br>FDA recognized standard, applicable to all medical devices. |
| ANSI/AAMI/IEC 62366-1: 2015, *Annex C: Evaluation of a User Interface of Unknown Provenance (UOUP)* | Annex C in this version of the standard was created so manufacturers can apply the tools defined in 62366-1 to user interfaces or parts of user interfaces that have already been commercialized prior to the publication of this edition of this standard, thus the devices were not developed using the processes of 62366-1 and as a result are of unknown provenance with respect to these processes.<br><br>FDA recognized standard, applicable to legacy medical device products that are being changed. |
| ANSI/AAMI/ISO 14971-1:2007/(R) 2016 - *Medical Devices — Application of Risk Management to Medical Devices.* | This standard details the process to identify medical device hazards, estimate and evaluate the associated risks, and monitor the effectiveness of the controls.<br><br>FDA recognized standard, applicable to all medical devices. |

**TABLE 1.6** Examples of international human factors medical device standards that may apply to U.S. submissions.—cont'd

| Title/version | Description |
|---|---|
| ANSI AAMI HA60601-1-11:2015 *Medical Electrical Equipment — Part 1—11: General Requirements for Basic Safety and Essential Performance — Collateral Standard: Requirements for medical electrical equipment and medical electrical systems used in the home healthcare environment.* | This prescriptive standard contains collateral requirements for medical electrical equipment and medical electrical equipment and medical electrical systems used in the home healthcare environment.<br><br>FDA recognized standard, applicable to medical devices with electrical components. |
| IEC 60601-1-6:2010 - *Collateral Standard: Usability* | This standard provides a bridge between IEC 60601-1 and ANSI/AAMI/IEC 62366. It specifies a process for a manufacturer to analyze, specify, design, verify and validate usability, as it relates to basic safety and essential performance of medical electrical equipment.<br><br>FDA recognized standard, applicable to medical devices with electrical components. |
| IEC 60601-1-8 Edition 2.1 2012-11 - *General requirements, tests and guidance for alarm systems in medical electrical equipment and medical electrical systems* | This prescriptive standard contains design requirements for alarm systems in medical electrical equipment and systems.<br><br>FDA recognized standard, applicable to medical devices with alarm systems. |
| IEC 60601-1-11:2015 - *Requirements for medical electrical equipment and medical electrical systems used in the home healthcare environment* | This prescriptive standard applies to the basic safety and essential performance of medical electrical equipment and medical electrical systems for use in the home healthcare environment. It applies regardless of whether the medical electrical equipment or medical electrical system is intended for use by a lay operator or by trained healthcare personnel.<br><br>FDA recognized standard, applicable to medical devices with electrical components used in the home healthcare environment. |
| ISO 15223-1:2016 *Medical devices — Symbols to be used with medical device labels, labeling and information to be supplied — Part 1: General requirements* | This prescriptive standard identifies requirements for symbols used in medical device labeling that convey information on the safe and effective use of medical devices. It also lists symbols that satisfy the requirements of this document.<br><br>FDA recognized standard, applicable to medical devices with symbol labeling. |
| ISO 14937:2009 *Sterilization of health care products — General requirements for characterization of a sterilizing agent and the development, validation and routine control of a sterilization process for medical devices* | This prescriptive standard specifies general requirements for the characterization of a sterilizing agent and for the development, validation and routine monitoring and control of a sterilization process for medical devices.<br><br>It applies to sterilization processes in which microorganisms are inactivated by physical and/or chemical means and is intended to be applied by process developers, manufacturers of sterilization equipment, manufacturers of medical devices to be sterilized, and organizations responsible for sterilizing medical devices.<br><br>FDA recognized standard, applicable to medical devices requiring sterilizing procedures. |

*(Continued)*

**TABLE 1.6** Examples of international human factors medical device standards that may apply to U.S. submissions.—cont'd

| Title/version | Description |
|---|---|
| ISO 9186-1:2014 *Graphical symbols – Test methods – Part 1: Method for testing comprehensibility* | This prescriptive standard specifies a method for testing the comprehensibility of graphical symbols. It provides a measure of the extent to which a variant of a graphical symbol communicates its intended message. The purpose is to ensure that graphical symbols and signs using graphical symbols are readily understood. The intention is to encourage the development of graphical symbols which are correctly understood by users when no supplementary (i.e. explanatory) text is presented. |
| | ANSI/AAMI recognized standard, applicable to medical devices with graphical symbols included in labeling. |
| ISO 9241-210:2010 *Ergonomics of human-system interaction – Part 210: Human-centered design for interactive systems* | This prescriptive standard provides requirements and recommendations for human-centered design principles and activities throughout the life cycle of computer-based interactive systems. It is intended to be used by those managing design processes, and is concerned with ways in which both hardware and software components of interactive systems can enhance human–system interaction. |
| | ANSI/AAMI recognized standard, applicable to medical devices with user-software interactions. |
| ISO/TR 16982:2002 *Ergonomics of human-system interaction – Usability methods supporting human-centered design* | This standard provides information on human-centered usability methods which can be used for design and evaluation. It details the advantages, disadvantages and other factors relevant to using each usability method. |
| | ANSI/AAMI recognized standard, applicable to all medical devices. |

## 3.5 Other standards

In 2017, the Medicines & Healthcare products Regulatory Agency (MRHA) in the UK worked collaboratively with stakeholders to produce guidance on HF aspects of design for medical devices and include drug-device combination products (MRHA, 2017). Its titled, "*Human Factors and Usability Engineering - Guidance for Medical Devices Including Drug-device Combination Products*" and is intended for manufacturers of all device classes, drug-device combination products as well as notified bodies responsible for assuring the quality of those devices.

Although the guidance aims to clarify regulatory expectations of medical devices marketed in the UK, it does not represent a compliance requirement (MRHA, 2017, p. 4), it applies to the design of future products and changes in user interfaces of existing products rather than those already approved.

This guidance references U.S. FDA Human Factors Guidance and is intended to be consistent with both FDA Guidance and international HF standards.

## 3.6 Staying current with standards

Periodically, regulatory agencies will update or create new standards. The links below can be used to check for updates or new guidance published by the respective agency:

- AAMI updates: http://www.aami.org/newsviews/content.aspx?ItemNumber=2704
- Recent FDA final medical guidance documents: https://www.fda.gov/medicaldevices/deviceregulationandguidance/guidancedocuments/ucm418448.htm
- Recent FDA draft medical guidance documents: https://www.fda.gov/MedicalDevices/DeviceRegulationandGuidance/GuidanceDocuments/ucm407274.htm

Once an IEC or ISO standard is purchased, any updates to that standard will be alerted to the purchaser in the form of an email. Alternately, the links below can be used to subscribe to newsletters and notifications of new or updated standards:

- IEC Updates: http://www.iec.ch/subscribe/
  - Subscribe to the "Just Published" feed
- ISO Newsletter: https://www.iso.org/news_index.html

# 4. Why might we want to do more

Agencies require that devices must be safe and effective and that all of the above must be demonstrated with evidence in submissions. However, there are additional benefits to adopting a user centric approach, wherein the design of the device use interface aligns with the users wants, needs and preferences.

Fig. 1.2, originally presented by Dr. Molly Story (Story 2017), illustrates the difference between meeting agency requirements and optimizing the user experience. While some might argue that FDA also requires usability, as long as a device is adequately safe and effective, it's usability (e.g. efficiency, user satisfaction) involves business risk, which is the

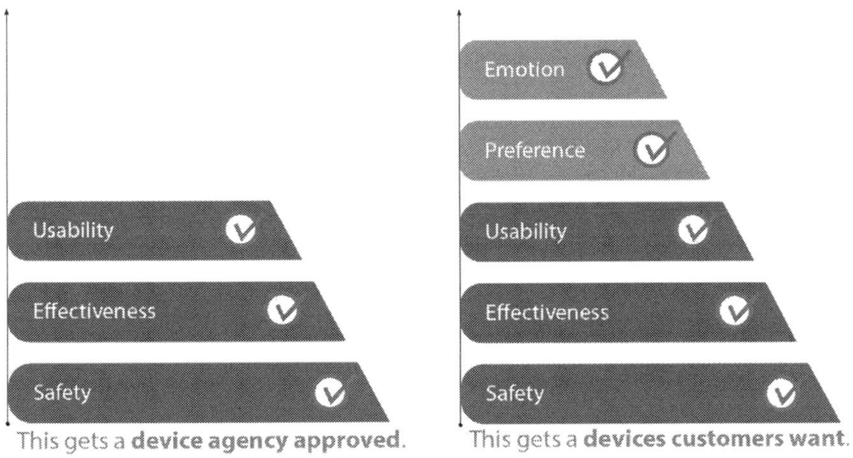


**FIG. 1.2** The difference between meeting agency HF requirements and a user centric approach: not much more effort required. *Adapted from Story (2017).*

manufacturer's responsibility but beyond FDA/s purview. Manufacturers, however, should care a lot about their device's usability but also users' preference for their device over their competitors' devices and also the emotional satisfaction users get from using their device/ Optimized medical device design can provide significant competitive advantage to manufacturers who follow good human factors practices.

## 5. Summary

This chapter introduces and defines the application of human factors for the purposes of medical device design. The application of HF is integral and necessary to ensure patient safety is promoted throughout the development and design process. In fact, HF is required to meet agency and international standards. To incorporate and maximize the value of HF, device designers and manufacturers must not only make a strong commitment to a user-centric approach, they must also continually incorporate a thorough understanding of HF methodologies in their work.

## Acknowledgments

Special thanks go to Elissa Yancey for editing.

## References

Cafazzo, J. A., & St-Cyr, O. (2012). From discovery to design: The evolution of human factors in healthcare. *Healthcare Quarterly (Toronto, Ontario), 15*(Spec No), 24–29.

Carayon, P., Wooldridge, A., Hose, B.-Z., Salwei, M., & Benneyan, J. (2018). Challenges and opportunities for improving patient safety through human factors and systems engineering. *Health Affairs, 37*(11), 1862–1869. https://doi.org/10.1377/hlthaff. 018.0723.

Desai, K., Medvedev, S., & Yanco. (2016). *Human factors*. Human Factors Time.

FDA. (2016). *Applying Human Factors and Usability Engineering to Medical Devices Guidance for Industry and Food and Drug Administration Staff Preface Public Comment*. Retrieved from http://www.regulations.gov.

ISO/IEC. (2007). *International standard international standard 14971 medical devices-application of risk management to medical devices* (Vol. 2007-10-01).

Joint Commission. (2015). Human factors analysis in patient safety systems. 13(4). Retrieved from http://www.jointcommission.org/assets.

Karsh, B. T., Holden, R. J., Alper, S. J., & Or, C. K. L. (2006). A human factors engineering paradigm for patient safety: Designing to support the performance of the healthcare professional. *Quality and Safety in Health Care, 15*(Suppl. 1), 59–65. https://doi.org/10.1136/qshc.2005.015974.

Kohn, L. T., Corrigan, J. M., & Donaldson, M. S. (2000). *Institute of medicine. Committee on quality of health care in America*. To Err Is Human: Building a Safer Health System.

MRHA. (2017). *UK notified bodies for medical devices*. GOV.UK. Retrieved from https://www.gov.uk/government/publications/medical-devices-uk-notified-bodies/uk-notified-bodies-for-medical-devices.

Russ, A. L., Fairbanks, R. J., Karsh, B. T., Militello, L. G., Saleem, J. J., & Wears, R. L. (2013). The science of human factors: Separating fact from fiction. *BMJ Quality and Safety, 22*(10), 802–808. https://doi.org/10.1136/bmjqs-2012-001450.

Sawyer, D. (1996). *Do it by design*.

Story, M. (September 2017). *The human Touch : Develop a patient-centric injection device*. Retrieved from https://www.wesrch.com/medical/paper-details/pdf-ME1XXF000KSCW-the-human-touch-a-patient-centric-injection-device.

Yurkiewicz, I. R. (2018). Complicated: Medical missteps are not inevitable. *Health Affairs, 37*(7), 1178–1181. https://doi.org/10.1377/HLTHAFF.2017.1550.

CHAPTER

# 2

# Overview of a human factors toolbox

*Mary Beth Privitera*

HS Design, Gladstone, NJ, United States

OUTLINE



*Knowledge is a tool and like all tools, its impact is in the hands of the user.*

## 1. Introduction

Human factors methodologies adds important value in all phases of device development, including generating new opportunity and advancing device design in order to improve the user experience which ultimately contributes to value as perceived by the user/customer. This chapter introduces commonly used human factors methods for the purposes of medical device design. Methodologies described in this book are not intended to provide exhaustive

Applied Human Factors in Medical Device Design
https://doi.org/10.1016/B978-0-12-816163-0.00002-5

descriptions of all methods found across the broad practice of human factors in all industries. Rather it is intended to provide examples of best practices for the development of a medical device and includes those methods which have been most helpful in meeting regulatory requirements.

## 2. Contents of a human factors toolbox

The practice of human factors in medical device design often emphasizes scientific validity and reliability of data in human performance through usability testing. In addition, HF methods in healthcare include opportunity definition through observational research techniques and the design of the aesthetic characteristics of a user interface with relation to their function and task. This is accomplished with careful examination of anthropometry, biomechanics and widely known usability design principles, thus reducing risk of injury to users, patients as well as reducing business risk.

Methods that comprise the human factors toolbox for medical device design (Fig. 2.1) include contextual inquiry, task analysis, applied human factors in design (hardware/software device design, design of instructions and training materials), heuristic analysis, cognitive walkthroughs, expert reviews, simulated use studies, use focused risk analysis, root cause analysis, known use error analysis, post-market surveillance, and usability validation studies.

Each method is briefly described below:

### 2.1 Contextual inquiry

Contextual inquiry is the study of people, tasks, procedures, and environments in the "real world." It applies methods from the social sciences (e.g., cultural anthropology/ethnography) to better understand real-world conditions; for example, how devices are

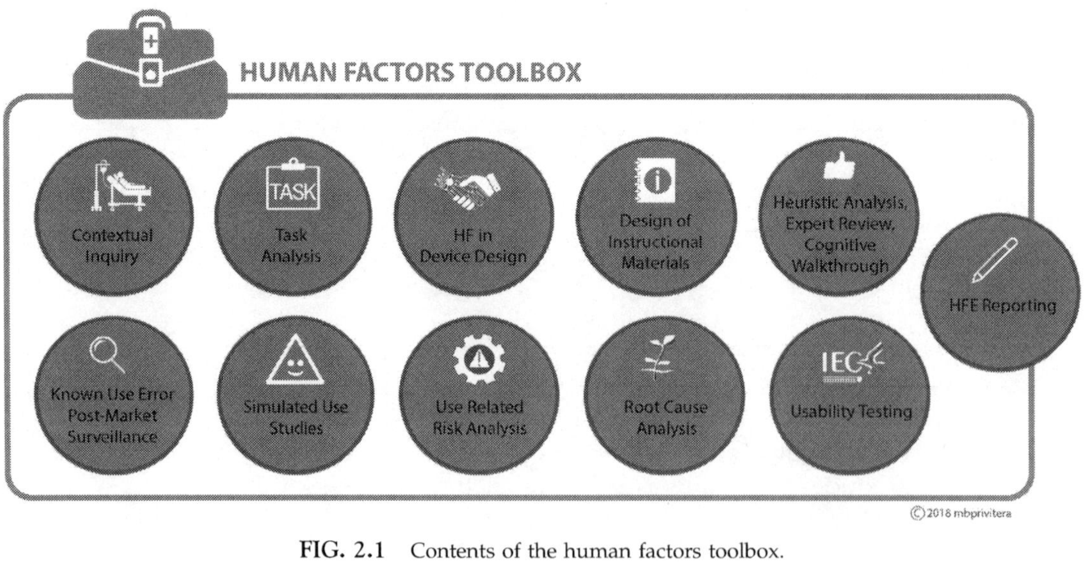


FIG. 2.1 Contents of the human factors toolbox.

actually used. It involves interviews and observations, often supported by video documentation, in the actual environment of use for a given product or system (AAMI TIR 51).

Full details can be found in Chapter 5.

## 2.2 Task analysis

Task analysis determines detailed descriptions of the user-device interaction with sequential and simultaneous manual and intellectual activities of people operating, maintaining, or controlling devices or systems (AAMI, 2009). For the purposes of producing design-related considerations, each step is studied in regards to how a user would likely perform each task.

Full details can be found in Chapter 6.

## 2.3 Applying human factors in design

Human factors data is used to inform the design of the device, instructional materials, and training programs. This involves gathering HF information regarding human capabilities, limitations, and needs (coupled with detailed descriptions of context and environmental considerations—) in order to assure the resulting design is optimized for usability in the context of the user.

A user-centered design process is encouraged by agency HF guidance's and requires a definition of the user, including characteristics such as: gender, age, experience, physical strength, dexterity, limitations, culture, training, and education. It also requires knowledge about the environment and context of the use situation. This information can assist in determining visual, auditory, and tactile constraints. It may also include other factors, such as: vibration, atmospheric pressure, radiant energy, biological contamination, injury risk, electrical/magnetic interference, and cleaning/maintenance/repair needs.

Full details can be found in Chapters 7–9.

## 2.4 Heuristic evaluation, cognitive walk throughs and expert reviews

These benchtop techniques define how experts evaluate the device's user interface against design principles, and identify challenge areas and potential risk for use error. These are often completed by a team of experts.

Heuristic Analysis is an evaluation during which a small set of evaluators examine the user interface against recognized usability principles (the heuristics) (Nielsen, 1995; Nielsen & Molich, 1990; Zhang, Johnson, Patel, Paige, & Kubose, 2003). The process consists of identification of problem areas, after which an assignment of severity is provided.

A cognitive walkthrough evaluates the design of a user interface with attention to whether or not a novice user can easily carry out tasks within a given system (Cognitive walkthrough |Usability.gov, n.d.). It is a method in which people work through representative tasks and ask questions about the tasks as they complete them. It focuses on the tasks, the interface, and the learnability (Blackmon, Polson, Kitajima, & Lewis, 2002; Spencer, 2000).

In expert reviews, a usability interface expert/s inspect a system in order to identify possible issues within the use interface. An expert review expands on heuristic evaluations by assessing the design against known heuristics, and principles of usability, as well as the reviewer's expertise and past experience in the field. The reviewer should have deep knowledge of usability best practices, a large amount of past experience, and not have been involved in creating the design to be reviewed in order to avoid bias.

Full details can be found in Chapter 10.

## 2.5 Simulated use study

This powerful method assesses, at one or more stages during the device development process, a user interface or user interactions with the user interface to identify its strengths and weaknesses and to identify potential use errors that would or could result in harm to patients or users (FDA, 2016, IEC 62366). It requires users performing actual tasks with prototype design/s.

A simulated use study involves testing with representative users completing realistic tasks with a device user interface in a simulated and/or representative use environment. Objective (i.e., performance) and subjective (i.e., information from the user's point of view) data are obtained, analyzed then used to inform design decisions.

Full details can be found in Chapter 11.

## 2.6 Use focused risk analysis/risk management

This is the identification and analysis of use-related hazards. Use focused risk analysis/ risk management is integral to ANSI/AAMI/ISO 14971. Eliminating or reducing design-related problems that contribute to unsafe or ineffective device use is written in to the requirements of the risk management process (FDA, 2016). HF in risk analysis/management are derived from aspects of the user interface that cause the user to fail or adequately or correctly perceive, read, interpret, understand, or act on information from the device which may result in potential harm to the user or patient.

Full details can be found in Chapter 12.

## 2.7 Root cause analysis

In device-user interface testing, this method is used to determine weaknesses in design or information provided as part of a device user interface (e.g. training or IFU). It is a requirement of all HF validation (summative) studies to ascertain the root causes of all use errors and problems that result in harm and to determine the priority for implementing additional risk management measures (FDA, 2016). Root cause analysis includes defining the use error, identifying provisional root causes, analyzing anecdotal evidence, inspecting the device for user interface design flaws, considering other contributing factors, developing a hypothesis, and reporting the results (Wiklund and Dwyer, 2016).

Full details can be found in Chapter 13.

## 2.8 Known use error and post-market surveillance

Known use-related problem identification uncovers use-related problems (if any) that have occurred with similar devices based on their use, the user interface or user interactions (FDA, 2016). When found, these problems should be considered during the design of a new interface, regardless of manufacturer. This process utilizes database-searched, in-house customer complaint logs as well as the knowledge of in-house sales staff familiar with use-related problems.

Post-market surveillance is required by law to monitor the performance of medical devices that have been approved by competent authorities. It may include regulatory vigilance systems, customer knowledge management, and market observation (Zippel & Bohnet-Joschko, 2017). Data from these efforts can be mined for use-related issues.

Full details can be found in Chapter 14.

### 2.9 Human factors engineering (HFE) validation/summative usability study

This type of testing is conducted once the user interface is completed, assesses user interactions with device-user interfaces and identifies use errors that would or could result in serious harm to the patient or user. Human factors validation testing is used to assess the effectiveness of risk management measures. Human factors validation testing represents one portion of design validation (FDA, 2016, IEC 62366).

Full details can be found in Chapters 15 and 16.

### 2.10 Preparing an HFE report for agency submission

This report includes summary information regarding device use safety and effectiveness. The level of detail is sufficient to describe the identification, evaluation, and final assessment of all serious use hazards and is intended to facilitate FDA review (FDA, 2016).

Full details can be found in Chapter 17.

## 3. Purpose of each tool

Safety in the use of medical devices is dependent upon them being used as intended as well as their reliability (MRHA, 2017). Throughout the development process, human factors methods can provide valuable information about: the discovery of new opportunities; advancing device design through utilizing proven HF principles in design; conducting formative usability evaluations; evaluating use related safety, risks, and hazards; and conducting the final use validation study. In order to accomplish this, human factors methods can be broken into two categories: those that assist in generating user interface design and those which are used to evaluate designs.

Generative methods (Fig. 2.2) are typically aimed at the discovery of a new opportunity, challenge areas of existing devices or like devices, and the actual design of the user interface. As Fig. 2.2 illustrates, these include the following: contextual inquiry, known use error/ post-market surveillance for discovery, and applied principles in the design of the device, IFU, and training programs. Task analysis can also be used to generate new workflows as the method illuminates details of product use.

Evaluative methods (Fig. 2.3) are aimed at providing a critique of the use of a device for an intended purpose or task through formative usability testing, determining risk and root cause

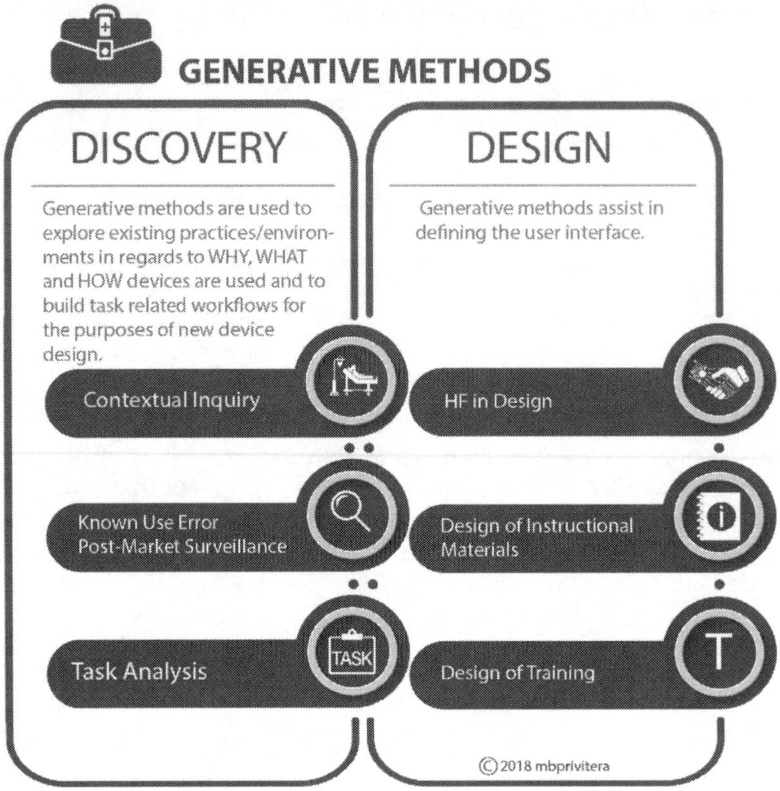


FIG. 2.2 Generative HF methods.

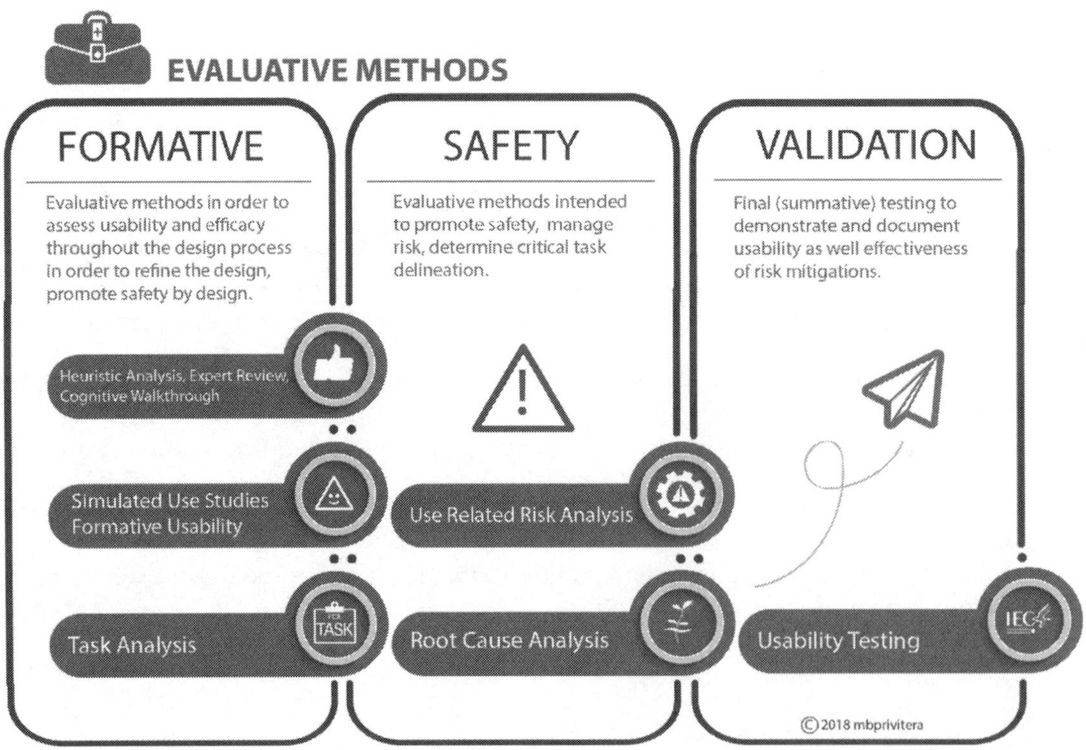


FIG. 2.3 Evaluative HF methods.

of use error as they relate to safety, and in order to provide final usability validation. These methods include: heuristic analysis, cognitive walkthroughs, expert reviews, simulated use studies, task analysis (PCA), risk analysis, root cause analysis, usability validation studies, and post-market surveillance.

In some instances, a method may be used more than once and with varying aims (e.g, the identification of a need for a control vs. detailed design of the control; or, in the case of task analysis as a generative method, to determine steps of use for a novel product, or, as an evaluative method, to determine the perception, cognition and action at each step as an input to risk analysis).

In addition, it is important to note that each method builds on other methods and some methods may be used in combination. For example, a task analysis method may be integrated within a contextual inquiry program in order to achieve a specific goal of understanding use behaviors within a given step in a task. Knowing the output of human factors information needed (e.g., assess usability or inform design), an appropriate method can be placed in a human factors strategy or plan.

## 4. Summary

This chapter provides an overview of common human factors methods used in medical device design, which include: Contextual Inquiry Task Analysis, Applied HF in Design, Heuristic Evaluation, Cognitive Walkthroughs, Expert Review, Simulated Use Studies, Use Focused Risk Analysis/Risk Management, Root Cause Analysis, Known Use Error, Post-Market Surveillance and Usability validation Testing. Each of these methods may be used in combination with others toward the common goal of improving device usability and reducing risk.

## Acknowledgments

Thank you to the HS team, Audra Wright and Catherine Jones of Avanos, for their diligent review; to Elissa Yancey for editing.

## References

AAMI. (2009). *ANSI/AAMI HE75, 2009/(R)2013 Human factors engineering—design of medical devices* (USA).

Blackmon, M. H., Polson, P. G., Kitajima, M., & Lewis, C. (2002). Cognitive walkthrough for the web. Proceedings U6 - Ctx_ver=Z39.88-2004&ctx_enc=info%3Aofi%2Fenc%3AUTF-8&rfr_id=info%3Asid%2Fsummon.Serialssolutions.Com&rft_val_fmt=info%3Aofi%2Ffmt%3Akev%3Amtx%3Abook&rft.Genre=proceeding&rft.Title=Confe. In *Conference on human factors in computing systems*. Retrieved from http://uc.summon.serialssolutions.com/2.0.0/link/0/eLvHCXMwtV1LS8NAEB4ivQgefFStjxJBvEikbpPaHLxYLEIrCFY8hjy2Gs2jkITaf-_sbjYv7cGDlyXswm6YGWa_HWa-AdglVz2t4RPq3XsUGV8r5_5VsTiHqmWFsn9QbrEpTuA3qhhHVDKODfz761VTFuyxqL-IyI_zFjq8no_1bOCk24KZ_FITpQF8n6RqL6Mij2hpB5-y.

Cognitive walkthrough | Usabilitygov. (n.d.). Retrieved October 21, 2018, from https://www.usability.gov/what-and-why/glossary/cognitive-walkthrough.html.

FDA. (2016). *Applying human factors and usability engineering to medical devices guidance for industry and food and drug administration staff preface public comment*. Retrieved from http://www.regulations.gov.

MRHA. (2017). *UK notified bodies for medical devices*. GOV.UK. Retrieved from https://www.gov.uk/government/publications/medical-devices-uk-notified-bodies/uk-notified-bodies-for-medical-devices.

Nielsen, J. (1995). *How to conduct a heuristic evaluation*. Nielson Norman GroupNorman.

Nielsen, J., & Molich, R. (1990). Heuristic evaluation of user interfaces. In *Proceedings of the SIGCHI conference on Human factors in computing systems* (pp. 249–256). https://doi.org/10.1145/97243.97281.

Spencer, R. (2000). The streamlined cognitive walkthrough method, working around social constraints encountered in a software development company. In *Proceedings of the SIGCHI conference on Human factors in computing systems - CHI '00*. https://doi.org/10.1145/332040.332456.

Wiklund, & Dwyer, D. (2016). *Medical device use error cause root analysis*. CRC Press.

Zhang, J., Johnson, T. R., Patel, V. L., Paige, D. L., & Kubose, T. (2003). Using usability heuristics to evaluate patient safety of medical devices. *Journal of Biomedical Informatics, 36*(1–2), 23–30. https://doi.org/10.1016/S1532-0464(03)00060-1.

Zippel, C., & Bohnet-Joschko, S. (2017). Post market surveillance in the German medical device sector — current state and future perspectives. *Health Policy*. https://doi.org/10.1016/j.healthpol.2017.06.005.

CHAPTER

# 3

# Strategy, planning, documentation & traceability for human factors

*Mary Beth Privitera*

**HS Design, Gladstone, NJ, United States**

OUTLINE



*If it's not written down, it didn't happen.*

## 1. Introduction

All classes of medical devices and device-drug combination products are required to have a usability engineering file for the purposes of agency approval (IEC, 2007; FDA, 2016). Conveniently, all regulating agency bodies are closely aligned in their respective guidance and international standards. This makes the job of putting together an HF dossier for multiple global submissions a little less burdensome.

It's no surprise that strategy, planning, documentation, and traceability for human factors is of great importance to device development because the goals of human factors in

Applied Human Factors in Medical Device Design
https://doi.org/10.1016/B978-0-12-816163-0.00003-7

healthcare are to: (1) support the cognitive and physical work of healthcare, and (2) promote high-quality, safe care for patients (Russ et al., 2013). This chapter presents best practices for developing a human factors strategy, developing a plan of action for HF, integrating HF within design control, documenting HF efforts, and ultimately enabling traceability.

## 2. Developing a human factors strategy

Human factors addresses problems by modifying the design of a system to better aid people (Russ et al., 2013). As such, considering developing a human factors strategy early in the design process will be beneficial to users. A human factors strategy includes the ethos and methods needed to meet specific usability goals. Best practices include an integrated approach wherein human factors methods were applied in conjunction with the conception and ideation phases in design development, then verified through formative usability tests, refining the user interface as necessary and finally include validation with near production devices or prototypes which represent final, production-quality design.

In some instances, the step of developing a strategy is either not completed or it is accomplished by non-human factors personnel, which can lead to poor results. The reasons for this are varied; however, anecdotal evidence suggests that manufacturers often believe only a minimum of HF input is required based upon what they believe to be the agency requirements. Manufacturers then provide a rationalization that doing more is not necessary because they, in fact, "know their users." In instances where manufacturers lack a HF analysis procedure, they may rely on hearsay without an optimized or documented approach. This becomes problematic in developing the required HF submission documentation for agency review. Lastly, personnel without additional education in HF, may not ask the right questions necessary to determine appropriate human factors strategies required to optimize the device interface. Thus, developing a HF strategy at the onset of a device program assures that all User Needs are met and that the manufacturer has a firm commitment to quality from the users' viewpoint.

In developing a HF strategy, it is important to assess the complexity of the device as well as that of predicate devices. A user study, at the onset, may not be required in order to optimize usability in the design process. For example, design guidelines (ANSI/AAMI HE 75, 2009) can be referenced and the design principles implemented, or a Heuristic Analysis/ Cognitive Walkthrough (See Chapter 10) could be used. The decision to involve users frequently or infrequently depends upon the following (ANSI/AAMI HE 75, 2009):

- The overall complexity of the user interface. The more complex the device, the more users should be involved in evaluating the user interface design at an adequate frequency leading up to design validation.
- If the intended user/s include broader user groups, then this equates to additional formative studies.
- If the device includes a new technology that is unknown to users, then it may require several studies to optimize usability.
- If the device requires a procedure change that impacts a tenet behavior to be modified, more formative usability studies may be necessary in order to lessen the burden of behavior modification.
- User studies are always subject to budget requirements and require careful delineation in regards to purpose and impact of a formative study.

- Like budget, timing is almost always critical. In order to get the correct evaluation with meaningful results that optimize usability, formative studies require consideration of the time required to execute. This includes time spent recruiting, building prototypes or materials with which to test, as well as time for data analysis and final design recommendation preparation.
- Access to users may also determine the number and frequency of formative studies. While some manufacturers have clinical staff on contract and in alliance, other smaller organizations may struggle to gain the necessary access to users.

In summary, assessing accurately what requirements need to be met that will ensure optimal use safety is key to developing a successful HF strategy, as opposed to blindly applying HF methods. Teams can start by examining assessments of similar devices and known use problems (See Chapter 14). Once those are known, then it is possible to assess whether performing extensive human factors activities will improve use scenarios and final validation studies. Using standards such as ANSI/AAMI HE 75 to inform design can provide a strong background for applied HF. Questions regarding the number and type of formative evaluations should be discussed with a focus on meeting the use expectations of stakeholders. Lastly, it is imperative to understand the regulatory requirements. For example, Class I devices may not have any critical tasks associated with them and therefore may require less HF efforts.

### The relationship of design validation and human factors validation

Among HF professionals there is some debate about the difference between design validation and use validation. Both are required with initial production units and in specific environmental conditions replicating actual use. Validation tests are required to demonstrate the medical device functions as expected and meets the User Needs. They may include inspection and analysis as part of validation in addition to testing. To meet this goal, in design validation there is a specific marking of pass/fail which is traceable to a user need or user dependent risk control. The test must provide objective evidence that is specific enough to be used for validation purposes and in determining conclusions.

The differences of design validation and HF (use) validation include the following:

- Testing methodology: HF always requires the User to be involved, whereas design validation may not.
- Acceptance criteria: HF validation does not have pass/fail criteria per se, rather it assesses critical tasks as determined by risk and the level to which a User was able to accomplish the task without assistance. Design validation is pass/fail testing.
- Disposition of findings: HF validation is typically maintained in separate files from design validation. The findings of a Use Validation Study (Summative Usability) are required as an integral part of the HFE dossier to be submitted to the FDA. Design validation documents are not part of the HFE submission and as such should not be included.

Both activities are part of the validation process. However, a formative study toward the end of the design process may be used to validate use requirements and the effectiveness of risk control. In contrast, a HF validation study (study) must be performed on a production equivalent device, post design freeze, at the end of development.

## 2.1 Considering previous knowledge

Agency guidance (FDA, 2016) indicates the usefulness of identifying known use errors and issues prior to the beginning of product design. This effort is especially helpful on product line extensions or when designing a competitive device. It can assist in providing an understanding for the range of harm associated with use error for 'like' products as a result of use error. This requires assessing the intended user interface design related to consistencies and differences between existing devices and a new product concept. By conducting a brief search, the design team may be able to incorporate safety through design and incorporate protective measures in order to reduce the probability of occurrence or harm associated with use error.

This knowledge, along with a complete understanding of User Needs, is fundamental toward developing a strategy of applying human factors in medical device design. In order to understand User Needs, it is important to define the continuum of users and environment/s: this includes defining all people (stakeholders/users), places (use environment/s), tasks and touchpoints along with the ability (human performance) of all users involved. In order to assist building this knowledge, a wide range of human factors methods (e.g., contextual inquiry, known use error analysis) can be applied, depending on the level of confidence the design team has about previous knowledge.

Further, device companies typically have established a wealth of information regarding their customers, their devices and device uses (Privitera, 2015). Typically marketing departments will have data available which can help define the User and User Needs. There may be scientists, clinical specialists, or technologists who may be able to offer initial insights regarding the product concept that can help determine the overall human factors strategy.

Another possibility is to conduct a literature review of published clinical trials and research (Privitera, 2015). Journal articles, research papers, other publications (e.g., books, product manuals, procedure protocols), and available demographic and usage data may assist in developing a foundation of knowledge (Privitera, 2015).

### Marketing studies are not human factors studies

Marketing studies are extremely helpful for the application of human factors; however, there are distinct differences (Fig. 3.1).

Market research includes gathering, analyzing, and interpreting information regarding a product or service to be offered for sale for an intended market. It may concern previous, current, or future customers and their characteristics, opinions, location, size of the opportunity, or competitors within the market. Market research provides data relevant to the challenges the business will face as an integral part of business planning purposes. The research methods incorporated in order to define the above details is similar to some types of human factors studies; however, marketing studies tend to be broader in nature.

Human factors studies always focus on usability; however, they may include the reactions, opinions, and preferences of a variety of stakeholders in relation to a stimuli or concept variance. The purpose of a human factors studies in medical device design is entirely dependent on the objective relative to the stage in the product development process. For example, a common element in human factors' studies is examining how humans behave physically and psychologically within a specific environment and/or with a specific product. In assessing the ways and means of reducing error, increasing efficiency, and improving the user experience, human factors studies provide relevant design information or validation. The research may be conducted informally or formally, and the direct participation of users in human factors studies is not always required.

Finally, human factors studies are intended to provide depth of understanding and empathy of the use experience. They require depth within challenging areas (root cause analysis) in order to mitigate risks through design and ultimately improve the customer experience. As a result, they increase value to the customer, result in fewer CAPA's (corrective and preventative action), and lead to more efficient uses and mastery of the market.

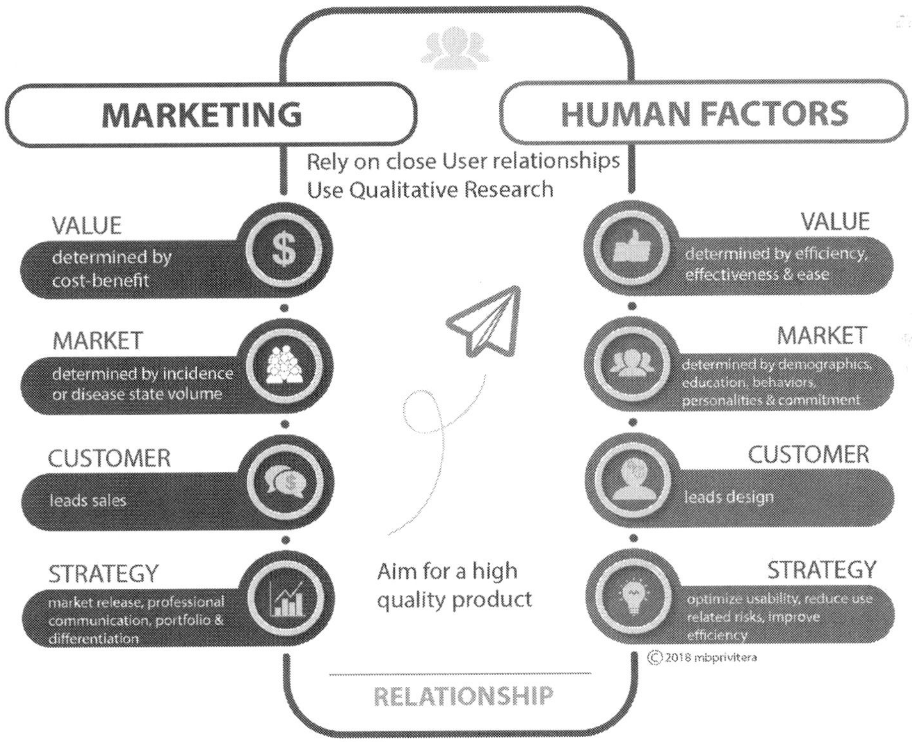


FIG. 3.1 Relationship of marketing and human factors studies.

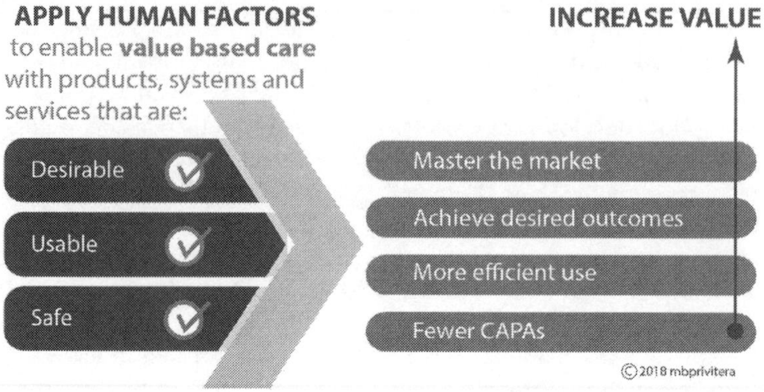


**FIG. 3.2** Apply human factors to increase value. *Adapted from Torenvliet (2018).*

## 2.2 Considering risk

While it is impossible to uncover all possible use errors in a risk analysis, a realistic approach of capturing errors which are reasonably foreseeable with an expectation of rationale and prioritization of risk mitigations is expected by agency reviewers (ANSI/AAMI 14971, 2010; FDA, 2016).

Best practices call for starting use error risk control when evaluating concepts in order to move from concept phase to full product development. If there is no effort regarding usability and the application of HF, it might be worth starting, since the goal of human factors is to mitigate as much risk as possible through product design rather than through labeling, instructions for use, quick reference guides, or training. Informing design early can save valuable time and resources.

In summary, applying human factors early will minimize risk and maximize opportunity (Fig. 3.2).

## 2.3 Identifying HF activities

The table below highlights human factors activities which are typically used in medical device development. The purpose of this table is to serve as a quick reference guide that can be used when planning in more detail and accounting for resource allocation and budget considerations.

| Human factors activity | Definition | Purpose |
|---|---|---|
| Chapter 5 | Contextual inquiry is the study of people, tasks, procedures, and environments in the "real world." It applies methods from the | This process is used to identify opportunities for device design improvements, process improvements and identify key detail design |

| Human factors activity | Definition | Purpose |
|---|---|---|
| | social sciences (e.g., cultural anthropology/ ethnography) to better understand real-world conditions; for example, how devices are actually used. It involves interviews and observations, often supported by video documentation, in the actual environment of use for a given product or system (AAMI TIR 51) | requirements throughout the design process. This is considered an empirical method used to inform the design and should be included in the Human Factors Report as a reference. |
| Chapter 6 | An analysis to determine the user-device interaction systematically into discrete steps. Once completed, each step is studied based on how a user would likely perform each task. | This process is used to determine use flow and usability. It is fundamental to all HF programs. This is considered a user interface evaluation technique and can be referenced in the Human Factors Report. |
| Chapter 10 | These are benchtop techniques wherein experts evaluate the device's user interface against validated heuristics- design principles, identify challenge areas, and potential risk for use error. These are often completed by a team of experts. | This process is used to evaluate a novel design's overall usability. While not required by agencies per se, it is a recommended best practice. This is considered a formative usability test. |
| Chapter 12 | This is the preliminary identification and analysis of use-related hazards and is integral to ANSI/AAMI/ISO 14971. It is required to delineate critical and essential use tasks. | This process determines task delineation according to risk. It is required for agency approval. |
| Chapter 11 | This powerful method assesses, at one or more stages during the device development process, a user interface or user interactions with the user interface to identify the strengths and weaknesses as well as to identify potential use errors that would or could result in harm to the patient or user (FDA, 2016, IEC 62366). It requires users performing actual tasks with prototype design/s. | This process requires prototypes and typically walks a user through a device use for the purposes of assessing usability. This is considered a formative usability test. |
| Chapter 13 | The process includes defining the use error, identify provisional root causes, analyzing anecdotal evidence, inspecting the device for user interface design flaws, considering other contributing factors, developing a hypothesis, and reporting the results (Wiklund et al. 2016). | This process determines why a use error may have occurred. It is required for agency approval. |
| Chapter 14 | This process uncovers use-related problems (if any) that have occurred with similar devices. It utilizes database and in-house customer complaint logs. | This process determines known areas of challenge. It is required for agency approval. |

(*Continued*)

| Human factors activity | Definition | Purpose |
|---|---|---|
| Chapter 15 | This testing is conducted at the end of the device development process to assess user interactions with a device user interface to identify use errors that would or could result in serious harm to the patient or user. Human factors validation testing is also used to assess the effectiveness of risk management measures. Human factors validation testing represents one portion of design validation. (FDA, 2016, IEC 62366) | Protocol, data, and report are required to meet Human Factors Report Requirements for FDA approval submission. |
| Chapter 17 | This report includes summary information regarding device use safety and effectiveness. The level of detail is sufficient to describe the identification, evaluation, and final assessment of all serious use hazards and is intended to facilitate FDA review (FDA, 2016). | The purpose of writing a human factors report is to meet agency guidance as a formal record of human factors activities. |

## 2.4 Considering budget

There is a common misconception that conducting a human factors study is expensive. In reality, it does not have to be. Studies can and should be scaled according to their intended purposes and confidence level. For example, if the area of inquiry is known—such as uncovering unmet needs (open innovation) versus iterative design evolution (next generation)—the scope of a study can be tailored in scale/cost.

When applying human factors, if the ethos is simply a box-checking exercise in order to seemingly fulfill agency requirements, then typically the costs are increased because the risks of use error are often higher and may result in increased need for customer service, training, or increased risk of CAPA's.

The value of a successful HF program is derived by uncovering unmet needs, refining the product design for optimized usability, clearly defining the design inputs, and ensuring design outputs meet the market opportunity.

The human factors effort involved should include, at a minimum, a plan for an expert review, a use risk analysis, at least one if not more formative studies, and summative usability study. Consider the overall market size, and if the device is intended for broad consumer use, it warrants more preliminary studies and broader participation in user studies. If the device manufacturer is planning a product line extension, often HF is overlooked and not budgeted. If there is no change to the user interface from the existing approved device, then there should be no problem. However, a budget should be considered for resources to assess the need for any human factor's application.

Listed below (Table 3.1) are cost ranges. Please note these are only intended to provide a reference and will vary depending upon the supplier, as well as the complexity and/or the uniqueness of the device.

TABLE 3.1 Cost estimates for human factors activities.

| Human factors method | Cost range | Considerations |
|---|---|---|
| Contextual inquiry | $50,000–$250,000+ | # of site visits, costs associated with preparation, recruiting, travel, access, honoraria |
| Task analysis | $5,000–$15,000+ | # of iterations/revisions, level of detail required, complexity of device |
| HF in design | $5,000–$15,000+ | Uniqueness of user interface, complexity, # of iterations or variations |
| Formative studies | $5,000–$75,000+ | User involvement, # of formative studies, costs relating to preparation, recruiting, location, travel, honoraria |
| Use risk analysis | $5,000–$40,000+ | # of iterations/revisions, level of detail required, complexity of device |
| Known use error analysis | $2,500–$60,000+ | # of incidents reported, # of varying devices inquired, complexity of findings |
| Summative study | $60,000–$250,00+ | # of distinct user groups, costs associated with preparation, recruiting, location, travel, honoraria, residual risk analysis |
| Agency submission preparation | $5,000–$25,000+ | Human factors documentation completeness, device complexity, human factors application complexity |

## 2.5 Developing the human factors report or usability engineering file along the way

As a best practice, human factors activities should be included within design control, including a procedure that contains the required level of documentation per phase leading up to the development of the package for agency submission.

Successful human factors engineering (HFE) or usability engineering (UE) analysis includes the following three steps:

- The identification of use-related hazards and unanticipated use-related hazards derived through preliminary evaluations and the determination of the degree of hazard and frequency of these situations
- The development and application of measures to eliminate or reduce use-related hazards that could result in harm
- A demonstration of the final user interface design which supports safe and effective use by conducting human factors testing

In order to expedite the development of a human factors report, plan on maintaining a reference to all activities throughout the development process. In some instances, human factors professionals will start the submission documentation at the onset of a program and continue to write the submission file at key milestones. For example, at the end of each formative, write a paragraph regarding the subject of the test, its outcome and any recommended design modification(s). This greatly reduces the documentation burden at the end of a development program.

**A human factors strategy and plan should contain the following:**

1. An Introduction-describing the purpose (including ethos, e.g., Keep it Simple), the scope, all human factors guidance, standards, and additional references expected to be used throughout the application of HF.
2. Description of the Organization involved in HF. This includes describing the roles, responsibilities, all key stakeholders, the sign-off matrix, as well as any relationships outside the organization involved in HF activities.
3. Description of any Preliminary Studies, such as Contextual Inquiry, Post-Market surveillance, literature reviews, Competitive Use Analysis, or Task Analysis, that may have already occurred or are required.
4. Description of Risk Management, including a discussion of the risks and benefits to the user with HF as an integral part of risk reduction.
5. Description of the role HF plays in the determination of the User Interface Design. This includes a description of influential aspects and key areas requiring the application of HF in design.
6. Usability Testing, which will be required throughout the design verification phase (formative usability) and for the final validation study (summative study).
7. A description of planned UI/UX Reviews, which may include heuristic analysis, cognitive walkthroughs, or expert reviews.
8. A description of Technical Design Reviews as an integral part of the design control process, anticipated integration of HF activities with regular checks.
9. Deliverable schedule and resource allocation, including target dates, responsible parties for each activity with a description of the anticipated deliverable, and required resources (personnel and materials).

## 3. Importance of documenting HF

For successful regulatory submissions, there is no other choice: human factors efforts must be documented.

"Documenting your risk management, HFE/UE testing, and design optimization processes (e.g., in your design history file as part of your design controls) provides evidence that you considered the needs of the intended users in the design of your new device and determined that the device is safe and effective for the intended users, uses and use environments" (FDA, 2016, p. 29).

Additionally, the MRHA indicates that Usability Engineering Files should be kept clear and concise. Documentation should be prepared with the reviewer in mind, to make it as easy as possible for them to access all the information they require (MRHA, 2017).

The human factors process should be recorded in a usability engineering file of the device technical documentation. Depending on the risk classification of the device, the file may be requested for review by regulatory bodies in order to understand how the process has been conducted and whether their particular use scenario has been considered. Simply stating "compliance with IEC 62366" is not sufficient without supporting evidence.

While the FDA encourages a separate human factors file where critical tasks are carefully analyzed, IEC 62366-1 supports that the usability engineering file can be part of the risk management file. As such, the standard further describes that the usability engineering file does not need to physically contain all the documents produced by usability engineering activities (see IEC 62366-1 definition 3.18). However, it should contain their references to the required documentation.

Regardless of agency, human factors activities are required for initial device approval and for approval of devices with modifications to the user interface, evidential documentation may be requested.

## 3.1 Incorporating human factors in design control

The purpose of AAMI TIR 59 is to describe the integration of human factors engineering/usability engineering into a manufacturer's design control process and quality system in accordance with Code of Federal Regulations (CFR) Title 21 Section 820.30 (AAMI Human Engineering, 2017).

In execution, large organizations which have a robust/defined Design Control System may have difficulty implementing the changes necessary to provide an integrative approach; however, in doing so, human factors are considered throughout all development phases and provide a single point of reference for the design history.

In order to expedite design control implementation, if there is limited amount of time, it is easiest to search for an expert who can 'proceduralize' staff members and provide them an example project. Alternatively, consider hiring an expert on staff who can help the company develop human factors procedures, and then train/hire a staff for implementation.

There is overlap functionality with HFE processes and design control activities (AAMI Human Engineering, 2017). Changing an existing Quality System to adopt comprehensive HFE processes might be disruptive initially. A phased transition plan may be more sustainable; however, care must be taken in order to prevent redundancy. Integrating HFE processes at one time allows for more efficient training and reduction of commercial risks related to redundant or inconsistent information and insufficient HFE inputs (AAMI Human Engineering, 2017).

# 4. Providing traceability

"On the traceability side, I think that that it is always helpful for understanding; especially if you're trying to map out how those use errors, and risks, and user needs were identified, kind of how they were carried through the process, mitigated, and finally tested and validated." Shannon Hoste, FDA Webinar.

"Traceability is necessary to demonstrate that the Usability Engineering Process has been applied (ANSI/AAMI/IEC, 2015, p. 21)"

Best practices are to enable digital traceability with a link other documents that assisted in the product development effort. An ideal option would be to build a trace matrix using Excel and available through both digital and paper documentation. However, these are sometimes very difficult to maintain in regards to traceability.

Software such as Greenlight Guru takes an electronic view of product specifications in order to tie user needs to individual elements. With built in checks and balances, it provides one way to minimize the deficiency gap which can occur for an agency submission.

## 5. Summary

A commitment to integrating human factors processes in medical device development is an agency requirement globally. But there are even more reasons why human factors are essential to effective device design. Adopting human factors methodologies assures a user-centered design process, reduces risk, improves quality, and is a fundamental customer requirement as all customers want a device which is "easy to use." This chapter offers an overview of standard HF activities in order to support developing a strategy, plan, and budget with resource allocation for this critically important work.

## 6. Further reading

- AAMI TIR 59
- FDA Human Factors Guidance (2016)
- IEC 62366-1, IEC 62366-2
- MRHA Human Factors and Usability Engineering (2017)

## Acknowledgments

Thank you to Gerard Torenvliet for presentation and discussion on the value of human factors; to Kate Cox for developing integration of HF processes within HS Design; to Elissa Yancey for editing.

## References

AAMI Human Engineering. (2017). *Technical information report AAMI TIR59* .

AAMI HE 75. (2009). *ANSI/AAMI HE 75:2009, Human factors engineering-design of medical devices* (United States) www.aami.org/publications/standards/he75.html.

ANSI/AAMI 14971. (2010). *Medical devices— application of risk management to medical devices*. American National Standard, 2008.

ANSI/AAMI/IEC. (2015). *ANSI/AAMI/IEC 62366-1*.

Commission, I. E. (2007). *IEC 62366:2007, Medical devices- Application of usability engineering to medical devices*. Geneva: Switserland.

FDA. (2016). *Applying human factors and usability engineering to medical devices guidance for industry and Food and drug administration staff preface public comment*. Retrieved from http://www.regulations.gov.

MRHA. (2017). *UK notified bodies for medical devices*. GOV.UK. Retrieved from https://www.gov.uk/government/publications/medical-devices-uk-notified-bodies/uk-notified-bodies-for-medical-devices.

Privitera, M. B. (Ed.). (2015). *Contextual inquiry for medical device design*. Academic Press.

Russ, A. L., Fairbanks, R. J., Karsh, B. T., Militello, L. G., Saleem, J. J., & Wears, R. L. (2013). The science of human factors: Separating fact from fiction. *BMJ Quality and Safety, 22*(10), 802–808. https://doi.org/10.1136/bmjqs-2012-001450.

Torenvliet, G. (2018). Implementing hf into design and development cycle to what stands between you and beer ?. In *4th annual human factors excellence for medical device design conference* (pp. 1–6). Minneapolis, Minn: MarcusEvans.

U.S. Department of Health and Human Services Food & Drug Administration Center for Drug Evaluation and Research. (2016). *Applying human factors and usability engineering to medical devices guidance for industry and Food and drug administration staff* (301).

Wiklund, & Dwyer, D. (2016). *Medical device use error cause root analysis*. CRC Press.

CHAPTER

# 4

# How to use this book

Mary Beth Privitera
HS Design, Gladstone, NJ, United States

OUTLINE



*Human behavior flows from three main sources: desire, emotion, and knowledge.* **Plato.**

## 1. Introduction

This book is intended to provide a comprehensive guide to the application of human factors methodologies in the medical device development process. While there are standards and guidance's which cover agency requirements (IEC 62366, FDA Human Factors Guidance 2016, etc.) and books regarding key methods (Contextual Inquiry and Usability Testing, etc.), there is no resource which consolidates this information and includes the references.

The authors leading the development of this book serve on the AAMI Human Engineering Committee and are AAMI faculty with many years of involvement in the development of medical devices. The methods described are based on industry experiences and are provided in an effort to assist others in the development of safe, useable, and desirable medical devices.

The book also includes a comprehensive description of human factors methodologies with detailed instructions on how to conduct each one. It provides key references to industry standards are guidance as well as descriptions of situations and benefits to utilizing one method over another. This is important as international standards indicate that a developer must incorporate human factors in medical device development; however, they do not indicate a process that outlines how such incorporation should be done.


https://doi.org/10.1016/B978-0-12-816163-0.00004-9

## 2. Who should use this book?

This book should be of interest to a variety of medical device development professionals and others who may be involved in the design process. This includes:

**Human Factors professionals:** those who are responsible for executing human factors tasks within product development.

**Software/Hardware Designers:** individuals who are directly or indirectly responsible for the design of the user interface and who might be responding to the findings of human factors methods in order to implement human factors suggestions in design. This may include industrial designers, graphic or user interface designers or others in associated professions.

**Engineers:** individuals who are responsible for device functionality, including mechanical, electrical, or software engineers.

**Quality Assurance personnel:** individuals who manage organizational initiatives to assure safety, efficacy, and usability while meeting agency standards.

**Regulators:** individuals who work for regulatory review and enforcement entities such as the FDA or notified bodies in the European union.

**Regulatory affairs specialists:** individuals who manage an organization's compliance to human factors standards and prepare submissions for device approval.

**Students:** those who are preparing for careers in the medical device development industry and may be involved in disciplines such as human factors, industrial design, interaction design, or engineering.

## 3. How should this book be used?

The intention of this book is to serve as a reference for the application of human factors in medical device design. It provides a listing and deep understanding of current U.S. and international human factors standards. As such, it should be used in conjunction with medical device standards and agency guidance's in order to provide examples and explain effective methods.

This book aims to provide guidance in all aspects of developing robust HF activities aimed at improving device design and withstanding agency scrutiny in regards to human factors. It promotes and communicates best practices in the application of HF in design and pushes the boundaries of perception regarding HF from seeing the field as a tool of quality/risk analysis to illustrating its true value.

## 4. Limitations

While this book references human factors standards and guidance available, readers are advised to regularly check for updates to assure regulatory requirements are met.

As there is a range of medical device types (e.g. consumer, combination drug delivery, surgical tools, imaging equipment, etc.), so too is there a range of approaches to applying human factors principles and methods. There is no right or wrong way to utilize the information and methods presented in this book, other than not incorporating human factors in medical device design at all. The information presented is based on professional practice and judgements by the authors, and it is anticipated that there will be variability in the execution of human factors processes.

## 5. Disclaimer

Where possible, examples have been provided which are generic in product detail and not attributed to any device/s currently on the market.

Opinions presented reflect those of the authors in their personal capacities and do not represent any employer or organization.

## Reference

*Applying Human Factors and Usability Engineering to Medical Devices Guidance for Industry and Food and Drug Administration Staff Preface Public Comment* (2016). Retrieved from http://www.regulations.gov.

# PART II

# Discovery & input methods

CHAPTER

# 5

# Contextual inquiry methods

*Mary Beth Privitera*[a], *Ian Culverhouse*[b]

[a]HS Design, Gladstone, NJ, United States; [b]Rebus Medical, Bristol, United Kingdom

OUTLINE



*Trying to understand the needs of the user and context in which the design is used will foster a deeper understanding of design function and empathy with the user.* **Joel Katz (2012, p. 125)**


https://doi.org/10.1016/B978-0-12-816163-0.00005-0

# 1. Introduction

Contextual inquiry (CI) is methodology aimed at immersing a design team within the use environment in order to define a deeper level of understanding that can only be gained by direct exposure to users and their environment. The methodology is promoted as best practice by FDA (2016) and AAMI TIR 51 (2014). CI is also cited in the International IEC standard for Usability Engineering for Medical Devices, 62366-1 (ANSI/AAMI/IEC, 2015) and the 2018 MHRA HF Guidance released in the UK (MRHA, 2017).

# 2. What is contextual inquiry (CI)?

An immersive process involving a small team at the site of device use, observing and video recording procedures in order to identify areas of opportunity for improved usability e.g. location of controls, feedback (or lack thereof), and overall user interaction with devices. The results are analyzed by human factors specialists/device designers and intended to inform new product development. They are not published, nor are they used to determine efficacy, rather they are used to identify what changes could be made that would improve overall ease of use from the users' perspective.

There are two types of CI studies:

1. Broad studies aimed at identifying new product development strategy
2. Specific studies aimed at identifying new product use requirements, user descriptions and use context/environment information.

Beyer and Holzblatt (1999), originally described contextual research approach wherein a seemingly casual to the study participants but involving rigorous data analysis and robust determination of the social and physical environments of the workplace through the use of tools that dissect individual task and user behaviors. The specific focus placed on human-computer interaction highlighting unique user challenges through the process of observation and interview.

As far back as the 1970's the design industry started to hire ethnographers (Koshinen et al., 2011) in order to understand the micro-cultures or user behaviors relevant to a specific product design within their use context. Contextual inquiry is a form of ethnography and rooted in social science. Both have similar methods and processes of analysis however, ethnography is centered on developing an understanding whereas CI is aimed at informing design (Fig. 5.1). This method is widely used in consumer product design, service design, and industrial system design. Organizations ranging from global corporates including Starbucks through to heavy industry such as nuclear power stations have all benefited from applying contextual inquiry methods to their design and innovation strategies.

## 2.1 Purpose and rationale

According the FDA Human Factors Guidance (2016) contextual inquiry studies are considered best practices in order to inform the entire human factors dossier. This type of research

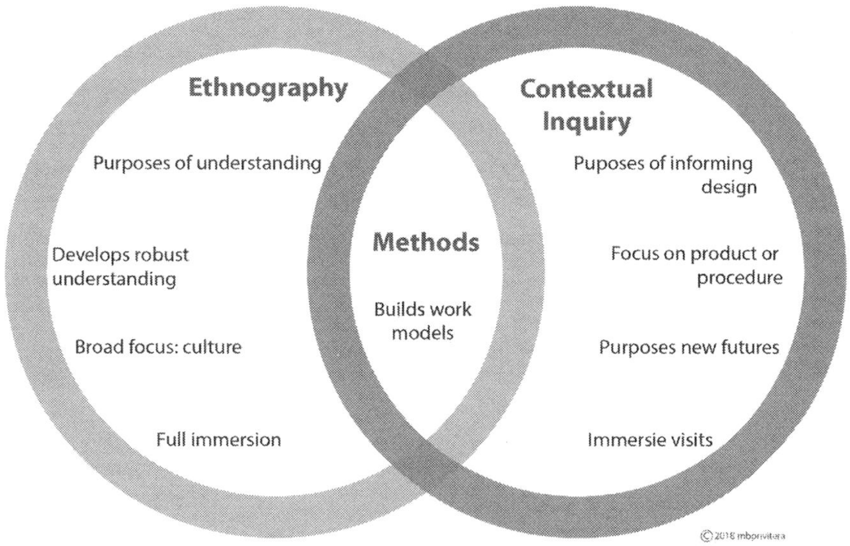


FIG. 5.1 Ethnography and contextual inquiry relationship.

can inform all aspects of design, all subsequent human factors analytical techniques as well as usability studies.

CI provides information about the constraints that a new device or system must operate within while providing additional information concerning unmet user needs. The process can uncover and identify problems with existing devices and systems that the new device can address. The information obtained in a CI study can be used to develop a robust task analysis (discussed in Chapter 6).

## 2.2 What information is yielded from a CI study?

The information gathered from a CI study is qualitative and rich with detail. In conducting a thorough study, device developers can expect to learn the following:

- Complete description of use environment and any changes throughout a procedure
- Description of the user, including demographics and training
- Challenges and mitigations with demonstrable evidence
- Insights for design and product development strategy
- Insights for clinical communication
- Insights into cultural bias
- Understanding of broader motivations for actions and decisions external to the device/ system user interface

Fig. 5.2 lists attributes of discovery that can be derived regarding the user, environment and task. Dependent upon the goal of the study, information regarding each one of these attributes can be more or less explored.

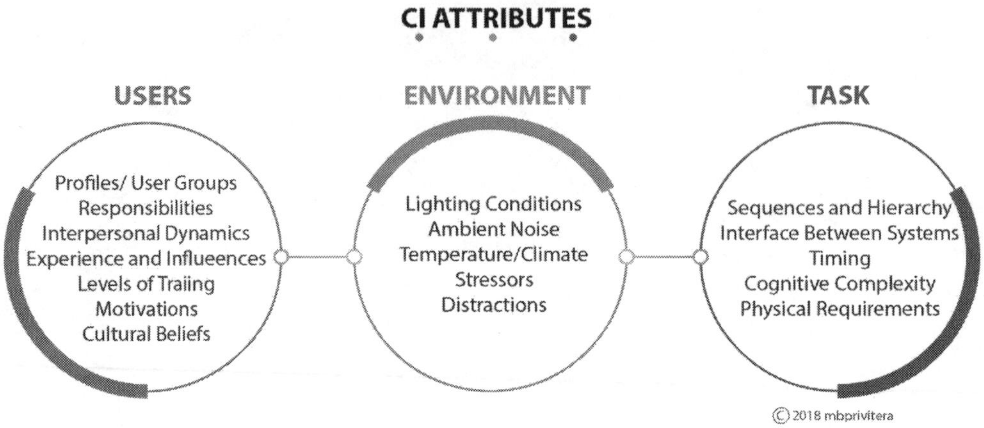


**FIG. 5.2** CI Attributes: potential areas of discovery.

## 2.3 Uses of CI in medical device development

As previously mentioned, there are two types of CI studies commonly used in medical device development; broad studies aimed at exploring new opportunities and focused studies aimed at exploring a specific device for the purposes of in-depth understanding of design improvements which enhance value at the viewpoint of the user.

Regardless of study type, Table 5.1 below highlights the impact of a robust CI study for each phase of product development including the development of an HFE report.

CI can assist in developing a business strategy regarding a new device by determining the most significant challenges and opportunities for improvement that are meaningful for users. These studies can assess perceived value through the discovery of culture in the workplace. In addition, and at the heart of CI studies, initial usability objectives can be determined through the in-depth study of behaviors. CI studies can be used to determine fundamental disciplinary tenets of healthcare providers or of a patient user group which should be considered in the product development process.

An agile process refers to an iterative process where requirements and solutions evolve through collaboration between cross-functional teams. The lean model of learn-build-measure (Fig. 5.3) is an example which aligns clinical and technical requirements with the value proposition. A CI study conducted early in the design process can assist all other research areas and assure user needs are identified.

# 3. Process

The process of conducting a CI study for medical device development focuses on the "why" and the "what" while in the field and then through analysis/interpretation the

TABLE 5.1 Impact of CI studies for each product development phase.

| Phase | Title | *CI impact* within each phase |
|---|---|---|
| 0 | Exploration | *Define new opportunities, define user behaviors, define use environment, define social structure around device use* |
| 1 | User needs | *Develop fundamental design requirements in the words of the user based on their values.* The market potential, use, intended user and use environment are fully described and the technical requirements are included. A need statement is a concise description of a goal (what the device needs to do) but does not indicate how to achieve the goal. It can also include qualitative targets and use descriptions. |
| 2 | Design input | *Inform storyboards of device, identify partners for preliminary formative evaluations, and other techniques which further design exploration and definition.*<br>During this phase best practices suggest that formal design verification and risk analysis begins with emphasis on preliminary testing of conceptual designs and identification of potential application risks that should be mitigated through design. |
| 3 | Detail design | *Inform design regarding desired user interface, inform risk analysis, identify partners for further formative evaluations* as the design is finalized into dimensional concepts.<br>Manufacturing considerations and further risk and usability assessments take hold. |
| 4 | Design output | *Provide root traceability of human factors* application at the onset of the program while conducting studies focused on assuring the design is robust<br>Preparing for regulatory submission and manufacturing pilots |
| 5 | Medical device | *Improve usability* and reduce likelihood of use errors, reduce issues submitted to post-market surveillance.<br>One important to note is the additional requirement of post market surveillance by regulating agencies in order to provide real-time assessment of the risks and benefits of a medical device. This requirement further emphasizes the need to focus early design efforts with the user in mind |
| Agency submission | HFE File | According to FDA Guidance (2016) the following sections can be determined from a CI study: *Section 2 Description of intended device users, uses, use environments, and training; Section 4 Summary of lnown use problems;* and included in *Section 6 Summary of preliminary analyses and evaluations* |

"so what?" that drives medical device design can be determined. The method is flexible adaptable however there are common steps for conducting a study. These include:

1. Determine study goals
2. Review existing research, literature, device use
3. Plan participants, # of site visits, # of users involved
4. Write study protocol
5. Determine need for IRB/Ethical approval, seek if required
6. Recruit participants and sites
7. Collect data in field
8. Upload data into system
9. Analyze data using annotations
10. Run specific queries
11. Generate report
12. Review and sign off

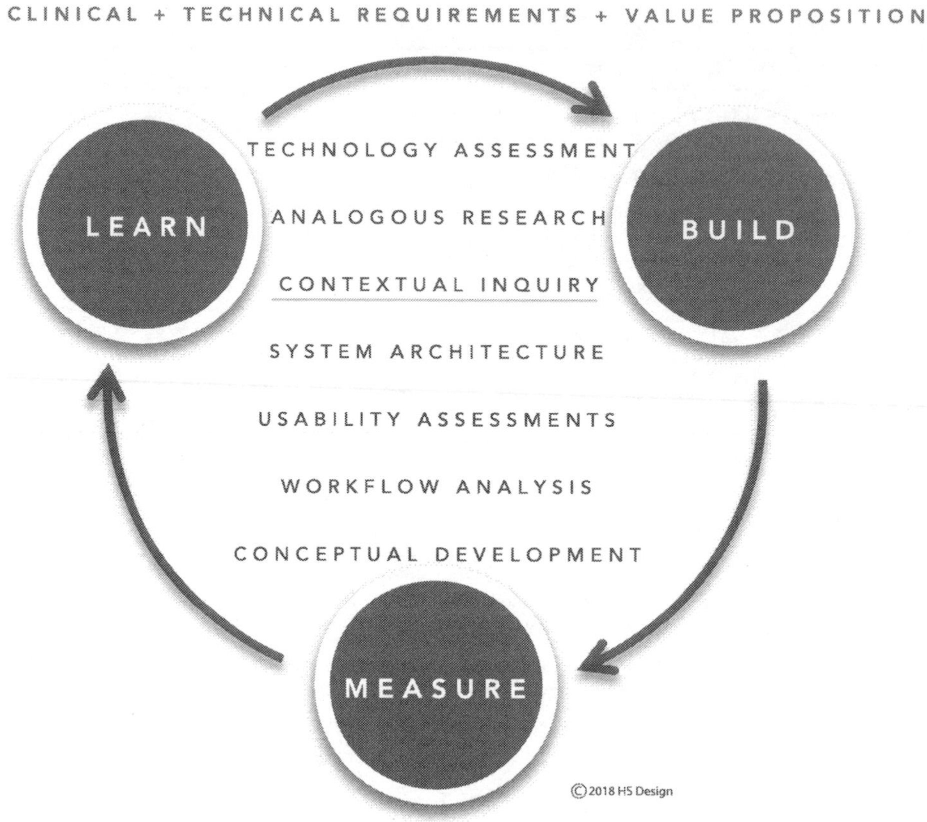


**FIG. 5.3** Contextual Inquiry in an agile process. *Used with permission from HS Design.*

Unlike an HF Validation study which typically follows an agreed upon format and methodology, CI studies will typically vary greatly in design, complexity and scale. CI studies are costly and time consuming to undertake, often requiring a multi continent perspective if the device(s) under investigation are intended for a global market. It is often the case that CI studies are scaled to meet the individual constraints of a particular project.

Vulnerable patient populations and user groups can sometimes make it impractical or impossible to structure a traditional approach to CI which also meets commercial budgetary constraints. An inspiring example of overcoming such challenges involving rare disease patients was demonstrated by (Bajars, Larson-Wakeman, Matchstick, & Biondi, 2016). Bajars et al. wished to gain a deeper understanding into the contextual factors influencin user needs related to patients with Cystic Fybrosis (CF). CF patients need to exercise extreme caution when coming into contact with others with CF due to the risk of cross infection between patients. Healthcare professionsal and caregivers working with CF patients also need to adhere to strict infection control protocols to avoid placing patients at risk. Notwithstanding the aforementioned risks of cross contamination, it is estimated that there are only 100,000 patients worldwide diagnosed with the disease ("Cystic Fibrosis Trust - What is cystic fibrosis?," 2018). To address the logistical challenges of working with this patient population, the team deployed a "remote contextual inquiry" approach to facilitate engagement with CF patients. Using an online video sharing platform over a longitudinal study, the team were able to learn about the needs of CF patients in relation to a novel Mobile Medical Application (MMA) designed to improve patient adherence.

## 4. Best practices

The essence of CI is to include focused observation and possibly having a conversation while a user is performing a task of interest, gathering artifacts, taking field notes, while conversing with the user in a seemingly informal manner (Privitera, 2015). At first glance, it may seem relatively simple to watch and talk to people as they use a medical device however unless the conversation and viewpoints are targeted and planned, data analysis becomes increasingly more difficult.

There are four approaches to use observations that are typically used in CI studies. These include the following: overt observation, think aloud approach, use simulation with reflection and a role-playing approach. Each of these is briefly defined below:

1. Overt observation: the research team is invited into a device use and with consent silently observes the user as they interact with a target device or procedure. All parties involved in the research and the use are aware that research is being conducted. This approach is useful when task being carried out by the users require concentration or is warranted by the situation. This is a 'fly on the wall' approach with little interaction between the research team and the user.
2. Think aloud approach: consists of asking a user to describe their actions and reasoning while they solving a problem or completing a task. It is a common and preferred approach for CI studies as it encourages the user to explain device use/procedure as they perform it in real time. This approach reveals aspects of opinions and behaviors that would be difficult or impossible to discern with observation alone. Using this approach, the research team can have concurrent discussions on what is happening or retrospective discussions on challenges at specific points in the device use which lead to new behaviors or compensatory behaviors.
3. Use simulation with reflection: involves the user demonstrating their processes, techniques and tool uses in the environment or simulated environment while the researcher ask questions and seeks explanation. This approach is best at the onset of a CI study in order to further education the research team, if there are challenges in recruitment or difficulties in observing in 'real life.' It is important to note that realistic conditions should be considered i.e. users should wear gloves, have wet hands, etc if that would be experienced. Additionally, a sense of urgency and/or critical life measures may not be present during simulated care.
4. Role-playing approach: involves direct interaction of the researcher as the user and potentially the device. In this manner, the researcher can experience the device use from the perspective of the provider or patient. The focus of this approach is to enable the researcher to encounter challenges under the guidance of the expert user.

Interviews conducted as part of a CI study are nearly always conducted as a semi-structured conversation. In this approach, each party has the freedom to explore new topics and dive deeper into specific topics that are viewed as important by either the research team or the user. This approach differs from a formal structured interview techniques wherein the researcher moves from one question onto the next without variance. In CI studies, the research team typically has a research field guide which assures consistency of inquiry throughout a site visit and assures efficient data analysis.

Another effective interview technique is a narrative interview approach wherein participants are asked to describe situations or use storytelling in order to describe experiences. This requires a line of questioning such as "can you tell me a situation in which you have experienced X and how it turned out?" or "Can you share any tips or tricks that you have found useful and why?"

Regardless of approach, care should be taken to establish a rapport with study participants and to know what questions to ask/when. To maximize input and further engage study participants, research teams should avoid leading questions and only ask one question at time then wait patiently until the participant has responded. An assessment of the questions to be asked might be necessary as asking questions which are too open ended may yield a challenging response from participants. Additionally, questions which include HOW, WHY, WHEN, FOR WHAT PURPOSE are generally the most important types of questions in order to generate conversations. Questions which use prior knowledge and are asked with the appropriate nomenclature will garner respect and deepen the conversation as a rule of thumb.

A minimum research team consists of two individuals for the conduction of a contextual inquiry study as there are challenges for one person to capture both the interview and observational data as indicated in a study protocol. Typically, at sites of care delivery the number of observers is often limited to three. As such it is recommended that each researcher have an assigned role and responsibility that makes the study execution run smoother. This includes the following examples:

- **Leader**: responsible for developing rapport with the study participant, following the study protocol including leading the interview and observation. This person is responsible for the majority of the interactions with the study participants while on site. They may also be responsible for gaining consent.
- **Wingman**: assists with the interview and asks follow-up questions, takes detailed notes and responsible for the majority of photography.
- **Observer**: responsible for assisting in photography and taking detailed notes. Typically a silent observer dedicated to capturing data and study oversight.

While in the field, overcoming the challenges of purpose -*"why are you here?"*- can be readily resolved with a solid, brief explanation of the study goals along with the name of the contact person who has assisted in facilitating the site visit. The research team should be prepared to explain themselves several times throughout the day as shift changes or care sites change and new personnel enter during an observation.

Below is a list of considerations to taken into account during a study plan and/or execution:

- If observing a procedure, where is the procedure performed most frequently? (Use Environments) what special equipment or location of the researcher needs to be accounted for?
- What devices or procedures are required to observe in order to meet the research objectives? (Research Competitive Products and Alternative Techniques/Procedures)

- What are the lengths of time for the procedures/work flows? (Schedule)
- Evaluate the ease of conducting CI during the proposed context.
- Is it appropriate to be in the room during the procedure?
- Should the procedure be recorded remotely and then reviewed in context in an Interview Format?
- What is the level of "stress" in the environment when the procedure is performed?
- How large is the Team – external – how many team members can participate in the CI data collection process?
- How comfortable will the participants or the research team be in context/onsite in use environments? If observing in context is too extreme other CI methods should be considered.

Note: If the objective involves comparing different techniques be sure that the study protocol does not influence clinical judgment or decisions. It remains that one can not push a particular course of treatment just because it is the basis of observation. It is a common edict from those involved in clinical studies that the 'best way to cure a disease is to study it.'

## 5. Importance of background information and protocol development

At the onset of a CI study, it is important to gain a complete picture of all the information regarding the design problem that is currently available within the company and within published literature. This assures the study builds on existing information and does not repeat what is already known. Often there are marketing studies, internal scientists or technologists within an organization who has been active in studying the design problem or has existing relationships with users that may me helpful in expediting a CI study.

Conducting secondary research by reviewing journal articles, research papers, other publications, product manuals, procedure protocols and available usage data can provide the framework for the protocol, field guides and analysis (Privitera, 2015).

In kicking off a CI program, specific discussions should be undertaken with the cross-functional development team which include (Privitera, 2015, p. 28):

- The purpose and desired outcome of the research
- Existing research within the organization or other sources
- Composition of the research team; specifically who will be involved
- Potential sites and/or environments
- Study participants i.e. work group or consumers to be observed
- The work/process/procedures to be observed

From this information a study protocol can be developed. A study protocol is best practice and will be required if the study is to undergo Internal Review Board/Ethics review, which is more often the case in order to gain access. A CI protocol will contain the following sections (Privitera, 2015):

- Background and Overview
- Objectives
- User Profile or Inclusion/Exclusion Criteria
- Methods for Data Collection

- Study Materials Required (if any)
- Data Management Plan including patient privacy provisions
- Target Schedule

Below are additional considerations for developing a study protocol.

## 5.1 Site selection considerations

Determining appropriate target sites and participant selection is amongst the most critical of decisions that can be made during the planning of a CI study. There are no hard and fast rules for determining an appropriate site however it is important that the following topics are considered carefully.

### *5.1.1 Research anticipated patient case load*

Before selecting the intended sites, an essential first step is to ascertain the number of patients treated in relation to the procedure you intend to observe. Determining a base-level criteria for site selection can save time, money and frus-tration later on. It is also worthwhile ascertaining the number of HCPs capable of performing the procedure of interest at each site to help inform the potential sample size for the study.

### *5.1.2 Access via a "friendly healthcare provider"*

Throughout any CI investigation it is crucial to gain the trust, respect and support of the healthcare professionals that the CI team will be working with and observing. This will inevitably require developing professional relationships with members of the hospital team. However, care should be taken to consider the potential bias that may be introduced if access is gained via a HCP who has a vested interest in the development of the medical device being developed or studied. It is often the case that Key Opinion Leaders (KOLs) and leading experts in their field have been involved in the development or conception of new medical technologies and techniques. Whilst persons close to the project may offer quicker access to hospitals for a CI study, it is recommended that the investigation focuses on observing HCPs who are not associated with the development to avoid only seeing 'the problems they want you to fix.'

### *5.1.3 RepTrax and vendor credential systems*

In decades gone by it was possible to gain access to clinical environments by registering as a company rep and gaining access through a RepTrax style credentialing system. Experience of conducting CI studies in the USA over recent years indicates institutions are placing increasing restrictions over the levels of access reps have within clinical environments. As such, the vaccination and training requirements of vendor credentialing are often always required however this does not guarantee access. In addition, in the USA hospital volunteers and workers are often required to get a two part tubercusous test whereas credentialing services only require a single part test. In the event a two part test is required, this may cause a delay in access as this test takes a few weeks to administer. Clear understanding of the requirements for observation, including administrative paperwork must be taken into account when planning a CI study.

## 5.2 International considerations

If the team plans to conduct a CI study across multiple countries it is important that to understand the process applicable for each country, as well as cultural influences that may affect the study. For example, studies in Japan need to carefully consider hierarchical influences which are deeply embedded in their culture. These can include being aware of who the team is allowed to observe, who is allowed to speak and when permission is required to speak. Planning CI study in Germany also brings with it a host of cultural considerations, such as the surgeons needing to personally seek permission from the hospital Chief before consenting to taking part in research.

While these are a few considerations, a detailed description of conducting a study in the UK is presented below.

### *5.2.1 Conducting a study in the UK*

When planning on conducting a CI study in the UK which requires access to hospitals, clinics and requires recruitment of patients with support by the NHS, expect a requirement Health Research Authority (HRA) approval. HRA is a non-departmental public body sponsored by the Department of Health who are responsible for coordinating and streamlining ethical reviews and the application process for conducting research within the UK NHS. An overview of the UK approval process is shown below (Fig. 5.4), covering the two key streams of focus by Research Ethics Committee (REC) approval and HRA Study Assessment.

HRA was established to provide harmonization to the review process across NHS Trusts within the UK. Prior to HRA it was necessary to obtain approval from individual Trusts for all aspects of the research study, HRA provides a single route for Ethics and Business case approval.

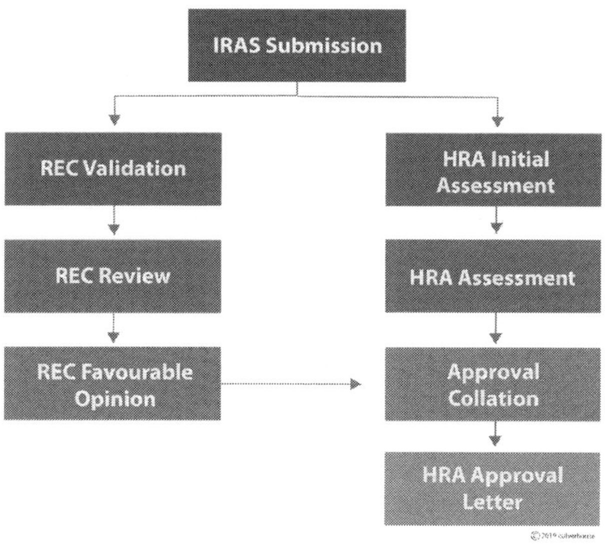


FIG. 5.4 UK IRAS study approval process.

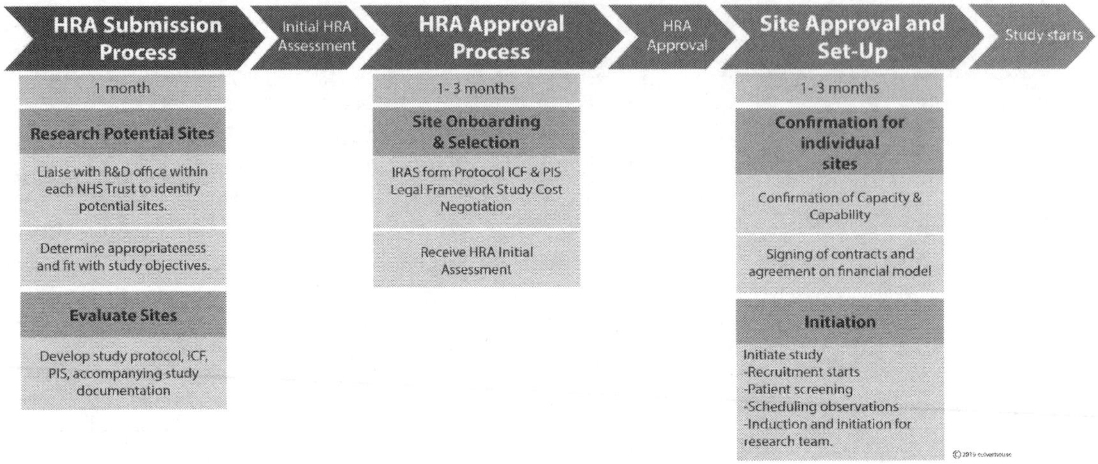


FIG. 5.5 Study approval process for the UK.

HRA has only been in place for a few years and many processes remain in relative infancy, constantly undergoing evolution and refinement. An optimized approach to gaining approval in the UK for CI studies is shown in the diagram below (Fig. 5.5).

### Conducting a study in the UK: Top tips

- Engage with potential sites as early as possible in the process, prior to submitting an application to HRA. Liaise with the local R&D office for the appropriate NHS Trust who will support the effort through the process.
- Plan for a 10%–20% site success rate. Include contingency sites in the initial HRA Application to avoid having to later submit amendments and go back through the review process to add additional sites.
- Appreciate that the protocol will be reviewed by the same Research Ethics Committee as those who review clinical studies. Expect a high level of scrutiny during the review process to ensure patient safety has been adequately consider.
- Establish communications with the NHS Clinical Research Network who will support identification of potential sites and Research Staff to support the study.

## 6. Clinical immersion best practices

Planning a CI study can take months and in some instances over a year to get all the necessary approvals in place, receive ethical clearance and meet the individual access requirements for each institution or site. That said, the hard work is only just beginning once approvals have been granted. Prior to stepping into the field it is vital that CI researchers know what to expect.

Researchers involved in CI studies need to develop the ability to blend seamlessly into their surroundings, often going unnoticed. This requires CI researchers need to be highly familiar with a multitude of guidance, legislation and red-tape all designed to protect patient safety, prior to being allowed on site.

University Hospitals can offer significant levels of support to CI teams to ensure policies and procedures are adhered to during a study. Teaching institutions will be more familiar with induction processes for trainees and students than private organizations who will typically only deal with qualified Healthcare Professionals.

Below is a short list of topics to enquire about during study setup and planning:

- Mandatory vaccinations, inoculations and blood test clearances (these vary from site to site so never assume they are the same)
- Induction into clinical hand washing technique
- Access to nurses scrubs and shoes
- Dedicated locker space and security passes
- Infection Control procedures
- If working in the EU, compliance with GDPR and the hospital data privacy policy

Once access has been granted, proof of vaccinations as well as any confidentiality agreements might be requested by clinical practitioners. As the CI team will be entering sites of care, patient privacy concerns such as HIPPA (Health Insurance Portability and Accountability Act of 1996) requirements/GDPR Data (General Data Protection Regulation) privacy are always monitored as well as policed, even stopping at a nurses station to check a cell phone may receive a reprimand from a nearby nurse.

Once in the field there are small but essential expectations of behavior that may assist with the research team 'blending in.' These include hand washing before and after leaving a hospital room, wearing short sleeves and minimal jewelry. Its important that CI investigators are burden free to the clinical team they are observing. These basic behaviors will improve relations and avoid being considered a nuisance.

### What to expect in the field of a CI study:

In a hospital:

- Lots of waiting around!
- Sporadic patient case load/be on call
- Importance of being mobile – no bags allowed in OR or wards

In a home:

- Animals of all sorts may be roaming around
- Patients may be sick or irritable
- The person the team wants to talk to may not want to talk or there will be interruptions by family members/small children

# 7. Analyzing data for optimum insights

After conducting all of the observations the real challenge begins, that is how to make sense of the data and generate meaningful output. Sharon (2012) states that the single biggest challenge in user experience research is creating the illusion that the research took place. Sharon goes on to state that whilst written reports hold some value, they fall short of adding true value for the following reasons:

- Reports take time to write
- Too few people take the time to actually read them
- Reports have a limited shelf life (e.g., they are filed away and forgotten about)
- Readers and writers alike get lost in the detail

A successful CI research team has a duty to deliver meaningful insights to a broad audience which is likely to involve Marketeers, Engineers, Sales Teams, Regulatory Affairs, Industrial Designers. It is necessary to make each of these functions understand the 'so whats' that have been identified in a manner whereby they could have been stood next to you during the fieldwork.

Edited video clips are amongst the most compelling mechanism to bring home the realities of how a device is used in real life. Short clips can be embedded into task analysis diagrams to help communicate the true context of use, whilst reducing the burden on having to read reems of text. Additionally, taking the time to edit video clips which include augmented labels and annotations can help the audience to understand the insights and focal points to observe.

The section below provides a brief overview of data analysis techniques and the process in developing meaningful insights.

## 7.1 Data analysis

> Data analysis actually starts in the field when the research team selects where to aim the camera/s and decides what notes are important enough *to write down* (Privitera, 2015, p. 117).

As the types of data collected in a CI program can vary widely e.g. field notes, recording, still photography, sketches or documents, etc. and some of this information may contain confidential patient information, data management requires a careful system and can assist in the data analysis process. Once the data has been collected and a robust database developed, analysis can begin and involves the following steps (Privitera, 2015, p. 117):

1. Condense data into themes and generate data analysis codes.
   a. Use debrief notes to look for surprises and common elements
2. Analyze all data for evidence of the same theme or code.
   a. This may involve conducting a task analysis with the data (see Chapter 6).
   b. Select images, quotes, video snips that best represent themes and codes
   c. Develop data displays; affinity diagram sketches or visuals

3. Condense themes and codes
   a. Eliminate unsupported themes
   b. Combine redundant themes
   c. Synthesize data to uncover connections and causality of behavior
4. Review the data and select top 2–3 representations
5. Develop design insights or need statements

In this model, a theme can be defined as a dominant behavior, idea or trend seen throughout an observation or interview. A code is a label applied to the data under a particular theme that can assign further meaning or category (Privitera, 2015). This analysis can be completed by hand or with readily available spreadsheet software or specific qualitative research software such as NVivo. See Chapter 5 Data Analysis in the book Contextual Inquiry for Medical Device Design (Privitera, 2015) for further details on data analysis methodology.

## 7.2 Developing insights

The most important tool for developing insights from a CI study is an open mind with the ability to imagine a different reality while recognizing the behaviors, actionas and thoughts of study participants. This requires putting aside personal bias. CI insights must answer WHY? users are currently doing what they are doing then collectively assessing SO WHAT? In this instance the 'so what' assigns meaning and by understanding WHY things work or do not work, a design team can then identify WHAT things might need to change in order to improve the situation.

The results from a CI a study can include design insights, user needs, design recommendations or considerations, design specifications and ideas (AAMI, 2014; Privitera, 2015).

Each of these are defined below:

- Design insights can be a statement made by a clinician or an observation around an issue relating to a particular part of the procedure or device use which may not be readily actionable.
- User needs are descriptions verbalized or observed which are straightforward and relate directly to the primary functionality of the device. In best practice, these should be positive statements as they result in better design criteria. Negative statements e.g. "preventing" or "not" doing something are very difficult in design as they are nearly impossible to measure. Need statements should include a direction (increase, ensure), a target (comfort, ease of use) and a context (activity, location).
- Design recommendation or considerations are high-level attributes that establish a general need however lack adequate detail in order to be a measurable outcome of the design.
- Design specifications are measurable attributes of the design.
- Ideas are concepts or opportunities to improve healthcare through design.

Generating useful design insights can happen at anytime throughout the conduction of a CI study. Throughout the process of data analysis team members should share information, seek alternative perspectives and ask further questions concerning a particular topic or interpretation (Privitera, 2015).

## 8. Visualization and communication

Developing graphical representations of data enable further analysis, can simplify observations and distill data into critical elements for the purposes of improved communication. As the collected data is in various forms, preparing to communicate the most important examples of evidence to demonstrate a particular theme may also be accomplished in the various forms e.g. still image, video, sketch or illustration.

Fig. 5.6 below is an example of using illustration to further communicate a complex medical situation during interventional radiology wherein the catheter being controlled by the physician has a build up of potential energy and winds on itself causing a negative situation. Without the illustration, it would be impossible to identify what's really going on with the patient.

Communication of CI findings may include still images, video snips, illustrations with the most extensive being entire procedure maps. Procedure maps include all details regarding the study as well as all of the findings. The purpose of the map is to serve as a reference for product development teams and generate discussion regarding relationships of the procedure itself, the environment or the user. Each of these are critical to the design of the device (see Chapter 7).

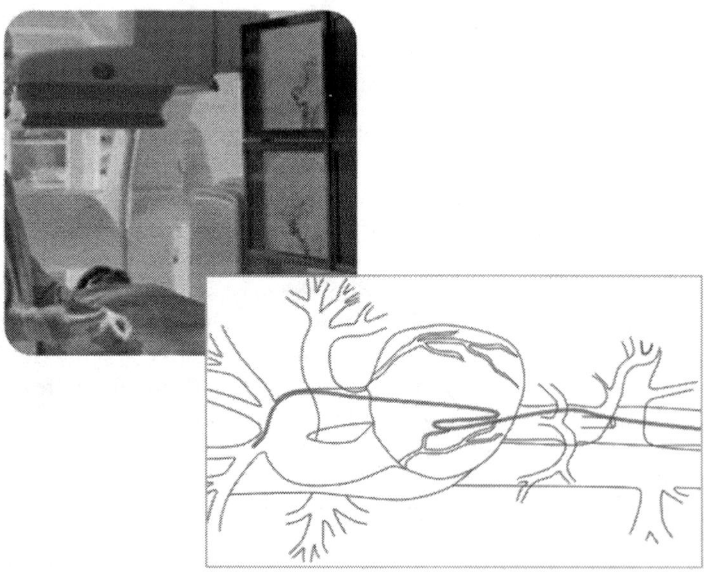

**FIG. 5.6** Illustration of catheter with redundancy inside the aorta explains what can happen inside the patient during interventional radiology.

## 9. Summary

CI studies add value to medical device development in that they clearly identify opportunities for new product development, clarify user needs, the user, the environment and the task. The impact of conducting studies intended for global market release or multiple environments of use can be invaluable as the learnings will include key differences between cultures. In addition, CI studies have a direct impact on device design, the HFE dossier for agency submission and are recommended as best practice.

## 10. Further reading

- ANSI/AAMI TIR 51 Contextual Inquiry
- Privitera, M. B. (Ed.). (2015). *Contextual inquiry for medical device design*. Academic Press.

## Acknowledgments

Thank you to the HS Design team and Rebus Medical for continuing to advance the practice of conducting CI studies. Special thanks to Juliana Privitera for the generation of graphics. Thank you to Elissa Yancey for editing.

## References

AAMI. (2014). *AAMI TIR51:2014 Human factors engineering-Guidance for contextual inquiry*.

ANSI/AAMI/IEC. (2015). *ANSI/AAMI/IEC 62366-1*.

Beyer, H., & Holzblatt, K. (1999). *Contextual Design Defining Custoner-Centered Systems*. Morgan Kaufmann.

Bajars, E., Larson-Wakeman, M., Matchstick, C. F., & Biondi, S. (2016). Novel ethnographic/contextual inquiry techniques for understanding connected device users in their native environment. *Proceedings of the International Symposium on Human Factors and Ergonomics in Health Care, 69*(1). https://doi.org/10.1177/2327857917061034.

Cystic Fibrosis Trust - What is cystic fibrosis? (n.d.). Retrieved from January 28, 2019. https://www.cysticfibrosis.org.uk/what-is-cystic-fibrosis.

FDA. (2016). *Applying Human Factors and Usability Engineering to Medical Devices Guidance for Industry and Food and Drug Administration Staff Preface Public Comment*. Retrieved from http://www.regulations.gov.

Katz, J. (2012). *Designing Information: Human factors and common sense in information design*. John Wiley & Sons.

Koshinen, I., Zimmerman, J., Binder, T., Redstrom, J., & Wensveen, S. (2011). *Design Research Through Practice From the Lab, Field, and Showroom*. Waltham, MA: Elsevier.

MRHA. (2016). *UK notified bodies for medical devices*. GOV.UK. Retrieved from https://www.gov.uk/government/publications/medical-devices-uk-notified-bodies/uk-notified-bodies-for-medical-devices.

Privitera, M. B. (Ed.). (2015). *Contextual inquiry for medical device design*. Academic Press.

Sharon, T. (2012). *It's our research: Getting stakeholder buy-in for user experience research projects*. Morgan Kaufmann/Elsevier. https://doi.org/10.1016/B978-0-12-385130-7.00001-8.

CHAPTER

# 6

# Task analysis

*Ashley French, Leah K. Taylor, Melissa R. Lemke*

Agilis Consulting Group, LLC., Cave Creek, AZ, United States

OUTLINE



*"Task analysis started by understanding the demands of tasks that need to be performed by an operator, so that for each task the key human capabilities and limitations could be derived."* **Bisantz and Burns (2008)**

## 1. Introduction

A task analysis is the process intended to identify a user's goal, what steps must be done in order to achieve the goal, what experiences (personal, social and cultural) users bring to the tasks and how the environment might influence the user. It is an important design tool that can be effectively used early in the design process (i.e., before the creation of device

*Applied Human Factors in Medical Device Design*
https://doi.org/10.1016/B978-0-12-816163-0.00006-2

prototypes) to help inform use-related design of the user interface (UI) components. Throughout the design process, a task analysis can also be used as a fundamental framework to the development of use-related risk documentation (e.g., uFMEA or Fault Tree Analysis), human factors protocol development, and usability evaluation. It can highlight elements of the user-device interactions that could be problematic for users, which provides designers opportunities early and throughout the product development process to implement risk mitigations proactively, ultimately saving time and manufacturing costs.

Additionally, a task analysis is a critical input for instructional designers while creating the device's instructional materials (e.g., Instructions for Use (IFU), quick reference guide (QRG), user manual or training materials). The instructional materials accompanying a device aim to guide accurate, safe and effective user performance. As the task analysis defines and describes what constitutes that performance, ultimately, good instructional materials cannot be developed without a comprehensive task analysis.

As defined by HE75 3.94 (ANSI/AAMI HE75, 2009), a task analysis is, "a set of systematic methods that produces detailed descriptions of the sequential and simultaneous manual and intellectual activities of personnel who are operating, maintaining, or controlling devices or systems." The FDA recognizes that a task analysis is an important type of preliminary analysis that can provide insight into the following questions (CDRH guidance, 2016):

- What use errors might users encounter for each task?
- What circumstances might promote users to make use errors on each task?
- What harm might result from each use error?
- How might the occurrence of each use error be prevented or made less frequent?
- How might the severity of the potential harm associated with each use error be reduced?

A complete and accurate task analysis requires a systematic process to break down necessary device operations into a hierarchical sequence of definable tasks that are intended to be completed by the device user.

This chapter describes the process of conducting a task analysis, provides examples and promotes its use as a fundamental requirement in the application of human factors in medical device design. It is one of the most important tools in the Human Factors Toolkit and is relevant to virtually all aspects of the design process including the following:

- Verification of function allocation is appropriate
- As input to the risk analysis process
- Analysis of errors, potential errors and critical incidents as part of post-market surveillance
- Highlighting potential design problems and crafting solutions
- Developing formative and summative usability evaluations
- Designing instructions for use, quick start guides and training materials

The methods and techniques for conducting task analysis have been devised in almost every field of endeavor involving human performance. Further, many techniques known by other names (e.g., Fault Tree Analysis, Link Analysis, Failure Modes and Effects Analysis (FMEA)) are often considered special types of task analysis by human factors professionals with the main reference text being Kirwan and Aimsworth's "A Guide to Task Analysis" (1992).

## 2. Overall process

Task analysis development is a systematic process that produces a coherent "road map" to device-user interactions. A use case is the first input necessary for developing a task analysis. Use cases are representative situations that describe how intended users will use the device in actual use environments. These use cases provide input to and guide the development of the task analysis by identifying specific tasks that users need to complete in order to successfully use the device in those specific situations. The next steps for developing a task analysis involve task identification, sub-task breakdown and application of the PCA model (i.e., Perceptual, cognitive, and manual action requirements model). The PCA model is recommended in CDRH guidance (2016) guidance however, this model can be limited for devices and systems that have multiple simultaneous users where coordination and communication tasks are involved (e.g., complex surgical systems or electrophysiology technologies).

Once the tasks and sub tasks have been accurately described, potential use errors are identified along with associated harms and severity. Fig. 6.1 illustrates the task analysis process starting from identifying use cases (Step 1 in the process) and in some cases (uFMEA) ending with determining potential harms and severity of harm (Step 6 in the process).

The terminology used in a task analysis may vary across device teams. The terminology used in this chapter includes terms such as; "use cases," tasks," and "sub-tasks." Although terminology may change slightly or have somewhat different meanings across manufacturers, it is important for a manufacturer to use a systematic and consistent hierarchy when creating and utilizing a task analysis and associated terminology. Table 6.1 helps illustrate the differences between use cases, tasks, and sub-tasks for an infusion pump.

### 2.1 Step one: use case identification

The first step in developing a task analysis is use case identification. Use cases are an important input to the task analysis as they provide context for the various, sometimes overlooked, uses of a device. Most medical devices involve numerous use cases. While some use cases are related to the main purpose of the device such as drug delivery or running a diagnostic test, other commonly overlooked use cases for some devices that should be considered in the task analysis if they are applicable include procedures such as maintenance, calibration, reprocessing, cleaning, disposal, shut down, and storage.

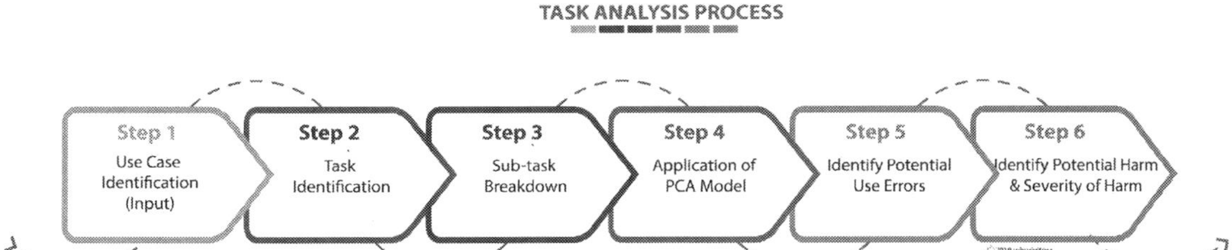


FIG. 6.1 Process map of an example task analysis development process.

TABLE 6.1 Example use case, task identification, and sub-task breakdown for an infusion pump.

| Terminology definitions | Examples |
|---|---|
| **Use Case:** High-level set of user-device interaction functions | **Use Cases:**<br>• Device set-up<br>• Troubleshooting<br>• Programming Therapy<br>• Monitoring Status<br>• Maintenance<br>• Shut-down |
| **Task:** Individual task (s) for the high-level functions<br>or<br>Action or set of actions performed by a user to achieve a specific goal task (CDRH guidance, 2016) | **Tasks:**<br>• Enter patient data<br>• Enter dosage concentration<br>• Enter start time |
| **Sub-Task:** Interaction steps (i.e. individual steps for carrying out a task in the order of their performance) | **Sub-Tasks:**<br>• Press power button<br>• Verify pump on<br>• Select dosage entry screen |

*Adapted from HE75, Section 5.6.3.*

Different users of the device may encounter the same or different use cases. Consider the previous example of an infusion pump. A health care professional may have a use case to program the pump, and a lay patient may have a use case to initiate a pre-programmed bolus of medication. When identifying use cases it is important to consider the different users throughout the life cycle of a device. When user-device interactions differ across intended user groups, the task analysis should include all tasks and specify the users who complete specific tasks in the analysis.

## 2.2 Step two: task identification

Upon definition of the use cases, the tasks associated with each specific use case can be established. Tasks may be determined from knowledge experts (e.g. clinical stakeholders) input, FDA feedback on similar devices, observations of users with similar devices, human factors experts interacting with the device, critical thinking, investigation into videos of device use or similar device use, instructional material, risk documentation, reverse engineering and/or experience with similar devices. When identifying tasks, it is important to analyze the user performance requirements, referencing that the device must support the user(s) toward meeting their intended goal. This analysis may include a heuristic review or cognitive walk through of the design for various use cases with the intended user in mind.

As a caveat, remember that relying solely on the IFU and risk documentation to develop a task analysis will not produce a comprehensive task analysis. Instructional materials and risk analyses are often created by engineers or the device developers and may not always account for all aspects of user-device interactions. Instructions that do not include all of the necessary information may cause preventable use errors or difficulties for users who choose to use the instructions. Additionally, a task analysis constructed solely from the instructions may inaccurately identify use errors. For example, the instructions may direct a user to hold injection for 15 seconds, but the medication may be administered after 7 seconds. The Task Analysis needs to identify the 7 seconds, not the 15 seconds to avoid inaccurately identifying the interaction as premature removal. Another example is if the instructions state to hold the needle perpendicular to the injection site, but in reality it is safe and effective use if the needle is at a slight angle. The user can technically insert the needle between a 75–105 degree angle, which is not evident from the instructions alone. Risk documentation that is incomplete may lead to unanticipated problems and unacceptable risk during clinical trials and human factors testing, as well as post-market issues that may not have been assessed or mitigated in the design and development process. Sometimes the manufacturer develops the task analysis and instructional materials at the same time. In these situations, the task analysis provides an input for the instructional materials (see Chapter 9 Applying Design Principals to Instructional Materials and Training for more details).

A crucial strategy in the task analysis development process is observation of use of the device to understand in detail how tasks are performed, the order they are performed in, and if the user is able to safely and effectively complete the task. A thorough and accurate sequencing of tasks is crucial, if a specific order of tasks is necessary for a user to successfully use a device. For example, consider an infusion pump that requires a user (i.e., a healthcare professional) to setup the device prior to use by first positioning the tubing set in the tube holder-clips that lock the tubing into place before closing the pump's access door. If the user closes the access door prior to ensuring that the tubing set is properly positioned in the holder-clips, the tubing could become clamped and result in an under-infusion for the patient.

Another strategy for task identification is for a stakeholder, who is not familiar with the device, perform simulated use, cognitive walk through, and/or ethnographic observation of the device. These can be done with and without using the instructional materials. If simulated use is used, a mock user can note what they have trouble with, their first impressions of the design, if the device works as anticipated and if he/she hears, feels or sees different cues while working with the device. A cognitive walkthrough (See Chapter 10, Section 3 for full details) can be a quick and inexpensive tool to understand the device's usability from the intended user's perspective.

While creating the task analysis, the mock user can input the "obvious" use errors based on these task identification techniques. It may also be helpful for the mock user to try to do tasks incorrectly with the device to see what types of responses the device has, and what additional potential use errors are induced. Similar observations can be made during an ethnographic observation. These data should be incorporated into the task analysis. Below are some questions that may be helpful to ask while creating a task analysis and may be answered by observation or mock use:

- Is order of tasks important?
  - The order of tasks may or may not be critical to the safe and effective use of a device. If a task is completed out of order, it should be determined if the result would be a use error or pose a potential harm.

    - If the order of tasks is important, sequencing should be documented in the task analysis. For example, most devices must be powered on before any other tasks can be completed.
- What happens if someone skips a task? If the task is skipped, is the task optional?
    - There may be a task that is inadvertently missed or intentionally skipped by a user. Some tasks are critical to the safe and effective use of the device, while others are included as steps in the logical workflow. Consideration for tasks that fall into these categories should be documented in the task analysis. For example, an optional task may be to pinch the skin around an injection site to administer therapy.
- Is timing a critical component?
    - Some devices include timing as part of the safe and effective use of the device. The timing may be related to warm up time of medication, length of holding an injection, or wait time for a diagnostic result to appear. If timing is determined to be relevant to the safe and effective use of the device, the time requirements should be integrated into the task analysis in an observable and measurable way.
- Are there alternative ways/shortcuts to complete a task that are acceptable or unacceptable?
    - For some tasks, there may be multiple ways to accomplish the same end goal. An ethnographic study may provide insight into these alternative paths or shortcuts. If alternative methods are acceptable they should be reflected in the task analysis. If a shortcut is unacceptable, it should be noted as a potential use error and mitigations incorporated in an effort to eliminate the occurrence.

## 2.3 Step three: sub-task breakdown

During the sub-task breakdown stage, each task should be broken down into independent actions or processes. Table 6.2 below illustrates an abbreviated task list for an infusion pump with both tasks and sub-tasks identified. For example, the task of "Program Continuous Therapy" consists of nine sub-tasks that delineate the specific sequence of actions that a user must take with the UI to complete the task (i.e., "Navigate to Therapy Modes Menu," etc.). Table 6.5 shows an example of a full task analysis excerpt for an infusion pump.

Most often it is important to avoid combining multiple tasks into a single line item of the task analysis, as this introduces the possibility of misidentification or omission of associated use requirements and potential use errors as well as "double counting" errors and losing specificity in data reporting during usability testing. For example, in the infusion pump example from Table 6.2, it may seem logical to combine Tasks #6 and #7 to be "Select CONFIRM and Understand Confirmation Message." However, each sub-task addresses separate parts of the UI (i.e., on-screen option and popup confirmation message) and separate required actions on the part of the user (i.e., select the on-screen option and understand meaning of confirmation message content). Thus it is better to separate the two into sub-tasks. When task analyses are used during the design process as the framework for human factors testing, appropriately separated tasks and sub-tasks are also important because they guide moderator observations and the debriefing interview so as to yield insights into root cause analysis and reporting of any observed difficulties and use errors.

TABLE 6.2 Abbreviated task list for an infusion pump.

| # | Device sub-tasks |
|---|---|
| ***Program a Continuous Therapy (i.e., constant programmed rate of infusion)*** | |
| 1 | Navigate to Therapy Modes menu |
| 2 | Select "Continuous" |
| 3 | Enter prescribed drug concentration |
| 4 | Enter prescribed amount to be infused |
| 5 | Enter prescribed rate of infusion |
| 6 | Select CONFIRM |
| 7 | Understand Confirmation message |
| 8 | Select RUN to start continuous therapy |
| 9 | Wait for a minimum of 30 seconds before additional programming |

Additionally, when breaking down tasks into sub-tasks, it is important to integrate order and timing (when applicable) in an observable and quantifiable way. Verbiage should be concise, consistent (with no ambiguity) and measurable (which becomes extremely important when task analysis guides usability testing). For example, terms such as "after" and "before" should be included when order is important. When timing is critical, measurable metrics should be included. For example, instead of "hold until the medication has been delivered," provide a metric such as "hold for 5 s until the medication is no longer visible in the window." While the above provides limited examples, the text "Taxonomies of Human Performance" (1984) by Fleishman and Quaintance provides more comprehensive examples of task descriptions.

The precision of wording included in task analysis can influence the quality of the analysis. It is important to avoid using words that are ambiguous and will be difficult to observe or measure (e.g., "adequate," "well," or "enough"). Instead define what is meant by ambiguous words with as much detail as possible that describes the specific UI components and the expected user-device interactions. The more specific and stand-alone the sub-tasks can be, the more useful a task analysis will be as an early design tool as well as a framework for subsequent human factors data collection and analysis. Table 6.3 provides examples of original task phrasing and suggested improvements to task phrasing, which also includes the separation of tasks into more discrete sub-tasks. Note that the specificity and phrasing included in a task analysis might lead to development of instructional materials that provide the user with similarly concise instructions, although this is not always the case.

It is important to capture all tasks for a device, including tasks and sub-tasks that are associated with observable performance data and those that are not observable (i.e., knowledge tasks). For example, understanding instructions on proper storage and understanding an expiration date may not be possible to observe during a simulated use scenario. However, if the task is included in the task analysis and is identified as critical, it can be assessed

**TABLE 6.3** Example task identification improvement in specificity, phrasing, and separation into discrete sub-tasks.

| Original task phrasing | Suggested improvements to phrasing and sub-tasks to reduce ambiguity |
|---|---|
| Hold the device horizontal with the button on top, press button to load dose, and do not turn upside down when pressing the button. | **Separate into sub-tasks:**<br>**1.** Hold device horizontally with the button on top.<br>**2.** Press the button all the way down to prepare dose.<br>**3.** Avoid tilting the device more than 90° while pressing the button.<br>**Notes**: Separated original task into discrete sub-tasks, added additional details about unacceptable device positioning. |
| Take a strong, deep breath and continue breathing until indicator changes from green to gray (after click). | **Separate into sub-tasks:**<br>**1.** Take a strong, deep breath by inhaling for at least 3 seconds.<br>**2.** Continue inhaling after a click is heard and indicator changes from green to gray.<br>**Notes:** Separated original task into discrete sub-tasks, since some users may have issues with the first sub-task that requires the user to perform an action for a specific period of time (i.e., take a strong, deep breath for 3 seconds) and others may have issues with the second part which relies on a specific design element to provide feedback to the user (i.e., continue breathing until indicator changes from green to gray). In addition, breaking into discrete sub-tasks will allow more detailed observations of UI elements and user performance if usability testing is conducted. |
| Open the medicine. | **Define sub-tasks:**<br>**1.** Tear off strip of foil along perforation (bottom of packaging).<br>**2.** Tear medication pack along perforation to separate one capsule segment from the rest of the packaging.<br>**3.** Press capsule through foil to remove it from individual packaging.<br>**Notes:** Separated original task into discrete sub-tasks that focus on different aspects of the packaging design that may impact user-device interactions (e.g., perforations in the medication pack, individual package segment containing the capsule). |
| Mix suspension well. | **Identify precise wording:**<br>• Mix until no particles are visible in the suspension.<br>**Notes**: Updated wording to indicate how "well" is defined in the original task wording. |
| Adequately press and hold green button. | **Identify precise wording:**<br>• Press and hold green button for 5 seconds until the light stops blinking<br>**Notes**: Updated wording to indicate how "adequately" is defined in original task wording. |
| Decompress the guard deep enough. | **Identify precise wording:**<br>• Decompress the guard so that yellow guard is no longer visible<br>**Notes**: Updated wording to indicate how "enough" is defined in the original task wording. |
| Twist cap to close. | **Identify precise wording:**<br>• Twist until a click is heard and the cap can tighten no more<br>**Notes**: Updated wording to indicate how "closed" is defined in the original task wording. |

TABLE 6.4 Task type examples in the PCA model.

| Task types | | |
|---|---|---|
| **Perceptual** | **Cognitive** | **Action (Physical/Motor)** |
| Detecting | Calculating | Adjusting |
| Scanning | Analyzing | Aligning |
| Identifying | Comparing | Synchronizing |
| Locating | Estimating | Opening |
| Discriminating | Planning | Connecting |
| | Categorizing | Pressing |
| | Deciding | Bending |

through a knowledge task question. Knowledge tasks are represented in the task analysis and are often phrased as "Understand [critical information content]" (e.g., "Understand the device should be stored in the refrigerator" or "Understand the device should be stored in the original carton"). If these critical non-observable knowledge tasks are omitted during the task analysis process, gaps will exist with respect to use-related risk and the user interface.

## 2.4 Step four: apply the perception, cognition, and manual action (PCA) model

A complete task analysis is fundamental to user interface optimization, use error prediction and prevention, and determination of the user's interactions with the device relative to task requirements. These user task requirements include user actions, user perception of information and user cognitive processes (Sharit, 2006). A Perception, Cognition, and Manual Action (PCA) model is an FDA recommended strategy for task analysis that is used to identify user-device interactions and characterize user capabilities. Applying PCA to a task analysis adds specific user requirements that lend support for identification of potential use errors and root cause analysis during human factors testing. This model identifies user actions related to the perceptual inputs, cognitive processing, and physical actions involved in the task (CDRH guidance, 2016). Table 6.4 below highlights task types according to the PCA model and is intended to provide examples though it is not an exhaustive list of types of tasks.

In more complex systems, there may be a fourth category of tasks which is Communication (people to people). For example, this includes the following task descriptors: informing, requesting, directing, advising, or querying.

Once sub-tasks are identified, the PCA model is used to further break down the sub-tasks to identify user requirements to complete each sub-task. For each sub-task, user requirements

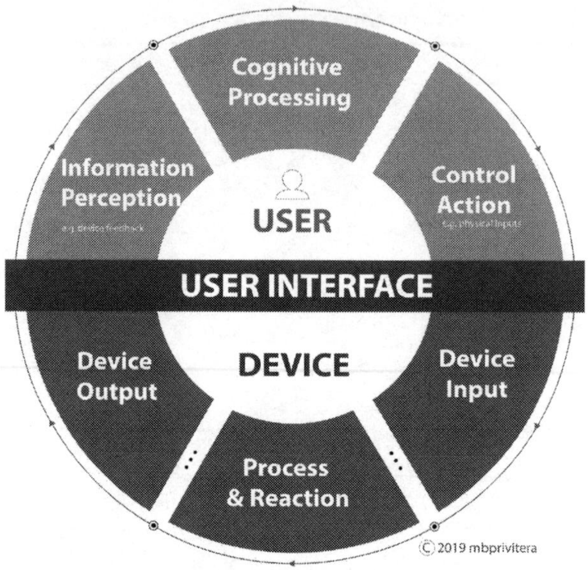


FIG. 6.2 Model of the operational context between a user (PCA components), device, and user interface. *Adapted from FDA HF Guidance (CDRH, 2016).*

related to perceptual inputs, cognitive processing, and manual actions necessary to perform the sub-task should be documented.

Fig. 6.2 below describes the user-device interface relationship. Risk occurs when any of the perception, cognition, or action tasks are difficult, confusing, or impossible for the user. For example:

- If perceptual information, such as warning labels or directions, are not seen or heard by the user (perceptual requirements), this information will not be available for the user to understand (cognitive processing).

Consider a wearable cardiovascular device that includes an alarm to warn the user to change the battery. If the user is unable to hear the alarm sound, he/she will not be alerted to take the necessary action(s) to change the battery, which could have fatal consequences.

- If a user correctly perceives the information from the device or labeling, but has difficulty understanding what that information means, the necessary action may be missed or done incorrectly.

For example, if a user with a wearable cardiovascular device hears the alarm but does not understand that it means to change the battery, the user may take no action or the wrong action in response to the alarm.

- If the action step requires users to act quicker or more forcefully than they are physically capable of doing, this interaction will not occur.

For example, if the user with the wearable cardiovascular device hears and understands the alarm to change the battery, the user may not be able to generate enough force to remove the old battery.

Fig. 6.2 demonstrates that the safe and effective interactions of the user-device interface are dependent on the user requirements outlined in the PCA process.

## 2.5 Step five: potential use error identification

The next step is to identify possible use errors and impacts of not meeting the use requirements. Use errors and difficulties observed during tasks can be traced back to the PCA requirements for each task, which can assist in determining the root causes for issues and the corresponding piece of the device user interface that may require redesign or additional risk mitigation evaluation. Potential use errors may occur if the user is unable to understand or carry out the user interface requirements. For example, refer to Table 6.2 Task 1, "Navigate to Pump Modes." If a user does not perceive, understand, or act on information, it could lead to the use error of navigating to an incorrect menu option. If during testing a user navigates to the wrong menu option, then the human factors moderators should probe the user to determine if the root cause is perceptual (i.e. user was unable to read the screen, or read the screen incorrectly), cognitive (i.e. user thought the menu option labels meant something different than intended), or manual action (i.e. user double-tapped the screen to select instead of a single tap).

Another part of use error identification includes leveraging data from prior formative evaluations, literature (e.g., peer published articles, FDA news flashes, device recalls), customer complaints, post-market data from similar devices and expert opinions of stakeholders (including clinical experts) on the potential use errors and the consequences of those use errors. If a user does not understand the tasks or is not able to complete the tasks, the device may be used incorrectly or may not be used at all. This could lead to decreased treatment efficacy or delay in therapy or treatment. Having input from multiple stakeholders with regard to use errors and associated consequences is critical to understanding all of the risks associated with the device use.

## 2.6 Step six: potential harm identification

Using task analysis as the foundation of risk analysis is a best practice. ANSI/AAMI HE75 defines harm as "(1) Physical injury or damage to the health of people, (2) damage to property or the environment," (ANSI/AAMI HE 75, 2009). A risk is defined as a, "Combination of the probability of occurrence of the harm and the severity of the harm" (ANSI/AAMI 14971, 2010). The probability of occurrence is the frequency with which the harm would occur. Severity is the measure of the possible consequences of a hazard. Although harm, probability of occurrence, and severity should all be considered in determining the risk profile for a particular device, probability of occurrence is not an FDA-accepted factor for determining the criticality and corresponding categorization of tasks in human factors testing.

TABLE 6.5 Example definitions of severity at the significant, moderate, and negligible levels.

| Example severity levels | Example description |
|---|---|
| 3—Significant | Death or loss of function or structure |
| 2—Moderate | Reversible or minor injury |
| 1—Negligible | Will not cause injury or will injure slightly |

The estimated probability of occurrence of a problem is not always accurate, and many use errors are not anticipated until device use is simulated and user interaction with the device is observed, or even later once the product is released and the manufacturer observes post-market problems. Therefore, severity and potential harm are preferred measures for determining if user interface modifications are required to reduce or eliminate harm (ANSI/AAMI 14971, 2010) and should therefore be included in a task analysis.

Severity levels are determined and justified by the manufacturer for a particular medical device under clearly defined conditions of use (ANSI/AAMI 14971, 2010). An example range of severity levels provided by ANSI/AAMI 14971 is described in Table 6.5.

Once severity levels are identified for all tasks or sub-tasks in the task analysis, tasks can be assigned a task category. CDRH defines a critical task as, "A user task which, if performed incorrectly or not performed at all, would or could cause serious harm to the patient or user, where harm is defined to include compromised medical care." (CDRH guidance, 2016). Although the only category that is defined by CDRH is a critical task, it can be beneficial to group tasks into a hierarchy to identify critical tasks and non-critical tasks. Task category should be linked to severity rating, and a definition of how tasks are categorized should be included in the task analysis. Note: CDER draft guidance defines a critical task differently than CDRH, specifically "user tasks that, if performed incorrectly or not performed at all, would or could cause harm to the patient or user, where harm is defined to include compromised medical care."

For the entire project team it is key to note that the task analysis is a living document. Throughout the development, testing, and re-design process iterations are likely. These iterations may include added, removed, modified, and/or re-ordered tasks. Additionally, revisions may involve modifications to the PCA analysis, risk severity, use errors, and task categorizations.

## 2.7 Example task analysis with risk and task category delineation

Table 6.6 below presents an example Task Analysis for an infusion pump including the sub-tasks, PCA elements, potential use errors, potenetial harms, severity of harm and task category.

TABLE 6.6 Task analysis example for an infusion pump.

| # | Device use sub-tasks | PCA elements | Potential use errors | Potential harm (outcomes) | Severity of harm[a] | Task category |
|---|---|---|---|---|---|---|
| *Program continuous therapy (constant programmed rate of infusion)* | | | | | | |
| 1 | Navigate to therapy modes menu | • P: See menu labels<br>• C: Read and understand menu labels<br>• A: Select menu label | User does not navigate to the correct menu | Inaccurate infusion, delayed therapy, or no therapy | 3 | Critical |
| 2 | Select continuous | • P: See menu labels<br>• C: Read and understand menu labels<br>• A: Select menu label to program continuous therapy | User does not select correct option to program continuous therapy | Inaccurate infusion | 3 | Critical |
| 3 | Enter prescribed drug concentration | • P: See editable field<br>• P: See measurement units<br>• P: See keypad<br>• C: Understand where to enter concentration value<br>• C: Understand to select the correct units of measurement<br>• A: Select editable field<br>• A: Use keypad to enter value | User does not enter the prescribed value correctly | Inaccurate drug concentration | 3 | Critical |
| 4 | Enter prescribed amount to be infused | • P: See editable field<br>• P: See measurement units<br>• P: See keypad<br>• C: Understand where to enter amount to be infused<br>• A: Select editable field<br>• A: Use keypad to enter value | User does not enter the prescribed value correctly | Inaccurate amount to be infused value | 3 | Critical |

(*Continued*)

TABLE 6.6 Task analysis example for an infusion pump.—cont'd

| # | Device use sub-tasks | PCA elements | Potential use errors | Potential harm (outcomes) | Severity of harm[a] | Task category |
|---|---|---|---|---|---|---|
| 5 | Enter prescribed rate of infusion | • P: See editable field<br>• P: See measurement units<br>• P: See keypad<br>• C: Understand where to enter rate<br>• A: Select editable field<br>• A: Use keypad to enter value | User does not enter the prescribed value correctly | Inaccurate rate of infusion | 3 | Critical |
| 6 | Select CONFIRM | • P: See option to CONFIRM<br>• C: Read and understand option labels<br>• A: Select option to CONFIRM | User does not select the correct option to proceed with infusion | Delayed therapy or no therapy delivered | 3 | Critical |
| 7 | Understand confirmation message | • P: See confirmation message<br>• C: Understand confirmation message | User does not recognize or understand confirmation message | Inaccurate infusion | 3 | Critical |
| 8 | Select RUN to start continuous therapy delivery | • P: See RUN button<br>• C: Understand that RUN button will start therapy<br>• A: Select RUN button | User does not select RUN to start delivering therapy | Delayed therapy or no therapy delivered | 3 | Critical |
| 9 | Wait for a minimum of 30 s before additional programming | • P: See Program Pump confirmation screen<br>• C: Know to wait a minimum of 30 s<br>• A: Wait 30 s before proceeding with any other programming | User does not wait a minimum of 30 s before proceeding with any additional programming | System error may appear | 2 | Essential |

[a]*Scale used for severity is as follow: 3—Significant, 2—Moderate, 1—Negligible.*

## 3. Hierarchical task analysis

An alternate method of task analysis is a hierarchical task analysis (HTA), which describes the activity or workflow to be analyzed in terms of a hierarchy of goals, sub-goals, operations, and plans (Stanton et al., 2013). An HTA may stand on its own or be integrated with additional task analyses. The end result of an HTA is a detailed description of a task or activity workflow. A benefit of a HTA is to further describe the relationships between the use case (parent task) and subtasks through a numbering scheme. This approach is extremely helpful in the design of complex software systems (see description in Chapter 7) wherein different approaches to describing user interactions can be broad and deep while maintaining a structured approach to the use case. For example, in HTA there could be two different paths to complete the same use case such as. If a new software introduced a login HTA, which could look like:

1. If a user is new to the system, complete Task 1
2. If a user has registered within the system, complete Tasks 1.1, 1.2 and 1.5

The benefits of HTA include the ability to compare different approaches to supporting the same task, assuring the development team uses the same consistent approach and language in order to compare the use case approaches. The results are an abstraction of the task which enable designers to capture multiple implementations of a design pattern by expressing interactions in a structured format.

An HTA is completed by focusing on a hierarchy of use cases, tasks, sub-tasks, and use plans, as described in a process for any application laid out by Annett (2004).

1. Define the use case(s) under analysis.
2. Collect data about the use case, such as tasks involved, interactions, and constraints, through observation, subject matter expert input, and walkthroughs.
3. Constructed the HTA with the starting point of the flow chart as the use case.

For the hand-held glucose meter example, the use case would be for the user to take a blood glucose reading. Tasks can then be identified that need to be completed to achieve the overall goal. For example, a task would be to turn on the blood glucose meter or load the test strip.

Tasks are then broken down further into sub-tasks, if required, so that the bottom level of each branch is a stand-alone functional operation. For instance, one sub-task of the HTA would be to press the power button or insert testing strip into BG meter. After all tasks and sub-tasks are in place, then use plans are added to dictate the order in which the goals are achieved. Use plans are numbered steps by which each task or sub-task will be carried out. If tasks and sub-tasks must be completed in a specific order(s), it is specified via a use plan. Fig. 6.3 shows individual use plans for different tasks. For example, for the task of turn on the glucose meter (Task 1, Use Plan 1), the batteries must be inserted first before pressing the power button, which would be dictated by a use plan.

The HTA can be completed in a hierarchical format and/or a tabular format. The hierarchical format is beneficial in visualizing interconnected tasks and the order in which they need to be completed. The tabular format is typically used as a supplement to the hierarchical format. Table 6.7 is the tabular format version of the HTA in Fig. 6.3. The tabular format (Table 6.7) structures the tasks into a more linear framework that may not suit certain

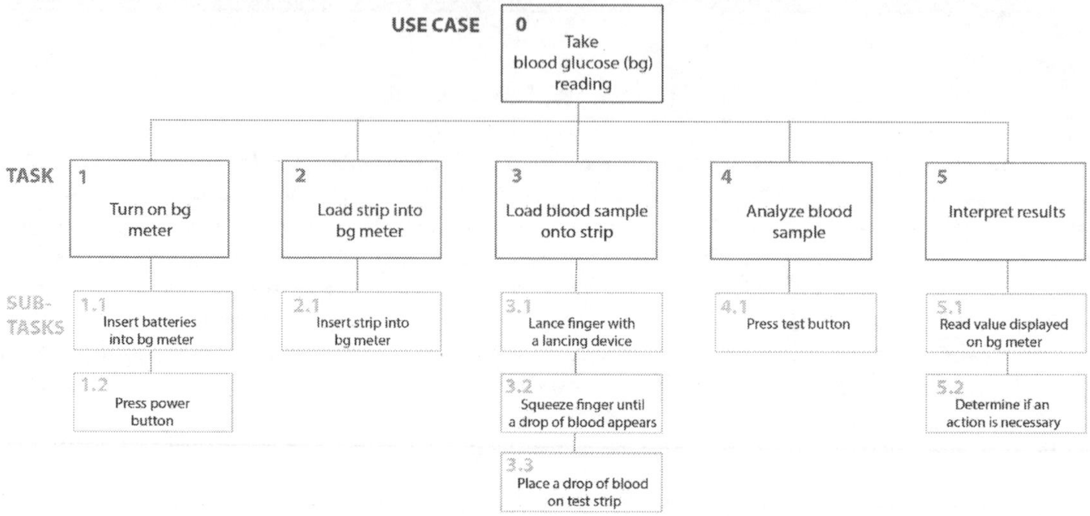


**FIG. 6.3** Hierarchical task analysis for a handheld blood glucose meter (hierarchical format). Example plans are included in Table 6.7 below.

**TABLE 6.7** Hierarchical task analysis for a handheld blood glucose meter.

**Hierarchical Task Analysis – Blood Glucose (BG) Meter**

0. Take BG reading

Plan 0 - User must complete tasks in the following order to successfully take a BG reading:
1 – 2 – 3 – 4 – 5, or
2 – 1 – 3–4 – 5

1. Turn on BG meter

Plan 1 - User must complete tasks in the following order to successfully turn on BG meter:
1.1–1.2

1.1 Insert batteries into BG meter
1.2 Press power button

2. Load strip into BG meter

2.1 Insert strip into BG meter slot

3. Load blood sample onto strip

Plan 3 - User must complete tasks in the following order to successfully load blood sample onto strip:
3.1–3.2 – 3.3

3.1 Lance finger with a lancing device
3.2 Squeeze finger until a drop of blood appears
3.3 Place drop of blood on sample area of the test strip

4. Analyze blood sample

4.1 Press Test button

5. Interpret results
Plan 5 - User must complete tasks in the following order to successfully interpret results:
5.1–5.2

5.1 Read value displayed by BG meter
5.2 Determine if action is necessary

applications but allows for easy conversion into a full task analysis (including PCA, use error identification, and determination of severity level).

Using HTA as an analysis method has both advantages and disadvantages. Advantages include that it is easy to create and implement and is a useful input to other analyses including task analysis, PCA analysis and use error identification. It provides a high level overview that can be beneficial as a reference while performing more detailed analyses. There are also disadvantages to using HTA as the only task analysis because the HTA contains mostly descriptive rather than analytical information and it might not be suited for complex systems and tasks (Annett, 2004).

## 4. Task analysis as a design tool

A task analysis serves as a tool throughout the design process. In addition to providing the foundation for task identification and task categorization, a task analysis can play a crucial role in improving device design. Device designers can utilize the task analysis and corresponding PCA to prioritize design features associated with critical tasks and implement mitigations to prevent use errors from occurring. Additionally task analysis can lead to a more efficient and effective design related to human performance, including safety and productivity. When a task analysis is used early in the design process, there is a two-way flow of information. Human factors requirements and limitations are fed into the design process and design decisions and constraints are then fed into the task analysis (Kirwan & Ainsworth, 1992). Examples of how a task analysis can be used as a design tool are below:

- A task analysis identifies the task of receiving a full dose from an auto-injector as a critical task. The associated PCA analysis identifies auditory and visual cues to inform the user of dose completion. This will inform the device designer that the auditory and visual cues are priority design features to be optimized for safe and effective use.
- A task analysis identifies the task of administering a dose using an auto-injector as a critical task and a green bar appears in the medication window when all medication has been administered. Because it was categorized as critical, early human factors testing was conducted. Users provided feedback that the medication kept streaming from the needle for a few seconds after the green bar appeared in the medication window and were concerned that they would not have received a full dose. Designers used the task analysis and testing results to revise the device design prior to validation testing.

## 5. Using task analysis for instructional design

The task analysis is a fundamental input to instructional material development. Instructional designers rely on the task analysis as an outline for the necessary steps, formatting and structure of the instructional material. The goal of the instructional material is to guide accurate, safe and effective user performance with a device, which is defined by the task analysis. This definition of performance includes identification of critical, non-critical, and knowledge tasks. Additionally, it may require that the task analysis include descriptions of the tasks in terms of the knowledge, skills and abilities (KSAs) the user must have in order to achieve successful performance of device use. It is important to note that KSAs are most helpful when developed by instructional designers or professionals who are familiar with performance-based training. Otherwise KSAs easily become an exhaustive list of, for example, required user knowledge that is content heavy and may not necessarily be relevant to performance descriptions or be useful in the development of instructions. Deficiencies in a task analysis, such as a lack of clear KSAs, can lead to poorly designed instructions, which can cause difficulties use errors, including incorrect performance and lack understanding and ability to find critical information.

Often, when human factors experts are analyzing use errors that have occurred in a usability evaluation, the instructional materials may be a potential source of the error. A robust and thorough task analysis can help pinpoint the actual source of the use error and provide insight into a potential mitigation.

Additionally, instructional designers may be required to look deeper and in more detail when developing the task analysis. They may ask questions related to the importance of timing and the consequences if a task is not performed correctly. These insights impact how the instructional material is presented including warnings, cautions and format as it is common for use errors to stem from knowledge comprehension aspects of a task. Much of the cognitive load on a user comes from the processing of the instructional material. The task analysis helps promote user performance expectations and clearly communicates KSAs. Chapter 9 provides more insight into the development and role of instructional design in overall device usability.

## 6. Summary

In summary, a task analysis is a fundamental process in early device design, instructional material creation, formative testing protocol development and final usability evaluation process. It is the development is a systematic process involving use case identification, task identification, sub-task breakdown. Providing key information in regards to device usability through PCA analysis, potential use error identification and potential harm/outcome identification. In creating a task analysis, it is important to involve input from knowledge experts (e.g. clinical stakeholders), MAUDE database research on similar devices, observations of users with similar devices, expert review, investigation into videos of device use or similar device use, instructional materials, risk documentation, reverse engineering and/or direct experience with similar devices. Task analyses can take the form of a hierarchical task analysis, which is beneficial in visualizing interconnected tasks and the order in which they need to be completed. When developed and used correctly, a task analysis is an invaluable tool for improving device design, serves as a fundamental input to instructional material development, provides insights into use-related risks and user-device interactions that could be problematic for users.

## Acknowledgments

Special thanks to Drs. Daryle Gardner-Bonneau, Deborah Billings Broky, and Jessie Huisinga for critical comments and suggestions. Thank you to Elissa Yancey for editing.

## References

Annett, J. (2004). Hierarchical task analysis. In E. Salas, H. W. Hendrick, N. A. Stanton, A. Hedge, & K. Brookhuis (Eds.), *Handbook of human factors and ergonomics methods* (pp. 329–337). CRC Press. https://doi.org/10.1201/9780203489925-9.

ANSI/AAMI HE75. (2009). *Human factors engineering – design of medical devices.*

ANSI/AAMI 14971. (2010). *Medical devices— application of risk management to medical devices.* American National Standard, 2008.

Bisantz, A. M., & Burns, C. M. (2008). *Applications of cognitive work analysis.* CRC Press.

Kirwan, B., & Ainsworth, L. K. (1992). A guide to task analysis. *University of Michigan, 16*(2), 417. https://doi.org/10.4324/9780203221457.

Sharit, J. (2006). *Handbook of human factors and ergonomics.* https://doi.org/10.1002/0470048204.ch27.

Stanton, N. A., Salmon, P. M., et al. (2013). *Human factors methods: A practical guide for engineering and design.* Ashgate Pub. Co.

U.S. Department of Health and Human Services Food & Drug Administration Center for Devices and Radiological Health (2016). *Applying human factors and usability engineering to medical devices.*

# PART III

# Human factors in design

CHAPTER

# 7

# Applied human factors in design


*Mary Beth Privitera[a], M. Robert Garfield[b], Daryle Gardner-Bonneau[c]*

[a]HS Design, Gladstone, NJ, United States; [b]Abbot, St. Paul, MN, United States; [c]Bonneau and Associates, Portage, MI, United States


OUTLINE




*Applied Human Factors in Medical Device Design*
https://doi.org/10.1016/B978-0-12-816163-0.00007-4

*Do it by Design! Applied human factors is about designing systems that are resilient to the unanticipated event.*

## 1. Introduction

Designing products for "everyone" is problematic as one person's experience will always be different than another's. This includes the fact that there is rarely a consensus of opinion and certainly a high degree of variability person to person. Each individual has a unique set of needs, perception and experience. In essence, environmental and human variability are perpetual. This reality must be considered when designing medical devices.

The role of the human factors engineers is to assist the design team to produce designs that better meet the capabilities, limitations, and needs of the user. In regards to capabilities, this means that the device fits the user's mental and physical constraints. That there is limitations in place which prevent injury; and that individual needs are considered to improve usability and efficiency based on context.

Human factors experts gather information regarding human characteristics and interactions with the work environment in order to design systems resilient to use error (Russ et al., 2013). The standard ANSI/AAMI HE75 (AAMI, 2009), is the most comprehensive resource in regards to human factors design guidance. This chapter provides examples of utilizing this guidance as well as providing further clarity of specific design processes for medical device design. For the purposes of design, key sections of ANSI/AAMI HE75 include:

- Human Skills and Abilities
- Anthropometry and Biomechanics
- Alarm Design
- Controls
- Visual Displays

- Software User Interfaces
- Medical Hand Tool and Instrument Design

This list is not exhaustive however represents the minimum references a design team should consider in their process. Other Sections such as Combination Products, Workstations and Packaging Design are examples of specific topics that are further detailed in the guidance however are not included in this chapter discussion.

This chapter discusses the importance of considering human factors in the product design for both the physical product embodiment as well as any controlling user interface or complimentary computer applications. This includes sections on how to know your users, know the use environment, specific elements of human factors in design including affordances, touch points, color, materials, and finishes with a case example. It also includes a section of software design or user experience design with detailed descriptions on design intent, constraints, graphic design including aspects of detection and discrimination. It concludes with the design of alarms with highlights from recent changes in the standards as a result of advanced research into the perception of alarms.

## 2. Understand your users

Applied human factors in design results in total solution for the user. It addresses human factors holistically including the physical interactions as well as the mental workload (perception and cognition) required. In order to do this an understanding of basic human skills and abilities, anthropometry and biomechanics, accessibility considerations as well as cross-cultural/cross-national design considerations must be explored in the context of the proposed medical device use (AAMI, 2009).

This includes exploring the "fit" of the device. Fit as defined by the overall shape and appearance of the device relative to the location of physical interaction. For example, the fit of a hand tool is the relationship of the tool itself to hand size and finger reach. Ultimately, fit is determined by the following:

- Anthropometry: determining the physical fit with the user/appropriate sizing
- Biomechanics: physical limitations for movement and control at a given joint which are required in order to use the device. It can be further broken down into:
  - Reach: access of a control
  - Operation: use of a control
  - Injury: reduction of pinch points, consideration toward repetitive use injury
  - Safety: inadvertent control activation

It also requires and understanding of the user's workload, stress, education and training. Specifically, the user's capacity for cognition and perception including:

- How our senses take in and process stimuli
- Working short term and long-term memory: tasks that are built upon tacit knowledge and those which are newly experienced

Understanding the user, (capabilities, limitations, anthropometry, biomechanics, etc) and the context of use (working and social environment) cohesively impact safe and effective device use.

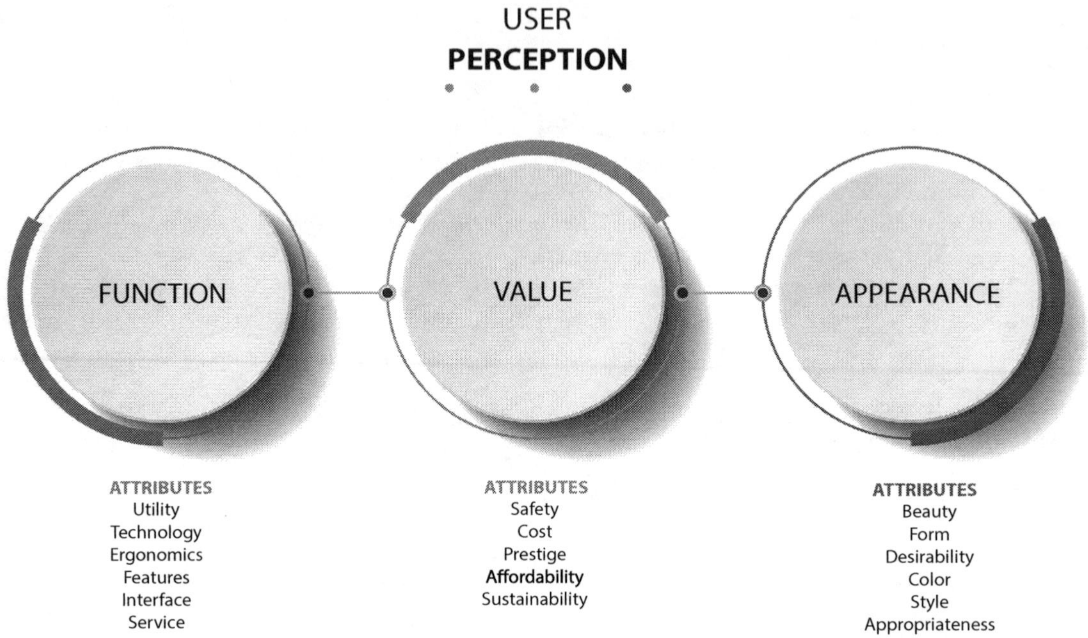


**FIG. 7.1** The relationship of function and appearance to user value (Privitera & Johnson, 2009).

Ultimately, the user judges' value and decides to select a device. This judgment is based on their perceived value as a result of function and appearance. Each of these are interconnected (Fig. 7.1) and related to one another. With enhanced aesthetic order, which is appropriate for the user, it will inherently incorporate human factors thus the usability will improve. Ultimately user perceived value will increase.

## 2.1 Using anthropometry and biomechanics to determine fit

Anthropometry refers to the measurements and physical attributes of the human body. Biomechanics refer to the structure and function of mechanical aspects of living organisms, most importantly for device design, it often refers to joint biomechanics. Each of these dimensions are important and directly influence design requirements as they determine how the device will "fit" the user. In other words, anthropometry equates static dimension and biomechanics equate dynamic dimensions. For example:

- Anthropometry = Static (structural) measurements are based on individual parts of the body in fixed positions (e.g., seated eye height or hand breath).
- Biomechanics = Dynamic (functional) measurements relate to the body in motion (e.g., arm reach envelope or angular hand movement).

Static anthropometric measurements can determine appropriate product size and shape details. Whereas biomechanics measurements can be used to determine the human limits for strength, endurance, speed, and accuracy.

### 2.1.1 Understanding percentiles

Anthropometric and biomechanics data is often presented in percentiles. Common examples include 5th, 50th, and 95th percentiles. Percentiles correspond with the frequency of measurements within the distribution of the data. For example:

- The 5th percentile is a value associated with the location within the data where 5% of data is below that value.
- The 50th percentile is the middle value where 50% of the data is less than and 50% of the data is greater than that value.
- The 95th percentile is the value where 5% of the data has is a larger value.

It is a common misconception that designing to accommodate a 5th percentile female and 95th percentile male will naturally lead to a good design. The trouble is there is no such thing as a complete 5th percentile female or a 95th percentile male in every aspect. The challenge is that each measurement is an individual characteristic. Any given percentile value is specific to its measure and is only useable for comparison purposes against similar data (e.g., percentile data can be used to compare hand strength data from one age group to another).

### 2.1.2 Deriving device form from anthropometry

Anthropometric data drives specific design requirements that ultimately determine the product form such as overall handle length or height of a workstation.

There are several steps for selecting the appropriate anthropometric data to use when designing:

**1.** Determine the intended user group(s).

It is important to define who the intended users are. Every user group has unique characteristics. Factors including nationality, age, and gender impact anthropometric data.

**2.** Identify the relevant dimensions (body attributes).

While this may be readily identified by one main dimension, often there are several dimensions that are important for a device design. In this case, a list of functional dimensions should be generated with prioritization and focus on the elements that are most influential to device performance.

**3.** Define the target percentile/size range.

An ideal design would accommodate everyone. Unfortunately, this is often not practical or possible. While there may be some anthropometric characteristics with narrow dimensional ranges, which can be entirely accommodated, many measurements include a wide range of data. There are a few paths to choose from that may help in selecting a target value. These include:

- Design using the average.

Using 50th percentile data is an approach that provides "average" inputs for a design. Although this path may not produce a good solution for any given individual user, average data can be acceptable depending on the situation.

- Design using 'edge' cases.

Selecting dimensions from either or both ends of the spectrum (e.g., 5th and 95th percentile) can be successful. This path can help accommodate the largest and/or smallest users for critical limiting dimensions. This path involves defining requirements that balances maximum and minimum limiting factors. For example, if a hand tool's grip span accommodates the smallest users and is acceptable to largest users, the overall design may be very inclusive.

When multiple connected dimensions are required, a different approach is necessary. As it is the case that when multiple variables interact, stacking errors occur since there is no "average", 5th percentile, 50th, or 95th percentile person. Therefore, it is not possible to select multiple variables and maintain inclusiveness. The use of two 5th to 95th percentile ranges results in a range that is far less than 90% inclusive. For example, a group of individuals that fall between the 5th and 95th percentile for stature and weight includes only 82% of the targeted population (Robinette, 2012). Solving for multi-dimensional non-interactive size/shape characteristics can be achieved by expanding the range for each measure (e.g., to the 1st to 99th range) (Robinette, 2012). If multiple dimensions interact other tactics are necessary including: 3D-CAD anthropometrics models, testing with live human subjects, and calculation of multi-variable sample boundary conditions (Robinette, 2012).

**4.** Consider constraints.

Anthropometric data can provide a robust framework to define requirements however there are limits to blindly using the data one-to-one. Common constraints include:

- Population changes.

The dimensions of the general population are ever changing; improving living conditions have resulted in a population that is taller and larger today than previous generations (Eveleth, 2001). Anthropometric percentiles within older textbooks and references may not reflect the exact measurements of the current population.

- Sample size limitations.

Available reference data may be based on limited sample sizes or narrow subpopulations and may not be generalizable. For example, a study on the strength and size of military age males and females conducted by the U.S. Army does not represent the broader population.

- Environmental factors.

It may be necessary to adjust anthropometric data to more accurately represent the intended user in their use environment. Adjustments for gloves, clothing or other personal protective equipment (PPE) are not included in the data. Anthropometric measurements are typically taken against bare or minimally covered skin. It is important to adjust the data to accommodate environment specific factors that were not included when the measurements were taken.

- Linear measurements versus 3D solutions.

Most anthropometric data is provided as 2D values. There are limitations to directly using this data to construct 3D designs. Designers may need to adapt 2D measurements to better represent users in 3D space. 3D-CAD anthropometric models assist with complex design problems where there are multiple size constraints and multiple body attributes to accommodate (e.g., automotive interior design). Software programs such as JACK (Siemens Inc., 2011) and RAMIS (Intrinsys Ltd., 2018) can be used to evaluate designs using interactive anthropometric models. These ergonomic CAD software programs are widely used in automotive, industrial, and defense applications and are helpful for complex workstation design (Blanchonette & Defence Science and Technology Organisation, 2010).

- The necessary data is unavailable.

Texts such as the Human Factors and Ergonomics Design Handbook (Tillman, Tillman, Rose, & Woodson, 2016) or HE75 (AAMI, 2009) provide common anthropometric measurements. If these or other resources do not provide the required data, it is possible to fixture a measurement tool in order to collect data. Tools such as dynamometers, force gages, or load cells to capture novel measurements. Using statistics, a correlation can sometimes be found between custom measurements and one or more standards measurements. This mathematical relationship enables project teams to predict a range of capabilities based on published databases of the standard measurements, instead of having to conduct a larger study for the custom measurements.

**5.** Test prototypes with users.

In all cases, anthropometric driven requirements should be confirmed via prototype evaluations with representative users. These studies ensure anthropometric design requirements fit the intended user.

Additional information on human abilities, anthropometry, and biomechanics can be found in ANSI/AAMI HE75 Basic Human Skills and Abilities and Anthropometry and Biomechanics.

## 2.2 Use related injury prevention

Injury prevention is for both the user and the patient. In some instances, the patient is the user, as devices can be used on oneself. The design of the device should prevent user injury from repetitive motion, vibration or from forcing users to exert themselves beyond their capabilities.

### *2.2.1 Nature of injuries*

The nature of injuries includes acute trauma, which can be traced to a single incident or sudden trauma (broken bones, strained ligaments, etc.) or cumulative trauma disorders (CTD). CTD develop over time as a result of "microtraumas" to the soft tissues of the body e.g. repetitive motion injuries. The following CTD risk factors are known to increase the likeliness of developing chronic injuries of the upper extremities: extreme postures, forceful exertions repetitive motion, contact stresses, low temperature, and vibration exposure Putz-Anderson (1988).

### *2.2.2 Using physiological measures to determine injury potential*

By using physiologic measures such as goniometers (joint angles), electromyography (EMG), pressure mapping or heart rate when assessing a potential design, injuries may be avoided.

For extreme postures, using a goniometer to record joint measurements can demonstrate postures which are out of range and could be injurious. This is best done if assessing repeatable task.

To measure how hard muscles are working, use EMG testing. EMG is a technique for evaluating and recording electrical activity produced by the muscles. The stronger the contraction, the more neurons firing the results are traces increase in amplitude with more activity.

To measure pressure distribution of tool-hand interaction, pressure sensors can be added to the users' hand then measured as an objective determinant of stress.

This can be an indicator of stress which is unavoidable. For example, if heart rate is used to measure stress during a complex life or death surgical case, it would be appropriate to have sharp increases at times. This information would have to be coupled with perceived stress in order to be maximally useful.

## 3. Know the use environment

Device designers must how a device will be used, what it must accomplish and all aspects of the use environment in order to ensure the design is safe and effective. Inherently, device designs are context dependent; it may not be appropriate for all applications or locations.

This requires exploration of the use environment in order to determine those constraints that the device design must accommodate such as ancillary equipment in the workspace, work height and orientation, lighting, noise, temperature, temperature as well as possible distractions. Each of these contribute to context specific design requirements, e.g. the design must provide thermal protection at all use interaction points due to extreme temperatures.

Specific environmental considerations include:

• Use work space

Devices are used in a variety of work spaces (e.g., hospitals, homes, and mobile scenarios) and may also be deployed in multiple environments. Each unique use scenario must be analyzed for unique characteristics. Multiple use environments may have conflicting factors which need to be balanced during development.

• Use workspace height and orientation

The height, orientation, and location of a device within its environment directly influences how it is used. For example, in laparoscopy users may have severe shoulder abduction-elbow flexion as a result of pistol grip designs, as a result of user position relative to the surgical table height and while this is not necessarily problematic, it will cause the users discomfort and reduced capabilities in forces required to activate controls. Device use is also influenced by other objects in the environment in regards to storage and access.

• Lighting

Lighting includes ambient illumination as well as direct sources of light. Dark or low light environments require special accommodations such as backlit screens and glowing touchpoints for easy identification and improved readability. Devices which reside within environments with extreme lighting often require display monitors that reduce/control for glare to ensure sufficient labeling contrast. Solutions for lighting are task and risk specific in the overall device design.

- Auditory noise

Ambient noise impacts all audible signals a device may present. Additionally, alarm con ditions can induce alarm fatigue and cause users to ignore or overlook important warnings. Devices that provide excessive auditory feedback are a distraction which can overpower vital information and/or hinder peer communication. The level of noise in the intended use environment must be considered when developing auditory features.

- Temperature and humidity

Temperature and humanity directly affect device and user performance. They also impact the clothing worn by users e.g. they may require protective clothing.

- Distractions

Distractions come from multiple sources and take various forms. Users may be distracted by ambient noise, background music, peer communication, multi-tasking requirements, fatigue, and stress.

To better understand the use environment in order to determine specific design requirements, contextual inquiry (Chapter 5) should be utilized as part of a robust human factors strategy.

For more information environmental considerations, see ANSI/AAMI HE75 Environmental, Organizational, and Structural Issues. Additional information on environment-specific design constraints can be found in ANSI/AAMI HE75 Design of Transportable Medical Devices and Devices Used in Non-Clinical Environments by Laypersons.

## 4. Device design

Physical products come in all shapes and sizes with varying degrees of complexity (Fig. 7.2). They range from simple tools (e.g., scalpels) and handheld devices (e.g., blood pressure monitors) to large workstations (e.g., CT scanners). The design process for developing these physical products is consistent: user research, requirements generation, design exploration, prototype, and test. This iterative process elicits feedback on ease of use and requirements fulfillment.

Design teams must solve for multiple human factors constraints to be successful. This section provides a brief overview of a human factors influences that when taken into consideration during development drive good physical product design. These range from universal factors such as the environment and workflow to hardware specific details such as color, materials, and finish.

### 4.1 Affordances and design cues

Users construct mental models of how the world around them functions. This extends to the operation of the products they interact with in their day-to-day lives. Common behaviors and design conventions associated with one product are often assumed to extend to another.

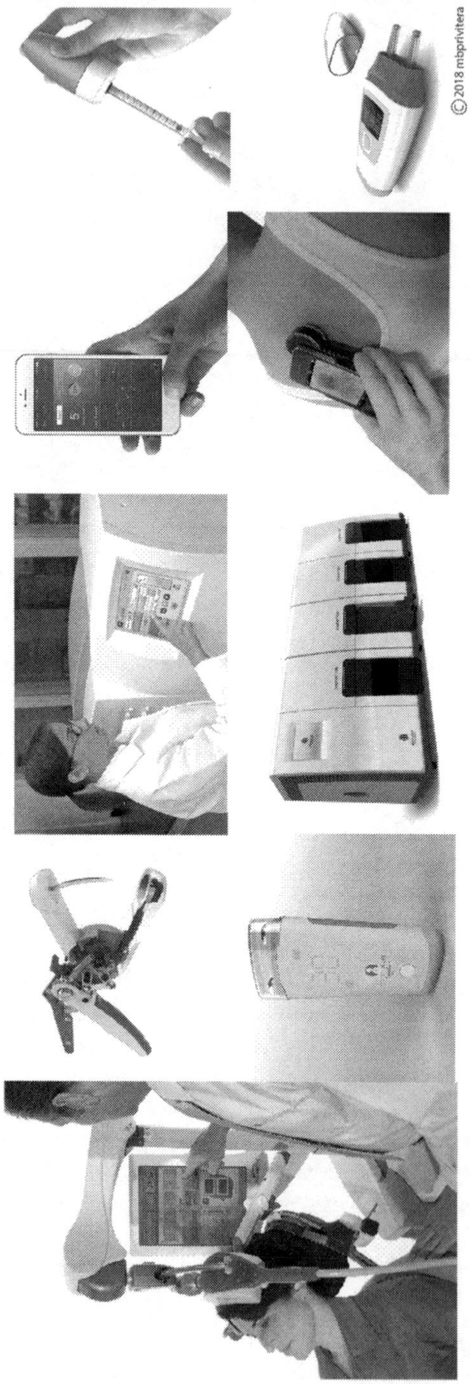


FIG. 7.2 Varieties of medical devices from hand held devices, software, to large lab equipment and workstations. *Image provided by HS Design.*

To assist users' understanding of proper device function, a physical product's form should embody how it is used. This is accomplished through affordances and design cues.

Affordances are design features that assist users in understanding how a device works; they can be real attributes or perceived properties of proper use (Norman, 1990). In physical product design, multiple elements of a design assist in communicating proper use. These cues are embodied by a product's visual form and the feedback provided to users.

Conscious and subconscious hints can be provided through simple design elements or expressed through the broader product embodiment. A product's surfaces (edges, ridges, styling lines) and textures (e.g., bumps for grip or ridges for friction) can imply directionality and movement. Labeling and visual indicators (e.g., arrows) also provide modest but powerful assistance. Likewise, handle location and form communicate use related information. For example, the provision of a power versus precision grip handle can imply the appropriate level of force to apply during use.

Affordances and design cues can also be utilized to prevent inadvertent action and error. Forced constraints and poka-yoke features help users avoid mistakes. In this regard, keying mechanisms stop misconnections from being made and steer users away from incorrect locations or orientations. Surface features can also be used to reduce error (e.g., physically recessed power buttons prevent inadvertent actuation).

Designers should provide cues for proper use as part of a physical product's exterior form and user interface. These cues should build upon common design conventions and/or communicate any variability from the norm. Additional information regarding common conventions, mental models, affordances, and design cues can be found throughout ANSI/AAMI HE75 including General Principals.

## 4.2 Aesthetic beauty as it relates to usability

Both aesthetics and affordances are considered to be measures of product success as designers use these two ostensibly distinct theoretical elements in order to provide effective ways of interaction (Xenakis & Arnellos, 2013). In combination, these enhance a user's ability to detect action possibilities (affordances) that allow the user to form an opinion of aesthetics. The aesthetic experience can be viewed as a complex cognitive phenomenon that constitutes several processes that emerge through interaction (Xenakis & Arnellos, 2013).

There is a stereotype that "beautiful is good and good is useable," while beauty is accessible immediately through visual presentation, usability reveals itself only through interaction (Marc Hassenzahl & Andrew Monk, 2010). There is a correlation of beauty and usability that implies devices are either both beautiful and useable or ugly and unusable (Marc Hassenzahl & Andrew Monk, 2010). Visual aesthetics do affect perceived usability, satisfaction and pleasure (Moshagen & Thielsch, 2010).

The formal parameters of aesthetic objects include simplicity and diversity, color and craftsmanship (Moshagen & Thielsch, 2010). Each of these parameters carry a unique meaning which are:

### *4.2.1 Simplicity*

Simplicity reflects the aspects that facilitate perception and the cognitive processing of layout. This includes clarity, orderliness, homogeneity, grouping and balance (Moshagen & Thielsch, 2010).

#### 4.2.2 *Diversity*

Diversity is the visual richness, dynamic nature, variety, creativity, and novelty of a design (Moshagen & Thielsch, 2010).

#### 4.2.3 *Colorfulness*

This includes the number of colors and their composition.

#### 4.2.4 *Craftsmanship*

Craftsmanship refers to the construction of the design itself. It can be characterized by skillful and coherent integration of all design dimensions (Moshagen & Thielsch, 2010).

The initial reaction a user has to a device is the functional role of aesthetically-oriented emotional values to detect interactive opportunities or threats (Xenakis & Arnellos, 2013). The second is to signal other functions which control our decisions and behavior regulation processes. This emerges in relation to success or failure in reaching a potential goal. Interaction aesthetics aid the user to construct meanings that make clearer the way (action) to goal achievement (Xenakis & Arnellos, 2013). It is the role of the designers to enhance devices with characteristics which embody simplicity, diversity, appropriate colorfulness with skillful craftsmanship. This ultimately leads to perceived value (see Chapter 1 Section 3 Why might we want to do more).

### 4.3 Use interaction touch points and control selection

User interaction touchpoints (e.g., handles) and controls physically connect the user to the product. They are the primary areas of interaction and can be defined as:

- Controls are input mechanisms that allow users to change the state of a product or feature. They take a variety of forms including thumb wheels, toggle switches, triggers, slide controls, pushbuttons, and rotary knobs. Alternatively, they may be interactive screen button representations or icons with which the user can control the device.
- Touchpoints describe areas of targeted user interaction and points of contact with a product. They are the locations of physical interaction such as door latches, input controls, and handles.

#### 4.3.1 *Use interaction touch points*

As described earlier, the use of affordances or cues can assist in the user interface design. For example, in Fig. 7.3 the overall shape of the device impacts how the device is held and is explored (Figs. 7.5, 7.7, and 7.8).

While this illustrates a handheld device, use touchpoints refer to any area that the user is expected to interact with and can include areas such as handles for pushing or carrying, push buttons, control levers, etc.

The shape, color and texture provide the affordance/cue to the user in device design. Designers should factor the obviousness of use interaction areas when developing the overall device architecture and detail design of the outer housing (physical design) or screen display (software design).

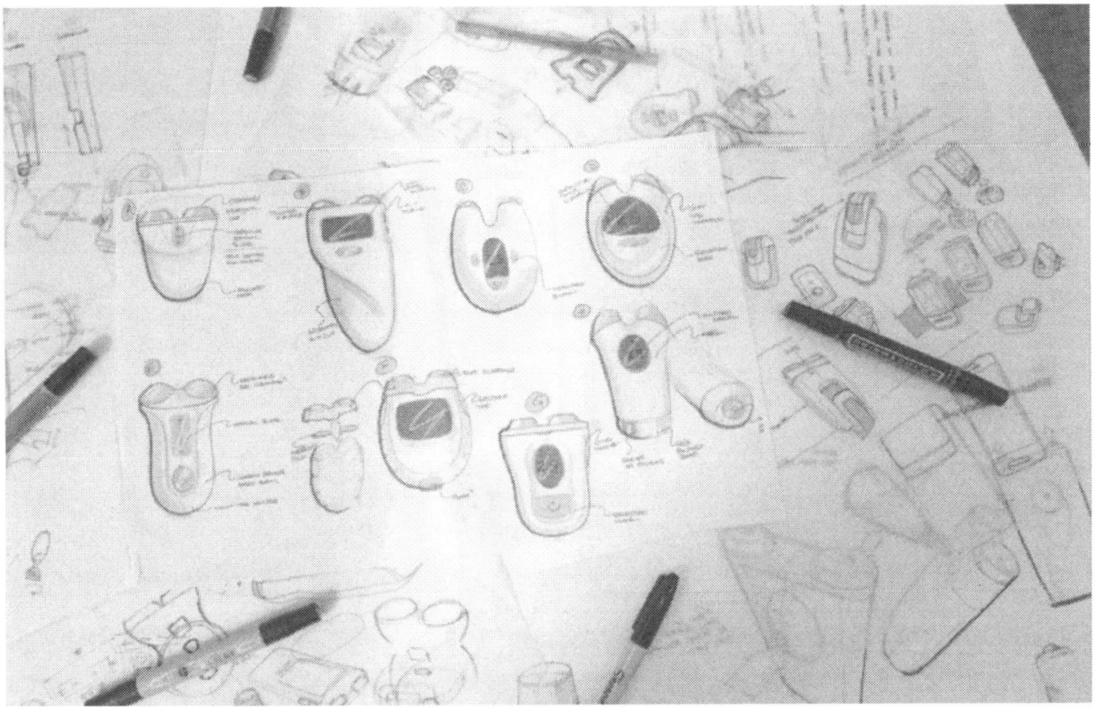

FIG. 7.3 Form sketches exploring use interaction touch points (how it feels in the user's hand). *Image provided by HS Design.*

### 4.3.2 *Control selection*

The selection of controls begins with mapping the functions that require controlling with how they should be managed. This includes the control type (e.g., button vs. switch) as well as control function (e.g., multi-state, dual-state, and continuous controls). Selection of control type and function is typically dependent on access requirements, frequency of use, as well as determining if control function is continuous or discrete in activation.

Recommendations and available options for control size, shape, positioning, travel distance, labeling, activation force, and feedback vary by control type and application. Each attribute should be tailored to a device's intended user, task, and environment of use. The design of physical controls should match common interaction paradigms, user preferences, and users' mental models of devices/control functions.

### 4.3.3 *Layout*

The position of controls and touchpoints on workstations, devices, and handheld tools can be determined based on their importance, type, or sequence of use. Control layout should consider risk and potential use errors such as inadvertent activation.

Each control type and functional option has different geometry and can be organized based upon access, use, and visibility. These attributes are driven based on the perception, cognition, and physical abilities of users. Anthropometric and biomechanics data is a primary driver for control placement and layout. It is particularly important for layout and placement of controls on handheld tools.

There are nuances for the implementation of controls on fixed input panels for devices and workstations versus handheld tools and instruments. For application specific information on touchpoints, controls, and layout see ANSI/AAMI HE75:

- Controls
- Medical Hand Tool and Instrument Design

## 4.4 Color, materials, and finish

The selection of color, materials and finish during the design process goes beyond brand guidelines and user preference. These visual and tactile design characteristics influence product perception, safety, and ease of use.

### 4.4.1 Color

Color is an influential attribute for function and aesthetics in a product's design. Color coding can be used to distinguish different elements from one another or associate discrete functions. Color plays a powerful role as a signifier for prioritization and order of operations. It is also influential as a mechanism to highlight safety critical elements, providing warnings for areas of risk. Additionally, it can serve as a status indicator, distinguishing between functional states (e.g., traffic lights).

The specific meanings derived from colors are influenced by psychological and cultural norms. U.S. specific color codes and conventions for medical applications are not applicable in all countries. Designers need to ensure the selected color conventions match the product's target market.

Color selection is influenced by human limitations. Color blindness affects 8% of men and 2% of women in the general population (AAMI, 2009). Designers should choose colors (including attention to tone, tint, shade, and hue) with these limitations in mind. Common forms of color blindness include protanopia (red blindness), deuteranopia (green blindness), protanomaly (reduced sensitivity to red light), and deuteranomaly (reduced sensitivity to green light) (AAMI, 2009).

Common conventions and multiple standards influence proper color selection. For example, IEC 60601-1-8 provides input on the use of color for the prioritization of alarm conditions: red reflects high priority alarms whereas yellow should be used for medium priority conditions (IEC, 2006). Additional influences can be drawn from guidance provided by HE75 (AAMI, 2009), the American National Standard for Safety Colors ANSI/NEMA Z535.1-2017 (NEMA, 2017) and the Occupational Safety and Health Administration (OSHA). These sources provide detailed descriptions for colors, their meanings, and their application. Designers should use these references and check for updates regularly.

Additional information on human vision and use of color for product development can be found throughout ANSI/AAMI HE75 including:

- Basic Human Skills and Abilities
- Cross-Cultural/Cross-National Design
- Software-User Interfaces

### 4.4.2 Materials

Material selection directly influences product appearance and functionality. Aspects including weight, appearance and tactile feel are influenced by material selection. In addition to manufacturing constraints, functional and economic considerations, the selection process must consider the tangible and intangible influences materiality plays on usability. There are several human factors attributes to consider as part of the material selection process, these include:

- Temperature

Materials can transfer temperature which may result in the material of handles can transfer the temperature from the internal mechanisms or the use environment to the users' hands. For example, the use of metal handles within a cold environment may negatively impact user comfort.

- Energy isolation

Materials can provide energy (e.g., electrical, or magnetic) insulation and isolation. These functional properties can help reduce the risk of injury to the user.

- Vibration

Materials can reduce the impact of vibration on the user by dampening its effect.

- Cleaning and sterilization

Some materials can more difficult to clean or during sterilization the material properties may change. Careful assessment of cleaning and sterilization needs must be considered. This process may be expedited by selecting materials with imbedded antibacterial properties or those materials with proven performance.

- Friction

Moving parts or connector interfaces are always subject to friction. Reducing friction can help lower activation forces and reduce injury risk.

- Weight and center of gravity

A product's weight and center of gravity is directly impacted by the materials used in manufacturing. By balancing the design for the use context and application, usability will be improved. This avoid situations like tipping in moving large equipment or awkward movements in hand tools.

- Visual and functional properties

Material selection affects the color and feature options available as part of an overall design. For example, plastic and glass provide options for transparent components which can improve visibility of internal mechanisms or provide the ability to check fluid levels. The material selection process should ensure that intended visual and functional properties of a given design are achievable.

#### 4.4.3 Finish

Material finish includes the visual appearance, tactile feel and surface treatment of the exterior of a product. These attributes can be selected independently from the underlying material. Various manufacturing processes can be used to achieve desired finishes including in-mold texturing and two-shot over molding, as well as secondary processes including milling, sandblasting, and paint.

Final surface selection should be resistant to damage and wear, textured to improved grip, and suitable for easy cleaning (e.g., limit crevasses). Surface texture can also be used as a signifier for touchpoints or other areas of interest.

Additional application specific information on materials and finish can be found within ANSI/AAMI HE75 including:

- Connectors and Connections
- Medical Hand Tool and Instrument Design
- Workstations

## 4.5 Case study: applied ergonomics for hand tool design

Ergonomic hand tools link the user, the device, and the use environment (Aptel and Claudon, 2002). To meet this goal, designers must control multiple characteristics to transform user needs and design inputs into finished concepts (Garfield, 2015). A parallel inside-out (internal mechanism) and outside-in (external housing) development approach is beneficial. Using this approach, engineers lead internal mechanism design while human factors and industrial design experts develop exterior handle concepts (areas of hand and tool interaction). These multidisciplinary collaborations balance relevant technical constraints along with users' capabilities.

For the purpose of brevity, this case will only describe the activities of the human factors and industrial design team in their concept development. This includes four main activities: (1) handle shape selection; (2) control selection and placement, (3) handle and controls size definition, and (4) form language and surface refinement.

#### 4.5.1 Step 1: handle shape selection

The initial step is selection of the handle shape (type), angle of use, and grip position.

The position of the hand tool relative to the user's body and the work area directly influences the recommended handle shape (Quick et al., 2003, Ulin, Ways, Armstrong, & Snook, 1990, Ulin et al., 1993). Based on information found in ANSI/AAMI HE 75:

- Power-grip pistol handle allows users to exert a high degree of force while maintaining a neutral hand position (e.g., linear laparoscopic staplers);
- Precision-grip inline handle provides users a direct axial connection to the tool's end effector.
- Inline hands are beneficial for steering and rotating devices within a fixed lumen (e.g., cardiac catheters).

In this case, an inline handle shape was selected based upon contextual information regarding user position relative to the target. Specifically, the hand tool is used in a minimally invasive procedure with a small thoracotomy approach (small incision between the ribs).

*Note:* both human factors references and information regarding the user and use environment were used to select device shape.

### 4.5.2 Step 2: control selection and placement

Controls connect the user to the device's intended function. It is important to consider multiple aspects when determining control type and placement including the underlying mechanism, similar hand tools, anthropometric constraints and the user's mental model of device function (Garfield, 2015). There are multiple control types and hybrid options to choose from (e.g., triggers, switches, rotary, slide, and push button controls) and each has unique attributes tailored to specific applications. Further compounding the control selection and placement process are devices with multiple controls. Multi-control layouts need to take into consideration factors such as task frequency, workflow, and elevated use-safety constraints (e.g., inadvertent cross-control activation).

Exploration of control selection and placement via a control's possibility matrix (Fig. 7.1) assists in answering the underlying questions of what controls to select and where to place them on the handle. A possibility matrix uses symbols (color dots) to represent a control type and a potential location providing a visualization of potential control layouts. The matrix starts with multiple outlines of the predetermined handle shape; and colored dots or other signifiers are placed to symbolize various control options. Using a possibility matrix as a discussion guide with engineering teams can expedite the design process and tradeoff discussion to best meet functional and usability requirements.

The human factors related to individual control type and constraints can be found in ANSI/AAMI HE 75 and provide the backbone of a possibility matrix. This information includes the maximum/minimum force capabilities and will be required by the technical team.

*Note:* Using published HF data regarding controls, different device configurations can be quickly generated that include the technical requirements. This enables robust conversation regarding tradeoff decisions required to maximize functionality and optimize usability.

### 4.5.3 Step 3: handle and control size

The process to determine size (e.g., cross-sectional size, control spacing, handle length, form elements) balances factors including anthropometric guidelines, the internal mechanism and feedback from users. It is important to assess basic environmental inputs (e.g., gloved hands) and critical hand dimensions (e.g., finger loop size minimums, grip span limits, and hand breath) to ensure the design is appropriate for indented users. User testing with prototype models further assists with detail design process (Fig. 7.2).

*Note:* both published literature and user testing can assist in further form refinement to assess "fit."

### 4.5.4 Step 4: form language and surface refinement

3D Surface refinement builds on the previous steps of the process and involves determining cross-sectional shape, form language, surfaces textures, and materials (Fig. 7.3). These attributes communicate critical interface elements to the user. They directly influence hand placement, touchpoint identification, and grip friction. Adding guards, removing sharp edges/creases, and eliminating pinch points are also important to increase safety of the final design. The visible features of the handle are not the only aspects of design refinement.

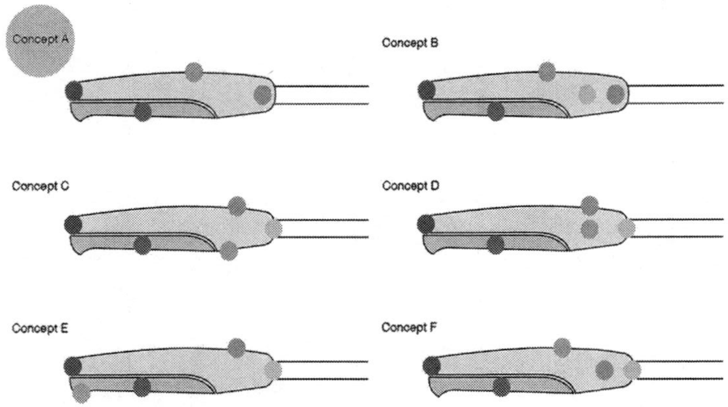


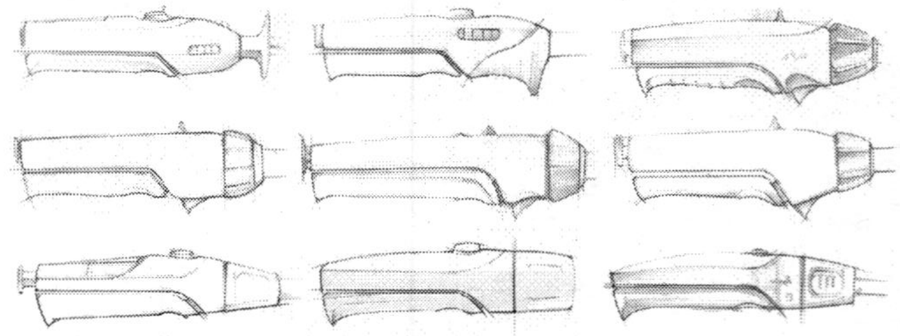

| Icon | Control | Shape | Order of Use | Critical Use 1 (Low) - 5 High) |
|---|---|---|---|---|
| ● | Jaw open/close | Lever | 3 | 5 |
| ● | Lock/unlock | Button/Trigger | 1&7 | 5 |
| ● | Articulation | Wheel/arm | 2 | 3 |
| ● | Needle Advancement | Slide | 4 & 6 | 5 |
| ● | Plunger | Plunger/Twist | 5 | 5 |

FIG. 7.4 Controls possibility matrix for an inline handle (Left) and corresponding hand sketches (Right) by Kevin Zylka.

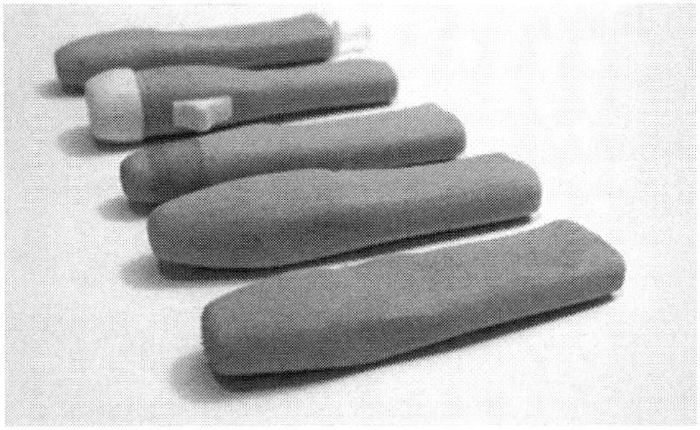

FIG. 7.5 Form handle prototypes by Kevin Zylka.

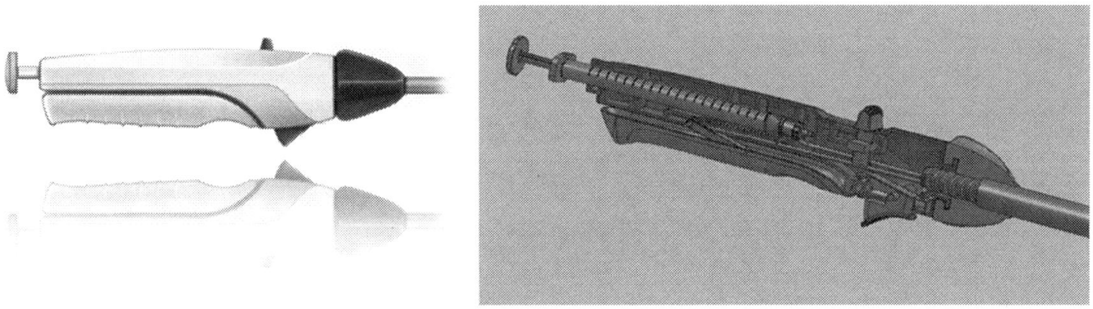

FIG. 7.6 Sketch Rendering of final visual direction (Left) and final mechanical design (Right) by Kevin Zylka.

Control activation forces and feedback, as well as weight, center of gravity, and tether location impact ease of use and use safety.

As with all product development, hand tool design is an iterative process. Models of increasing fidelity facilitate evaluation and refinement. The process involves a give and take between engineering and industrial design where the benefits of textbook solutions are weighed against technical constraints. Successful designs find an equilibrium that thoughtfully navigates these tradeoffs.

***Note:*** form refinements prevent pinch point injuries, provide cues for the gripping surface and the use of color and shape indicate controls.

For detailed information on ergonomic hand tool design refer to ANSI/AAMI HE75 Section 22 Medical Hand Tool and Instrument Design.

## 5. Software design: user experience (UX) design

The design of software either embedded within a physical product design or as a standalone product, has the same design process: requirement elicitation, system specifications, design, implementation to functional prototypes, unit and system testing, release, and maintenance.

The biggest difference between software design and hardware design are that updates and substantive modifications of software become readily available. This section describes considerations for improved design by incorporating key human factors principles at the onset of design and throughout the development process.

## 5.1 User experience design

User experience design is the art and science required to integrate all the elements that comprise an interactive system. This includes the programming, interaction design (means of input), interface design (interactive graphics), information design (relationship and user comprehension), motion, sound, graphic design, and written language. These elements combined comprise an interactive system as a user does not naturally distinguish individual elements of the system. Table 7.1 below provides an example of individual elements with further definition.

**TABLE 7.1** User experience elements.

| Element | Definition |
|---|---|
| Programming | Code for data input, processing, and retrieval |
| Interaction Design | Workflow, system flow/behavior and human comprehensibility provided by the user interface |
| Interface Design | Graphic information design utilized to indicate data control or manipulation |
| Information Design | Text style, graphics, aesthetic order (composition & hierarchy) for information structure, meaning, relationship and user comprehension |
| Motion | Animation, motion, changes, time, or rhythmic movements of elements |
| Sound | Audible signal, music, voice used to enhance experience as feedback or input |
| Graphic Design | Shapes, symbols, lines, color, texture, dimension, composition, and all elements of the visual screen representations |
| Language | Users natural language e.g., English, Spanish, Japanese, etc. |

## 5.2 Describing the design intent and constraints

In virtually every software design, the design intent is to be clear and concise to maximize usability and ease of use. By stating the design intent in general with increasing detail, better decisions can be made regarding the overall design aesthetic to optimize clarity and visibility of the information presented. A design intent statement will include specific details regarding the user and any training requirements. Below is an example design intent statement:

The ACME 2000 Graphic User Interface is designed to be clear and concise, maximizing usability and ease of use principles from published guidelines. The interface utilizes flat visuals, bright colors, and high contrast to optimize clarity and visibility of the screen. Users are expected to undergo training or in-service on the entire device prior to using.

There are several common design constraints that impact UI/UX, these include:

- Display

The physical screen size and the active area of display. The screen bezel often limits how closely a user can touch to the edges of a screen.

- Screen resolution

The resolution limits how smoothly curves can be rendered for a given size and must be considered when designing the screen visual elements.

- Viewing distance of the user

The viewing distance impacts the visibility of elements and can vary within various use cases e.g., a nurse may input control parameters on a screen while viewing from 20 inches from the screen, however this information is referenced by a physician standing 9 feet from the screen. There are often competing constraints regarding viewing distance.

- Input methods and response

Input methods are increasing in variety and should be carefully assessed regarding the use context. Input methods can be in the form of a: keyboard, mouse, touchscreen (see box), scan, pen/stylus, hardware button, or numerical keypad, etc. Entering specific details may be accomplished utilizing functions such as: scroll-select, virtual thumbwheels, gestural typing or constrained data entry using traditional input means. With innovative technologies available input methods expand to include voice, gaze, and gaze-dwell (see Chapter 21).

### Touchscreens: resistive or capacitive?

A resistive touchscreen is made of two thin layers separated by a thin gap. These are not the only layers in the resistive touchscreen however they are the focus for simplicity of discussion. When these two layers of coating touch each other, a voltage is passed which is in turn processed as a touch in that location. A stylus or other instruments can cause a "touch" to occur. Resistive screens are readily cleanable, typically cheaper, and less sensitive to touch. These screens also tend to be less durable, have lower resolution and if not properly designed can register "false touches" as they are not as accurate as capacitive.

A capacitive touchscreen makes use of the electrical properties of the human body. A capacitive screen is usually made of one insulating layer, such as glass which is coated by a transparent conductive material on the inside. Since the human body is conductive, the capacitive screen can use this conductivity as input. When a touch is completed there is a change in the screen's electrical field. Other devices can be used to input on this screen however they must be conductive. These screens are relatively expensive however are accurate, cleanable and have higher resolution. For medical applications, these can be slightly more difficult with gloves and can be sensitive to touch if the user hovers over the screen. As a result, usability testing early with contextual elements is recommended.

## 5.3 Communicating interactive conceptual design

Table 7.2 below communicates all the necessary details regarding the "general settings" screen design for a novel software system. The description includes the purpose of the screen and the functionality intended to be delivered by the software. The screen design and navigation is communicated as is a reference to developed visual guidelines for asset management. The term asset refers to each visual element within the user interface. There is opportunity to describe dynamic behaviors of elements and specific conditions that the operating system must accommodate. This type of communication can assure human factors principles are applied to the user interface and provide a traceable document for design modifications.

## 5.4 Graphic design: detection and discrimination

The visual channel has more bandwidth than touch, hearing, smell, or taste.

It is the strongest signal and reaches the brain first then dominates attention using contrast as the biggest determinant of signal strength. Visual elements such as borders, edges, color, and size create powerful contrasts. While visual elements may be clear in contrast, users in general are bad at conceptualizing differences and comparing things from memory. Rather, users will need to discriminate what is in front of them rather than rely on recall. This means that in user interface design, the overall composition of individual elements and navigation pattern will determine system usability. The sections below describe composition and comprehension in more detail.

### *5.4.1 Composition: grouping and organization – how does the mind group signals at a pre-attentive level?*

The brain finds patterns to reduce its workload, even if there was not supposed to be a pattern there. Cognitive science describes a step between seeing something and making sense of something called "pre-attentive sensing." The brain tries to make things easier for the attentive mind by grouping. Signals that share similar qualities are grouped together and sent along as a single chunk. In visual design, this includes:

- Proximity
- Alignment
- Symmetry
- Similarity in size, color, or shape

This leads into Gestalt principle wherein the sum is greater than the individual elements. In this case if someone squints until this text is unreadable, they will see pre-attentive groups and read words as shapes rather than putting individual letters together for words. When designing a user interface, it is reasonable to assume that when grouping elements and objects together:

- White space is least intrusive
- Common background color is next
- Use borders/frames as a last resort

**TABLE 7.2** Design communication table highlighting interactivity.

| | General settings design details |
|---|---|
| Purpose | The overall purpose of the settings menu is to provide the user with a way to navigate the various settings. The purpose of the general settings screen is to provide the user with an opportunity to modify the current date and time, and to adjust the brightness and volume of the device. |
| Description | The user must press the settings button on the home screen to enter the settings menu. The user must also be able to navigate back to the home screen directly from the Settings Menu.<br>The date, presented in two formats, can be adjusted via a numerical keypad initiated by selecting the text fields. The format is specified via a radio button. Formats provided are DD-MM-YYYY and MM-DD-YYYY.<br>The time is presented in a 12-h format by default, and is also adjusted via a numerical keypad initiated by selecting the text fields. The user can choose to use 24-h time format by selecting the checkbox. The device should default to the current date of system set-up. The time zone is selected from a drop down menu.<br>By using the (–) and (+) buttons, the user can adjust volume. The (+) buttons increase volume. The (–) button decreases volume. There are 10 settings for both brightness and volume.<br>Selecting "Save" will save all changes and selecting "Reset to Default" will return all settings on menu to system default. |
| Screen Design | 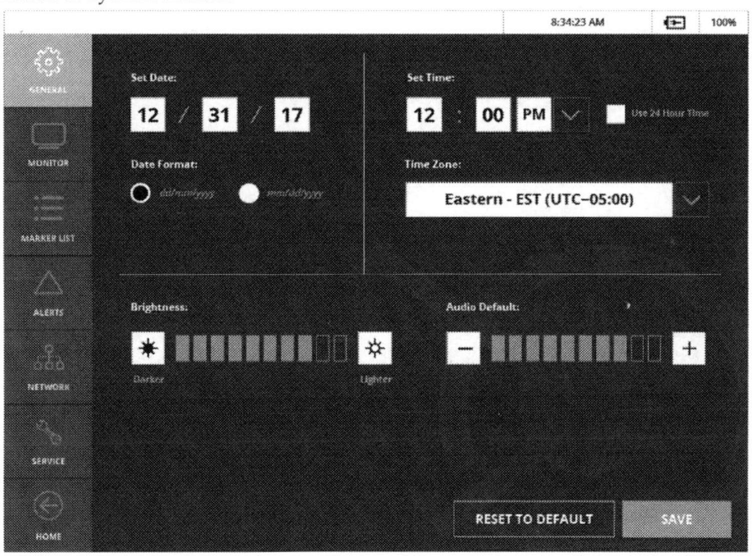  |
| Visual Guidelines | Refer to **Appendix A** for styling and visual guidelines. |
| Dynamic Behaviors | **Date and Time:** Text fields initiate a keypad for manual entry.<br>**Brightness and Audio Volume:** Button "pressed" state change appears upon selection. As each setting level is reached, the black boxes will fill with light gray to communicate the current level. User must receive a visual and single-tone audible alert indicating the change in audio volume.<br>*See* **"XXXXX-YYYYYY& Visual Guidelines.pdf"** *for specific asset descriptions.* |
| Conditions | The last saved entry will appear in the button until the entry is either reset or changed. |

*Used with permission from HS Design.*

### *5.4.2 Comprehension: meaning and working memory- can users find meaning at a cognitive level?*

Working memory has a limited capacity, limited duration is highly volatile an is affected by motivation. ANSI/AAMI HE75, describes three kinds of memory-sensory, long term memory and working memory (short term).

Working memory acts as a scratch pad where information is processed. It is where the thinking happens. Working memory gets information, either by sensing it from the outside world or by retrieving it from long term memory and then processes it. There are three important descriptors for working memory: it has limited duration, limited capacity and is highly volatile.

Research has demonstrated that the magic number of objects working memory can hold is $7 \pm 2$ (although this number is smaller if the objects are more complicated) and another study suggests working memory is limited to 4–7 chunks of information. Further that the capacity of working memory gets smaller with old age, learning disabilities, anxiety, or exhaustion.

Working memory has limited duration in that it only holds information for 20–30 s. The mind can keep it in working memory longer with effort such as repeating the information over and over. Which indicates that it is highly volatile; working memory can evolve or be corrupted as it sits there for its 30 s life. Alternatively, it may disappear entirely e.g. asked to remember something and then there is a sudden distraction of significance it will be gone forever.

Research also indicates that working memory is increased by motivation and decreased by anxiety. Reward, gratification, pleasure, and efficiency motivate users. They can extend their biological and cognitive abilities to reach complete use when they are motivated. Anxiety in small doses can also help us stay focused and fully engaged but too much anxiety overwhelms working memory. If designing something that deals with a sensitive topic, then users may have higher levels of anxiety resulting in use errors or inefficiencies which are unanticipated.

In designing a UI, considerations for a heavy cognitive load must be considered. For example, when the tasks require more cognitive processing/thinking than a person can give, there will be mistakes and abandonment of the software. Asking people to perform fine discrimination tasks between similar sounds/colors/shapes draws on working memory, which as described above is limited. This means that it is even possible for a person to detect two signals however fail to discriminate between them as different, like a nurse who is busy and confuses 2 similar sounding medications.

People are engineered to only give enough to get by. They are good at efficiency, including auto responses and unitizing. They sample bits and pieces; yet they do not devote their full attention to everything. People are also foragers. Information theory says that they will seek, gather, and consume the flux of information in an environment to understand and complete a task if the proper motivation is present.

## 5.5 Learning and long-term memory - can users retain and recall knowledge at a metacognitive level?

People expect new systems to mirror the ones they already know but can learn new ones with greater ease with help from cognitive scaffolding.

While working memory is limited, long term memory can hold an enormous amount of information for a lifetime. People arrange knowledge into semantic networks also known as schemas. When a schema stores information about systems, they are known as mental models. These mental models help people anticipate events, reason, and underlie the explanation. When a person interacts with a system for which they have no categorization, they 'thrash about' randomly. While the medical device loves new technology and innovation, their users love familiarity. When designing an interface, intuitiveness is accomplished by mapping what people already know (even if technology is groundbreaking) and examining the intended shifts the new technology will require.

The building of mental models and long-term memory retention is accomplished through the following exercises:

- Rehearsal-rote memorization where the muscle memory "wears a groove" into the mind
- Elaboration-where users build on understanding with a self-generated information
- Duration-longer time spent learning helps it stick; for example, 10 half day sessions are better than 5 full day sessions.
- Distribution of presentation-users need time to absorb, assimilate and accommodate information.

Learning new processes and methods can be aided by cognitive scaffolding where helpful guides are set up at the beginning and serve as a framework then reduced until the learning is complete. If the use environment is constantly changing or there is a long gap in time between uses, the cognitive scaffolding may require it to be permanently included. For example, software that is not readily used in day to day tasks however is critical to accomplish a complex surgical intervention may have an embedded user manual or training session to refresh a user on the workflow.

Legacy users have ingrained user behavior that makes innovation on workflow a challenge. Habits and known workflows can lead to negative transfer. Negative transfer from a diabetes user (extreme comfort with needles and self-injection) when using other forms of self-administered drug delivery – they may be careless, ignore the IFU, act out their routine even if it runs counter to the simple directions of the alternative product. Legacy users can give a design team the hardest constraint as new users may not have any preconcieved notions of the system and be readily able to pick it up and use it. Legacy users may have habits and ingrained behaviors that need to be taken into consideration.

## Developing a wireframe for navigation design of software

Ultimately, the user's workflow serves as a guide for the system architecture as well as the user interface design. Developing a workflow is closely related to a Task Analysis (see Chapter 6) however it includes the response anticipated by the system, decisions, and actions that the user must complete to complete the task and is typically represented in a block diagram with reference to software specifications and supporting documentation for additional details. Fig. 7.7 below provides an example wireframe for a single task. It does not include the Perception-Cognition-Action

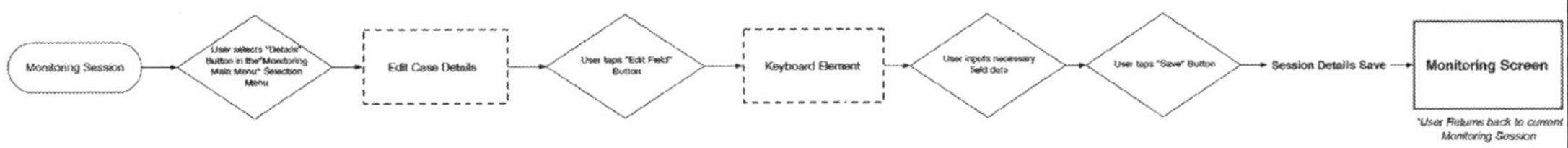


FIG. 7.7 Example wireframe based on a single task.

model of a task analysis, rather, it describes use case scenarios within the system constraints.

The process of developing a wireframe includes determining a high-level use navigation and interaction then visualizing it using block diagram format. A representative key is below in Fig. 7.8.

A complex software will have multiple wireframes each following a task and may have embedded relationships within the navigation which must be clearly communicated to both the software engineers, product designers and the human factors team. In the process of developing software, a team may elect to initially configure the visual user interface design then workout the wireframe. For the purposes of conducting preliminary usability analysis on conceptual software user interfaces, both the wireframe workflows and initial screen designs can be evaluated by users and documented as an early formative study. This testing is highly effective for assessing the overall layout of information and graphic element priority.

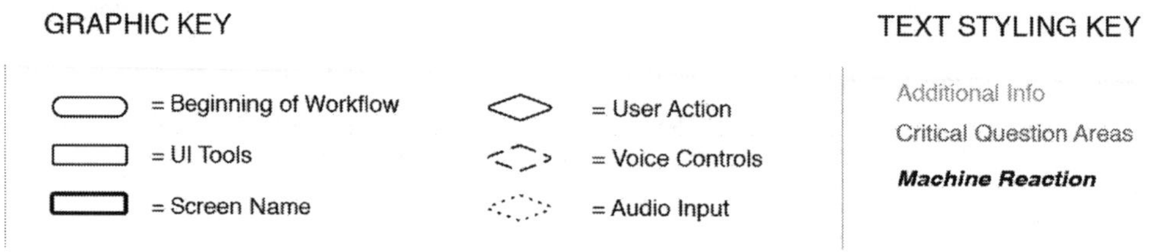


FIG. 7.8 Graphic and text styling key for block diagram generation.

## 6. Alarms (Daryle Gardner-Bonneau)

The design of effective medical device alarms and alarm systems poses many challenges. There are several reasons why medical device alarms are so problematic in practice. First, alarms and alarms systems are typically designed in isolation for a single piece of equipment, with little or no regard of the fact that, in practice, there may be many pieces of equipment in a health care environment, each with its own alarm system. There are, practically speaking, very few, if any, systems that can integrate the alarm conditions and alarm signals from multiple pieces of equipment operating simultaneously. Thus, an intensive care unit in a hospital is often filled with alarm signals coming from many pieces of equipment, attached to several or more patients. As a consequence, health care personnel can be overwhelmed by the shear number of alarms, and alarm fatigue is a serious problem that must be dealt with very far downstream from the actual design process of any given piece of equipment.

Further, up until recently, medical device auditory alarm signals consisted of patterns of, essentially, pure tone "beeps" that were very abstract, having no inherent meaning. They were, therefore, difficult to learn. Further, these sorts of alarms were also difficult to localize in space (e.g., which patient's heart monitor is alarming in the intensive care unit), and they were subject to masking effects. (Masking can occur when two auditory alarms occur simultaneously, and one cancels out or "masks" the other, perceptually.) Two simultaneous alarms can also combine, perceptually, in such a way that the ear "hears" an unidentifiable third signal that is actually a combination of the two that were presented. Up until, perhaps, 20 years ago, the reason for having such poorly designed alarm signals was technological; i.e., due to bandwidth issues, "beeps" were all the typical computer could produce. However, now that just about any sound can be digitized, we are no longer limited to designing with "beeps" (and have not been for some time), but the medical device industry has been slow to catch up to the fact that we can now use *meaningful* sounds that allow listeners not only to perceive that an alarm has occurred, but to identify what kind of an alarm it is.

The problem of too many alarms and alarm fatigue, described in the first paragraph, is a human factors problem, but not one that can be solved by device designers until such time, if ever, that it is possible to develop integrated alarm systems that are "smart" enough to pre-process sensed alarm conditions from multiple sources, in order to know what alarms, need to be presented to health care personnel. At present, human factors specialists can only work on solving the alarm fatigue problem through means other than device design, per se.

Improving alarm signals, on the other hand, is something that device designers and human factors specialists with a strong background in auditory pattern perception can do *today*. Luckily, a significant amount of work has already been done, during the revision of ISO 60601-1-8 — *Medical electrical equipment — Part 1—8: General requirements for basic safety and essential performance — Collateral Standard: General requirements, tests and guidance for alarm systems in medical electrical equipment and medical electrical systems*, to aid human factors engineers in designing and implementing better alarm signals.

The earlier version of this standard did include a set of "melodic" alarms in an annex, which were thought to be easier to identify than the set of "beep" alarms signals defined in the main body of the standard. These "melodic" alarms provided an option for developers who wanted to use them rather than the set specified in the main body of the standard. However, these "melodic alarms" were never validated, and those who tried to use them found them to be difficult to identify and confusable. The "melodic" alarms in that standard were, to some degree, based on melodic patterns that people with a music background or an "ear" for music might find easy to identify; however, for those with no such background, the sounds were difficult to identify, and still too abstract.

Judy Edworthy, whose work on auditory perception and alarm systems, specifically, is quite well-known, see Edworthy (2017), with support from the AAMI Foundation, was tasked with developing and validating a set of meaningful alarm sounds to replace the "melodic" alarms in the annex of the standard. Dr. Edworthy has conducted a number of studies over the past several years that tested this new set of alarm sounds, not only against the set in the annex of the standard, but against other sound sets, which varied in the extent to which the sounds were abstract or concrete. She also conducted studies with regarding to masking and sound localization of the set of new alarm sounds (Edworthy et al., 2017a,b).

The results of her studies showed that the alarm sounds that appear in the revised version of the standard were much better identified than any of the other alarm sound sets tested, including the one from the older version of the standard. The new alarm signal set employs auditory icons that sound like what they represent. For example, a "heartbeat" sound is used for a cardiovascular system alarm; the sound of pills in a pill bottle being shaken is used for a medication alarm. There are 14 auditory icons in all, representing low and high priority alarms for cardiovascular, perfusion, ventilation, oxygenation, temperature/energy delivery, drug or fluid delivery/administration, and equipment or supply failure. In addition, the alarm icons are preceded by an "auditory pointer" sound that serves solely to alert the user that an alarm is about to be presented. Because the auditory icons are meaningful, they are easy to identify. In addition, because they are complex sounds, acoustically, they are more easily localizable and less subject to masking effects. Finally, care was taken to ensure that the acoustic characteristics of the icons used differ enough so that one icon in the set is not confused with another.

The revised version of ISO 60601-1-8, in addition to providing this new set of validated alarm signals in a normative annex, has an additional annex which lays out a process for validating alarm signals, based on what Judy and her colleagues did in their validation studies with users. This annex should encourage human factors specialists to conduct work to improve upon the set of auditory icons in the standard, and to develop auditory icons for other auditory alarm conditions, as occasions arise. People are very good at identifying meaningful sounds in their environment, so the possibilities are almost limitless for the creative designer.

Although improving the alarm signals themselves won't solve all of the human factors challenges related to medical device alarms, they are likely reduce some of the alarm-related chaos that currently exists in hospital intensive care units and other places were multiple devices are used simultaneously for one or more patients. The healthcare provider will be able to identify much more easily than in the past what the nature of the alarm is, as well as the patient associated with the equipment from which the alarm is originating. The fact that the sounds are so highly associated with their meaning lowers the cognitive and memory burden on users. Even a lay person with no knowledge of the equipment could probably guess correctly what many of these alarm signals signify.

## 6.1 Designing auditory alarms

Öscan et al. (2018) recommend an audible alarm design process that is much more collaborative than what has been used in the past. It is one in which manufacturers, regulators, academic specialists, and users (clinicians, family members, and patients) are all involved with much more attention is paid to the broader context in which alarms will be used.

The auditory icons specified in the annex of the revised ISO 60601-1-8 may need to be modified, or new icons created, for several reasons. It may be that a particular use environment makes an icon less detectable than it typically would be (e.g., if there is an especially noisy environment, or there is other medical equipment running noisily that interferes with detection of the auditory icon). It is also possible that manufacturers will want to create more specific auditory icons, within the seven categories specified in the standard (e.g., a specific cardiovascular alarm), or may wish to create additional categories of alarms for which auditory icons will be developed.

There are particular criteria for the design of auditory alarm signals, generally, that should be considered during the design and validation process for auditory icons:

1. An auditory icon must be detectable in the environment for which its use is intended.
2. The association between the auditory icon and the alarm condition it represents should be as meaningful as possible, and easily learned by the intended user.
3. The design of the icons should consider the other auditory icons in use for the device to ensure that any new icons are not confused with existing ones by users (i.e., the icons should be as discriminable as possible by the users).
4. The auditory icon should be easily localizable in 3-dimensional space.
5. Designers need to ensure that no masking effects are occurring which would interfere with recognition of the auditory icon when it occurs simultaneously with another auditory icon or another auditory signal of a different type.
6. Auditory icons should not be displeasing or obnoxious to anyone in the environment (patients, health care providers, family members, staff) where the auditory icons will be used.

In addition to providing information about validating auditory icons, an annex in the revised ISO 60601-1-8 provides some information about the methodology that can be used during the earlier developmental stages for new icons. The standard recommends, for example, obtaining input from users about potential sounds that might be used to represent the alarm condition for which an auditory icon is being designed.

Similarly, if one or more new categories of alarm conditions are to be defined, it is recommended to use a card-sorting task with potential users in order to determine the categories to which users "naturally" assign particular alarm conditions. This will assist the designer in determining whether new categories are actually needed, and how those categories should be defined. Designers should not assume that categories that seem "natural" to them will be "natural" to the actual users. The revised standard also supplies specific details with respect to conducting the card-sort task.

Finally, it must be noted that auditory perception and auditory pattern perception are not, typically, strong suits for most human factors professionals, who tend to focus more on visual or physical user interface design. Aspects of design such as sound localization and masking effects can be very complex, technically; in addition, sound detection, recognition, and discrimination changes with age and various health conditions, and the designer needs to be aware of these effects, particularly when designing auditory icons for a broad audience of users (e.g., in the case of home care devices). Thus, it is highly recommended that if you are going to design new auditory signals for medical device alarms that you have an auditory perception/auditory pattern perception specialist on your design team so as to avoid mis-steps in the process.

## 7. Summary

As previously stated, medical device safety should be inherent by design as the initial risk control priority (ISO/IEC, 2007). A user will interact with devices with dynamic presuppositions of interaction from the initial experience to the detection of affordances ultimately leading to actions (Xenakis & Arnellos, 2013) that assist a user in reaching their goal.

A successful medical device design considers the users capabilities, uses human factors standards (to determine fit) and processes (to understand the context). This chapter communicates the importance of integrating human factors standards and the principals of design in order to optimize usability.

Design is the most effective means at improving safety, efficacy and usability.

## 8. Further reading

- ANSI/AAMI HE 75
- "Design of Everyday Things" by Donald Norman
- ISO 9241-210 Human centered design for interactive systems (2010)
- Healthcare Information and Management Systems Society (HIMSS), Selecting an EHR for Your Practice: Evaluating Usability (2010)
- ISO 60601-1-8 Medical Electrical Equipment
- Edworthy, J. E., Reid, S., McDougall, S., Edworthy, J. D., Hall, S., Bennett, S., Khan, J. & Pye, E. (2017). The recognizability and localizability of auditory alarms: Setting global medical device standards. *Human Factors, 42*(9), 1233–1248.
- Edworthy, J. (2017). Designing auditory alarms. Chapter 23. In A. Black, P. Luna, O. Lund, & Walker S. (Eds.), *Information design: Research and practice* (pp. 377–390). London: Routledge.
- Edworthy, J., Schlesinger, J. J., McNeer, R. R., Kristensen, M. S., & Bennett, C. L. (2017). Classifying alarms: Seeing durability, credibility, consistency, and simplicity. *Biomedical Instrumentation and Technology, 51*(s2), 50–57.

## Acknowledgments

Figs. 7.4 and 7.6 are courtesy of Kevin Zylka. Figs. 7.2 and 7.3, Table 7.2 are courtesy of HS Design. Special thanks to the engineering and software design team of HS Design for sharing their expertise in order to develop content for the boxes "Touchscreens: Resistive versus Capacitive" and "Developing a wireframe for navigation design of software." Special Thanks to Elissa Yancey for editing.

## References

AAMI. (2009). *ANSI/AAMI HE75, 2009/(R)2013 Human factors engineering—design of medical devices* (USA).

Aptel, M., & Claudon, L. (2002). Integration of ergonomics into hand tool design: principle and presentation of an example. *International Journal of Occupational Safety and Ergonomics, 8*(1), 107–115. Retrieved from http://archiwum.ciop.pl/790.

Blanchonette, P., & Defence Science and Technology Organisation (Australia) Air Operations Division. (2010). *Jack human modelling tool: A review*. Fishermans Bend, Victoria: Defence Science and Technology Organisation. Retrieved from https://apps.dtic.mil/dtic/tr/fulltext/u2/a518132.pdf.

Edworthy, J. (2017). Designing auditory alarms. Chapter 23. In A. Black, P. Luna, O. Lund, & S. Walker (Eds.), *Information design: Research and practice* (pp. 377–390). London: Routledge.

Edworthy, J., Schlesinger, J. J., McNeer, R. R., Kristensen, M. S., & Bennett, C. L. (2017a). Classifying alarms: Seeing durability, credibility, consistency, and simplicity. *Biomedical Instrumentation and Technology, 51*(s2), 50–57.

Edworthy, J. E., Reid, S., McDougall, S., Edworthy, J. D., Hall, S., Bennett, S., et al. (2017b). The recognizability and localizability of auditory alarms: Setting global medical device standards. *Human Factors, 42*(9), 1233–1248.

Eveleth, P. B. (2001). Thoughts on secular trends in growth and development. In P. Dasgupta, & R. Hauspie (Eds.), *Perspectives in human growth, development and maturation*. Dordrecht: Springer. https://doi.org/10.1007/978-94-015-9801-9_12.

Garfield, M. (2015). *Controlling the Inputs of Hand Tool Development through Design Research*. (Electronic Thesis or Dissertation). Retrieved from https://etd.ohiolink.edu.

Hassenzahl, M., & Monk, A. (2010). The inference of perceived usability from beauty. *Human-Computer Interaction, 25*(3), 235–260. https://doi.org/10.1080/07370024201050013

IEC. (2006). *IEC: 60601-1-8:2006. Medical electrical equipment – Part 1-8: General requirements for basic safety and essential performance – collateral standard: General requirements, tests and guidance for alarm systems in medical electrical equipment and medical electrical systems.*

Intrinsys Ltd.. (2018). *RAMSIS - the human touch to technology*. Retrieved from https://www.intrinsys.com/software/ramsis.

ISO/IEC. (2007). *International standard international standard 14971 medical devices-application of risk management to medical devices*. Vol. 2007-10-01.

Moshagen, M., & Thielsch, M. T. (2010). Facets of visual aesthetics. *International Journal of Human-Computer Studies, 68*(10), 689–709. https://doi.org/10.1016/j.ijhcs.2010.05.006.

NEMA. (2017). *ANSI/NEMA Z535.1-2017*. USA: American National Standard for Safety Colors.

Norman, D. A. (1990). *The design of everyday things*. New York: Doubleday.

Öscan, E., Birdja, D., & Edworthy, J. (2018). A holistic and collaborative approach to audible alarm design. *Biomedical Instrumentation and Technology, 52*(6), 422–432.

Privitera, M. B., & Johnson, J. (2009). Interconnections of basic science research and product development in medical device design. In *Conf proc IEEE eng med biol soc* Vol. 2009. (pp. 5595–5598). https://doi.org/10.1109/iembs.2009.5333492.

Quick, N. E., Gillette, J. C., R., Shapiro., Adrales, G. L., Gerlach, D., & Park, A. E. (2003). The effect of using laparoscopic instruments on muscle activation patterns during minimally invasive surgical training procedures. *Surgical Endoscopy, 17*, 462–465.

Robinette, K. M. (2012). Anthropometry for product design. In G. Salvendy (Ed.), *Handbook of human factors and ergonomics* (4th ed.). Wiley & Sons. https://doi.org/10.1002/9781118131350.ch11.

Russ, A. L., Fairbanks, R. J., Karsh, B. T., Militello, L. G., Saleem, J. J., & Wears, R. L. (2013). The science of human factors: Separating fact from fiction. *BMJ Quality and Safety, 22*(10), 802–808. https://doi.org/10.1136/bmjqs-2012-001450.

Siemens Inc.. (2011). *Jack: A premier human simulation tool for populating your designs with virtual people and performing human factors and ergonomic analysis*. Retrieved from https://www.plm.automation.siemens.com/media/store/en_us/4917_tcm1023-4952_tcm29-1992.pdf.

Tillman, B., Tillman, P., Rose, R. R., & Woodson, W. E. (2016). *Human factors and ergonomics design handbook* (3rd ed.). McGraw-Hill Education.

Ulin, S. S., Ways, C. M., Armstrong, T. J., & Snook, S. H. (1990). Perceived exertion and discomfort versus work height with a pistol-shaped screwdriver. *American Industrial Hygiene Association Journal, 51*(11), 588–594. https://doi.org/10.1080/15298669091370167 ISO 60601-1-8.

Xenakis, I., & Arnellos, A. (2013). The relation between interaction aesthetics and affordances. *Design Studies, 34*(1), 57–73. https://doi.org/10.1016/j.destud.2012.05.004.

CHAPTER

# 8

# Combination devices

*Jeffrey Morang*[a], *Mary Beth Privitera*[b]

[a]Sanofi, Cambridge, MA, United States; [b]HS Design, Gladstone, NJ, United States

OUTLINE



*Applied Human Factors in Medical Device Design*
https://doi.org/10.1016/B978-0-12-816163-0.00008-6

*Everything should be made as simple as possible, but not simpler.* **Albert Einstein.**

# 1. Introduction

Combination devices are unique as they are used as "diagnostic or therapeutic products that combine drugs, devices, and/or biological products." (FDA, 2018). Continual technological advancements in drug and biologic research and development coupled with those in engineering and manufacturing of drugs, biologics and devices have resulted in an increasing use of combination devices in nearly every environment. The FDA acknowledges that they expect "to receive large numbers of combination products for review as technological advances continue" (FDA, 2018). Furthermore, the European Medicines Agency (EMA) is currently working on creating guidelines on the assessment of device components in combination devices under the Medical Devices Regulation (MDR) (Haigney, 2018).

According to the National Institute of Biomedical Imaging and Bioengineering (NIBIB), a drug delivery system is defined as a system of engineered technologies for the targeted delivery and/or controlled release of a therapeutic agent (National Institute of Health, 2019). Historically, clinicians have directed their interventions to areas of the body that are at risk or affected by disease and dependent upon the medication, the way it is delivered, how our bodies respond and the side effects that sometimes occur (National Institute of Health, 2019). This includes designing drug delivery systems for both local (directly on area affected) and systemic (affecting the whole body) in order to decrease side effects. The routes of delivery vary and may include swallowing, inhalation, absorption through the skin, or by intravenous injection (National Institute of Health, 2019).

A few examples of drug delivery systems include hand-held injection devices, pulmonary inhalers, nasal sprays, infusion systems, ambulatory syringe pumps. As listed, these examples of delivery devices represent broad categories of delivery systems without regard to the limitations associated the drug agent being delivered, the patient, the disease state and/or comorbidities. Each of these must be carefully considered when applying human factors to the design of drug delivery systems.

Increasingly, patients are more responsible for their own healthcare and as a result of their medical condition may be enduring reduced cognitive abilities, memory and physical limitations. As such, this chapter discusses the evolution of healthcare, the uniqueness of combination devices, approaches to design, and the special considerations required when designing drug delivery systems. It is not intended to be exhaustive in regards to all types of combination products; rather it is focused on drug-delivery systems for use in the home or in care situations.

# 2. Health care (R)evolution

It is a fact that drug-device combination products play a vital role in the diagnosis and treatment of a wide range of disorders, including chronic diseases such as heart disease, cancer, respiratory diseases as well as diabetes (Masterson, 2018). They originate in a variety of industries including the pharmaceutical, biopharmaceutical, biotechnology and medical device sectors and draw on different conventions (Masterson, 2018). Manufacturers in this arena often face complex regulatory processes and all require the application of human factors processes.

Landers et al. (2016) describe the fact that America is experiencing a dramatic shift in demographics, and in 2019, people older than 65 years will outnumber those younger than five. As people age and live longer, increasing numbers of them will live with multiple chronic conditions (e.g., diabetes or dementia) and functional impairments (e.g., difficulty with the basics of life like mobility and managing one's household). One of the greatest health care challenges facing developed countries is ensuring that older citizens with serious chronic illness and other maladies of aging can remain as independent as possible. Our success with this challenge will help ensure that citizens age with dignity in a manner that meets their expectations, preferences and care needs. In turn, this also impacts the financial health of federal and state governments. Meeting this challenge will require envisioning the potential value of home-based health care, creating a pathway for home-based care to maximize its potential, and integrating it fully into health care systems globally.

Home healthcare has become an area of increasing focus for the pharmaceutical industry with anticipated growth in the United States of approximately 2–5% until 2022 is forecasted (IQVIA, 2018). This statistic includes the rapid growth of self-injection devices/systems which offer convenience by allowing patients to administer physician-prescribed medications in their preferred setting. Self-injection devices also continue to grow in outpatient care with the introduction of new injectable therapies for the treatment of chronic diseases, autoimmune disorders, infertility and many other diseases affecting large populations.

This longer life expectancy coupled with growth in home healthcare e.g., self-administration requires a comprehensive approach to the application of human factors. A thorough understanding of the potential users' behaviors, motivations and needs and for those with chronic conditions, the best designed delivery systems will meet the needs not only when first diagnosed but also in long-term care options. Further research into the environment of use wherein designers and engineers have the ability to observe users in the midst of daily distractions, caring for aging parents while interacting with children, pets, noise, temperature and lighting conditions to create patient and user centered designs are as much needed as required.

## 3. Designing useable combination products

For drug delivery systems in particular, the overarching design goal is to support users through the entire dosing process in a patient centric approach. This requires the design of clear alarms and alerts regardless of use environment; the need to connect patient use adherence (or lack thereof) with providers or provide instructional information for proper device use through designs that encourage use rather than deter as well as the use of mobile applications.

As mentioned in Chapter 7, understanding your users, knowing the use environment are key elements to successful device design in general but what happens when the conditions of device use change over time i.e., move from hospital administration to home administration. The section below builds on the foundation of Chapter 7 in order to support the design of drug delivery systems.

Described below are key elements and theories in support of AAMI HE 75 Combination Product Section. This section of HE 75 is a new addition and should be read in its entirety prior to undertaking the design of a combination product.

## 3.1 Support the user throughout dosing: patient centricity

At the center of every medical device design is a person, an end user. For most devices and systems the primary end user may be a health care professional. The uniqueness of a combination device is that the primary end user is commonly the patient. This simple fact means that designing with, or for, the patient as the central focus is paramount. Humans are individually unique, coming in all shapes and sizes with capabilities and limitations that are as exclusive as our personalities. This can be in stark contrast to the motivations and goals of device design teams who, likely, and usually, have the task of designing a one-size-fits-all solution. In such situations, taking the time to research, understand and define the end users possess much value; increasing knowledge and understanding reduces design risks that cause use errors (i.e., mistakes and difficulties related to use) decrease the time and costs of unnecessary design iterations (Stegemann et al., 2016).

Research has demonstrated that when it comes to drug delivery devices, the majority of patients are aware of options and willing to pay for advanced technology and user-friendly features. A well-designed drug-delivery device can impact three of the five dimensions that affect medication adherence as described by the World health Organization. These include therapy-related factors, healthcare team factors, as well as patient-related factors. Simply put, designing a simple and easy to use device can increase the patient's confidence in their ability to follow the treatment regimen. This is accomplished by ensuring that the device requirements are appropriately designed and the targeted support material compensates for the capacity, or lack thereof, of the healthcare system and within the constraints of the patient's abilities. Of equal importance is about the desire a person has to follow the treatment regimen. When combination device designs are excessively complicated, require additional steps to complete a task (dose preparation, dose administration) and/or perceived as unnecessarily complicated then users will be less likely to want, or actually use the device to adhere with their treatment regimen. Furthermore, the anxieties or frustrations of the user will decrease the pain threshold thereby compounding and worsening an already anticipated unpleasant experience. Adherence of the treatment regimen increases when users' experience is clear and pleasant.

Drugs are only effective if delivered as prescribed, in an accurate and consistent manner and a challenge for the design of drug delivery systems includes ensuring that the patient has received the correct dose of medication (McConnell & Ulrich, 2017). This is especially true during a medical emergency and can be difficult if the emergency happens when in a community with no assistance from healthcare professionals is available (Edwards, Edwards, Simons, & North, 2014). This requires that devices themselves provide clear indications of dose delivery and completion in order to ensure proper dosage is achieved and for some types of devices, this is readily available such as the case with syringes.

In other delivery systems, the steps to properly deliver a dose or the dosing status may be unclear and can lead to use errors. The most common example of this is the stopping of a nebulizer treatment too early or pulling an autoinjector away from the injection site prior to dose completion. In these examples, the user was not supported during the dose administration phase of the device use.

There are three phases to the use of combination devices:

1. Preparing the device for use
2. Dose administration
3. Disposing of device (potentially disassemble), cleaning and storage

The best way to support users is to provide "directed use" which incorporates design principles wherein the device itself indicates to users the proper operation (McConnell & Ulrich, 2017). The more users can rely on the device itself, the more likely dosing adherence is achieved (McConnell & Ulrich, 2017). For example, this can be achieved in design by incorporating a digital display, voice commands and/or clear device labeling, etc.

The absence of a "directed use" device requires a reliance on the users' mental model (see Chapter 7 Section 4.5 for more information on mental models). As Sir Francis Bacon said, "The human mind when it has once adopted an opinion draws all things else to support and agree with it." This translates directly to the development and importance of mental models. Mental models are basically the byproduct of learned reality based on individual perception and understanding. Simply, practice makes permanent. Ironically this reality can be invented, distorted and occasionally unstable. However, the purpose, and occasionally the benefit, of mental models are their ability to reduce cognitive load to free up those resources so they are available for use on other activities considered to be a higher priority; e.g., looking for and paying attention to the feedback given by the device for the status of dose delivery. They are pragmatic solutions to the complexity of life. In other words, using mental models has proven much benefit for accomplishing tasks as much as it has contributed to survival throughout human history.

Therefore, it should not come as a surprise that mental models are as much great allies as enemies to the practice and application of Human Factors. Researching and understanding existing mental models of the intended end users is extremely beneficial in understanding design improvement opportunities as well as potential use-related hazards to avoid. For example, it could be very valuable to know that the majority of the intended end users currently use devices that require them to push a button after they have inserted a needle into their body. That type of mental model could be leveraged in a new design or it could be disruptive because the new design requires un-learning and re-learning. Again, practice makes permanent and the new design is up against what has become permanent due to routine practice and use.

Training can be a valuable method in aiding with mitigating use errors and dose regimen delinquency by constructing, or reconstructing, a mental model through un-learning and relearning. It is important to note that often training might be considered by some device professionals to be the best, most successful option to mitigate use errors and increase adherence. However it is problematic and less effective than design aspects that support "directed use" because of its reliance on a person's perceptual, cognitive and memorization, both muscular and implicit, capabilities. Reliance on training to mitigate or prevent mistakes and difficulties should be a lower priority. This applies as much to a reliance on instructions as it does a demonstrative explanation from a person's doctor; the former is dependent upon an explicit willingness to open and read instructions whereas both are dependent upon cognition and comprehension translated into proper action.

Researching and understanding mental models help chart a design strategy that incorporates avoiding the use-related obstacles and pitfalls during the definition and design project phases before usability evaluations are conducted or the design is complete or "frozen."

### 3.1.1 *The power of predicate devices: known use-related problems*

There is a lot of value in knowing and understanding known use-related problems in order to identify what mistakes to avoid. Understanding the products or devices a drug-delivery device may replace or compete with is both best practice and recommended per FDA (2016) Human Factors Guidance (see Chapter 14 for more information).

Patients and caregivers will expect or, as previously stated in the previous section, apply their mental model, correctly or incorrectly, to similar/like devices to function in like ways i.e., the buttons, alarms, interfaces require the same interactions or steps of use. In considering this in the design, mismatches and confusion between the user expectations and the actual product function can be avoided. This starts with a search on common use-related errors associated with existing devices (Edwards et al., 2014). For example, search attributes in currently marketed drug-delivery systems may include inadequate product differentiation, unexpected device operation, complex device operation, and unclear messaging (Edwards et al., 2014).

To provide further clarity, this can include the following details.

Inadequate product differentiation, such as:

- Within a product line or similar products
- Between active product and demonstration/trainer product

Unusual or unexpected device operation, such as:

- Injuries due to users incorrectly holding the device

Confusing or complex device control activation, such as:

- Poor labeling
- Too many buttons
- Too much force to activate or manipulate controls

Electronic display illegible, such as:

- Lack of visual contrast
- Incorrect font size e.g., too small for sight impaired
- Confusion regarding the use of decimals related to dose

Outcomes related to "lost dose hazards" wherein patients do not receive the full dose due to unintentional use errors such as prematurely drawing the device away as in the case of autoinjectors.

In gathering this information, previous use errors can be analyzed in regards to root cause and used to inform the design of a new drug-delivery device for the optimization of design elements toward increased usability. It is best to conduct a root cause investigations by acquiring a device identified to have use-related hazards/use errors and performing the use tasks so that a more comprehensive understanding can be gained.

### 3.1.2 Considering dose features and feedback modalities

Dose windows may have different purposes depending on the type of product (see AAMI HE 75 Combination Products Section). For example, a dose window can indicate how many doses remain, confirm that a dose was given or allow flexibility in dosing by communicating a selected value as in a dial pen injector. However, a dose window may not be the best modality to support the user for using the device. For example, using a dose window to indicate that a dose has been delivered for only delivering into a leg when sitting may not best support the user to see the dose window and confirm the dose has been delivered. The chosen modality must be patient centric and based on the researched and defined capabilities and limitations of the users as well as the nature of the task(s).

The design of a combination product should provide feedback e.g., visual, tactile or audible cues that serve to reinforce and supplement instructions for use. In short, the use of the device should be obvious as a result of design. This includes:

- the shape indicates areas of use interaction e.g., form indents on gripping surface;
- visual (labeling) indicators being easy to see (based on the task) and remaining present until the task is completed or provide just in time information in the case of "remaining doses"
- audible indicators (clicks, tones, recorded tones) are incorporated appropriately to the context of use e.g., button to receive voice prompts on instructions or clicks/beeps upon dose completion
- tactile indicators that a device is assembled correctly or incorrectly

When designing feedback cues, careful attention to the use environment is required in order to ensure appropriateness e.g., if the device is intended to be used in a loud environment, the use of vibration plus a visual signal would be a better mode of feedback than an audible click. Fig. 8.1 below illustrates a digital interface intended to be used by patients who suffer

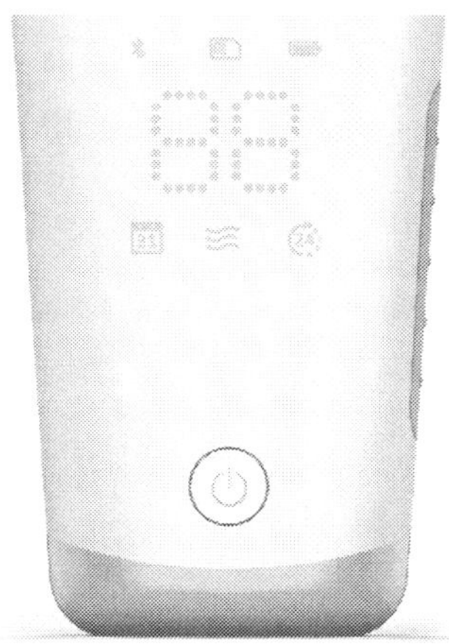

**FIG. 8.1** Digital display of user interface utilizing backlit transparent housing in order to dim brightness accommodating a specific disease state in which light sensitivity is increased in patient user. *Photo courtesy of HS Design.*

light sensitivity yet may be using the device in a dimly lit room. To accommodate these needs, it was concluded in usability testing, a backlit display with large dosing information was more acceptable to users than the use of a touch screen. It is best to use multiple, redundant cues to be sure that the feedback is perceived; e.g., visual and audible or audible, tactile and visual. Lastly, it is important to match the cue with the actual event as well as how that event is interpreted by the user. For example, an audible click that indicates the beginning of an injection followed by a second audible click that indicates the event is *almost* complete will likely confuse users resulting in use errors.

#### 3.1.3 Please, don't make me think

Cognitive tasks such as estimating time, wherein users estimate the time that has lapsed during a potentially discomforting experience, should be avoided as counting seconds will vary from one individual to another. In these instances, the device should provide the necessary feedback i.e., count on behalf of the user for the most effective approach.

Tasks that require a burden on memory, both working and long-term, should also be avoided. Where possible avoid relying on the user's memory to track uses, instead it is better to support this task by providing a means to help the user; e.g., automate this task, upload this information automatically and provide a clear and easy method for access. This includes the avoidance of any user performing calculations.

#### 3.1.4 Demonstration devices

In order to support the user it may be helpful or necessary to create a device replica for demonstration purposes and/or to allow end users an opportunity to practice using the device. Demonstration devices need to look and perform in a near identical manner to the actual device containing the drug product performs, but without hazards; e.g., an injection needle. This is particularly important for devices that are to be used infrequently such as in emergency situations or when the drug product requires infrequent dosing.

When including a demonstration device it is important to include these additional considerations:

- Define the intent for the demonstration device, i.e., is it for practice purposes or will it be used for training (see Section 6.2 for additional information);
- Identify what users will receive the devices (i.e., health care providers or patients) and how will access be granted; i.e., do they need to be ordered separately, are they delivered by a salesperson, etc.;
- Clearly indicate the difference between an actual device and a demonstration device (see Section 4.2.1).

### 3.2 Know the environment: considering the pharmacy, the home, storage, the cat and the TV

An environment is the sum total of all surroundings of the user, including natural forces, other living things (pets), and dwelling conditions such as ancillary objects (furniture, nicknacks and general clutter), lighting conditions and ambient noise (television, phone, people talking, other product alarms/noise). Use environments (Fig. 8.2) will have diverse characteristics that may impact the use of a combination device and the device might be used in more than one location including both indoor and outdoor use. This may require that the manufacturer design for all potential environments or place clear and obvious on-product warnings for special hazardous conditions that may result or that are critical to safe and effective use; e.g., overheating when placed in direct sunlight.

FIG. 8.2 Diabetic patient at home prepped with insulin pen for obvious reasons.

Additionally, manufacturers must evaluate whether the environment will potentially impact the physical, perceptual or cognitive abilities of the user when a device is in use as the majority of environments are not as see in television advertisements. For some drug-delivery devices the environment of use, as well as the user, may change throughout the use scenario (Fig. 8.3). For example, a PICC line will be placed in the hospital then accessed for extended periods of time in the home or as another example, a surgeon may place a pain pump for the care giver to administer when upon release from hospitalization would be intended for the patient to use. In this example the location of device use originated in the operating room (physician is user), transferred to a hospital room (nurse is user) and finally to the patient's home (patient is the user).

Additionally, adherence to the treatment regimen may dictate the use environment and require the user to use the device in uncommon, off-nominal environments; e.g., an airplane bathroom or a manufacturing floor. For these situations it is important to work with the team-mates focused on chemistry, manufacturing and control (CMC) to ensure that the device use interface, including the storage box and instructional materials, support the user using the device in any anticipated environment.

See Chapter 5 for more information on research methodology in the environment (context) of use and the user.

### *3.2.1 Design clear alarms and alerts*

Clear indications of device readiness and when the device needs attention are fundamental alerts in the design of drug delivery devices. For the home, in general, alerts such as "the dose is complete" and "refill or preparation" are always required. See Chapter 7, Alarms sections

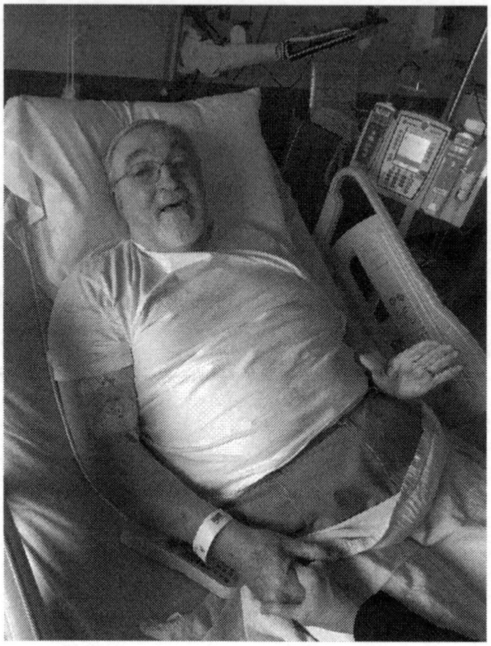

FIG. 8.3 Patient prepped with PICC line for home use while in hospital. In this image current drug delivery is managed by an infusion pump.

for more discussion on the design of alarms in general. Best practices for drug delivery devices include:

- Use alerts for urgent information only. Users will ignore alerts that come too frequently or include non-critical information
- Use multi-sensory alerts where possible (see AAMI HE 75 Alarms section)
- Use simple visual and/or auditory cues for alerts. Where possible, use universally accepted icons as recognized by the intended user group.

By providing a clear feedback of what is happening now (i.e., the state of the device/system), what needs to happen next, when the action needs to happen or how long the user has to complete the action and when the action is completed, the usability of drug delivery devices are improved.

Incorporating good Human Factors practices and techniques to design will reduce the amount of use errors committed by end users, but not completely prevent them. On the other hand, devices and systems do fail. When failures or use errors occur then users need to be supported in their efforts to troubleshoot and correct the problem. Therefore, when specifying feedback mode(s) it is important to design them to provide information that includes:

- What happened; e.g., what is the problem experienced by the device/system
- Why it happened
- How can the problem be corrected by the user; e.g., explain what steps need to be completed to fix the problem

Finally, when supporting the user with indications keep in mind that it is unhelpful to users when:

- Technical terms, e.g., "Syntax error 606", or acronyms, e.g., "the C-DMA is overloaded", are used, unless the acronym is a part of the users' common vocabulary, e.g., SCUBA;
- The indication tells the user to refer to the instructions or operator's manual for the error state definition
- Information for how to fix the problem is long and requires many steps because humans can only hold a limited amount of information in short-term/working memory
- The indication occludes or blocks access to a part of the user interface needing access to fix the problem; this frequently occurs with indications in software platforms

## 3.3 Design of connected devices: those that incorporate software and smartphone applications

The most critical aspect combination devices that are connected to a software application or system is the need to ensure that the system is safe, effective and serves its intended purpose (to deliver a drug) despite any challenges in software functionality (see AAMI HE 75 Combination Products Section). When combination products include software, the relationship between the drug delivery device and the software application or system should be assessed and optimized. In some instances, such as a smart wearable, the software is integral to the device whereas others, it may be an optional add-on e.g., an electronic dose tracking mobile application.

See Chapter 7, Section 4 for more information regarding the design of software and connected devices.

## 3.4 Design of packaging, labels and on-device labeling

Packaging is considered part of the user interface and is integral in facilitating safe and effective use of a combination product. When designing the package, considerations for whether to include opening/use instructions, device storage pre- and post-use, durability as well as any informational leaflets. Equally important are how the labels and labeling support the intended end user(s) and use environment(s).

The development of labels and labeling is product specific and the risk for errors will differ as a result. There is no "one size fits all" for labeling, in fact it may be vital for someone to clearly and easily see the difference between labels and labeling (see Section 4.2). Labels should be designed to be distinctive enough from other likely drug products which might be administered at the same time. In some instances, the use of a transparent label may be most appropriate to enable visual inspection of the device and/or the drug. Careful attention should be placed to assure the durability of the label is appropriate for the use context, user(s) and use environment and that there is no "look alike" label that can lead to medication errors.

Typography (i.e., font type, size and appearance) and legibility (i.e., readability) are important considerations as they determine how likely and how well the user will see and be able to read the information. For example, 10-point font is adequate both typographically and legibly for the general population; however a larger 12–14 point font better supports people who are older. Sans serif font types are commonly harder to read compared to Serif and Times Roman fonts (FDA, 2011). On-device labeling refers to the printed text or graphic labels which are directly applied on a drug container and/or device component as part of a combination product. This might include the following types of information:

- identifying information (drug product's name),
- drug strength/dose concentration,
- lot number,
- expiration date,
- warnings or precautions.

For labeling, in general, the devil is in the details. Considering more complex systems wherein there are components which are single use and some that are reusable, these elements should be clearly identifiable by the user and often labeling is used to accomplish this. Additionally, abbreviations and acronyms should not be used because they are easily misinterpreted. As is the case for trailing zeros and/or leading zeros which often go unseen by users. In cases where it is unavoidable to use decimals, include specific knowledge comprehension tasks pointing to numerical values in formative usability assessments.

Other labeling considerations include drug product strength and concentration, expiration dating, and storage requirements. Each of these elements is important to consider throughout the use scenario from product selection, use, disposal and/or storage. This information should be delivered in an obvious (to the user) location and at the exact moment of need.

## 4. Risk-based design approach

Frequently, conflict among design elements arises, and decisions must be made regarding the overall risk benefit impact a particular design element may bring. This approach to the combination device design can be extremely valuable in the conceptual or design phase of a new product.

Listed below are common attributes that may be unique to a combination product and may require additional testing feedback (see Chapter 16 for full human factors testing of combination devices).

Common design features that relate to use related risk can include:

- Removing from packaging or outer case
- Knowing how a device is to be removed from its outer case
- Meaning of icons (see Isherwood, Mcdougall, & Curry, 2007)
- Timing of voice or display prompts
- Control activation
- Specific instructions on an IFU/comprehension of illustrations and written captions
- Feedback on dosing status

In a study by Oray et al. (2017), patients together with healthcare professionals selected a device [concept] that best suited their personal needs, taking various factors such as control activation [manual or automatic] into consideration (Oray et al., 2017). In this study, it was described that an initial prototype was designed to inform the design through a risk and preferential human factors evaluation.

When developing combination products it is important for the Human Factors practitioner to form and maintain a tight partnership with whoever is responsible for risk management activities. This is helpful as proper use of the device is also commonly linked to the risks and efficacy of the drug being administered and to success of the treatment regimen. Having a tight partnership with risk management will also facilitate better cohesion between the risk and human factors documentation thereby making it easier to build a more comprehensive story of the human factors activities as related to the use-related risks. As such it also allows the Human Factors practitioner to demonstrate what improvements have been made to the user interface design to mitigate or prevent use errors and facilitate the path that ends with the completion of the Human Factors Validation Study and final report culminating the Human Factors activities.

## 5. Developing use requirements: the evolution of user needs

Designing for patient centricity ensures the product design is the right size and comfortable for the user with appropriate functionality to support use-related tasks. Design engineers need to consider how users need to and will interact with the product in addition to unintentional misuse. This is particularly important when establishing User Needs for combination devices due to the breadth of users. Furthermore, User Needs are best created in a manner in which they can be validated in the Human Factors Validation Study. It is recommended to omit statements containing use error mitigations, design specifications or qualitative information into a User Need. When any of those are included it will likely become difficult to properly validate the user interface.

To illustrate this point, assume a needle-based self-injection device user interface is being designed to be used by rheumatoid arthritis patients; a User Need may state:

The user shall be able to manually control needle insertion and retraction by a single user interface.

Note that the User Need example does not include addition use error mitigation, design attribute or qualitative statements like:

- *... without making safety critical errors.*
- *... shall be able to manually control with ease ...*
- *... and retraction by a cylindrical user interface* 4 *cm in diameter.*

It is best to include these additional statements in other, more specific documents related to design requirements and risk definitions. This allows design teams the flexibility needed to create and define the design requirements for the user interface.

For more details regarding the creation of User Needs and design input requirements consult the following publications:

- ANSI/AAMI HE75:2009/(R)2013 Human factors engineering – Design of medical devices
- AAMI TIR59: 2017 Integrating human factors into design controls
- Chapter 3: Strategy and Planning
- Chapter 5: Contextual Inquiry

## 5.1 Considerations for requirements

Anthropometrics, human performance and biomechanics, i.e., range of motion and strength, data for human beings of all ages and sizes are vital for defining device design requirements. Anthropometric data sets can vary significantly between user populations. Such data are particularly important and valuable when defining the requirements of combination device for people with a specific, unique disease state, chronic conditions, functional limitations and/or comorbidities. For combination devices, the importance of understanding and applying anthropometric, human performance and biomechanical data to design requirements translates to patient adherence to a given dose regimen; if the device is too large or small then the patient may not be able to complete the task(s) required to administer the drug. For example, a device intended to deliver a drug dose via the nasal passage by using a transfer of pressure generated by the user's own exhalation requires an understanding and application of nostril sizes as well as the distance from the nasal septum to the mouth.

Just like anthropometrics, human performance and biomechanics, the perceptual and cognitive capabilities and limitations of users will vary significantly between users within the intended user population as well as between groups within the intended user population. The reasons for these variations range from education and health literacy to the effects of a given medical condition and/or the combination of medical conditions including the side-effects of other medication regimens.

General design requirements for drug delivery device will likely include, but not limited to, the following considerations:

- The importance of anthropometrics and biomechanics
  - For combination devices, the importance of anthropometrics translates to patient adherence; if the device is too large or small then the patient may not be able to complete the task(s) required to administer the drug
  - Existing anthropometry and biomechanical data may be based on a healthy population and may not provide the correct measure of the user population and may require extrapolation and interpretation; e.g., obese or elderly data may not be readily available.
  - Considering biomechanics and the context of the use situation may require interpretation of existing data i.e., data readily available may not appropriately account for the dynamics of actual use.
- Design the device to be similar or familiar in appearance or function or alternatively, purposefully develop a design that is obviously dissimilar to indicate the device should be used in a new manner (McConnell & Ulrich, 2017)
- Support all phases of dosing (in general)
  - For example, provide a checklist or on-screen instructions for connected devices or devices with a digital user interface.
    - Provide different levels of information available for each step in order to provide shortcuts for advanced users.
    - Provide a visual indicator when dosing progress is not obvious
    - This may include status indicators visible from across the room (for hospital or care environments)
    - For example, provide a countdown (or timer) of drug administration remaining

    - Provide a clear indication of "dose complete"
- Support all phases of dosing for patient centricity
    - Indicate to the patient what state the device is in (ready to use, dose delivery, dose complete)
    - Provide direction on the next steps e.g., if using a mobile application the screen might read "ready" to indicate the device is adequately prepared to deliver a dose.
    - Use of lights and sounds in blinking patterns should use simple colors and a few audible combinations to indicate the dosing state. Care should be taken to avoid overly complex blinking patterns (see Chapter 7, Section 5 on the design of Alarms).
- Support unit dose packing- Your package may be the device.
    - Prevent accidental exposure, e.g., child-proof bottles that prevent ingestion by children.

## 5.2 Differentiation

For some users and use environments, it is likely that multiple combination products may be used and/or stored by a single household and that certain disease states may increase the likelihood a patient is prescribed multiple medications. The box containing the device(s) is considered to be a part of the user interface, just like the device itself, therefore it is important that developers of combination products consider what the clinical outcome would be if two products were confused. If the risk for an error exists, appropriate differentiation strategies aimed at minimizing risks of administering the wrong drug should be created. This includes considering:

- Packaging differentiation; including, if used, color and graphical elements
- Physical differentiation; including the device and/or the device box
- Label differentiation; also including, if used, graphical elements

Ideally, these may need to be combined in order for effective product differentiation.

In addition to the intended end users, commonly there are multiple development project/program stakeholders from different disciplines that are impacted by a chosen commercial differentiation scheme. Each of these stakeholders will have different goals and responsibilities that will drive different perspectives that should be considered. For example, the marketing stakeholder will likely have an interest in maintaining the brand identity whereas the one responsible for regulatory matters will have more interest in ensuring mandatory information is included.

When creating a differentiation scheme it is best to identify all of the stakeholders that will be impacted and include them at the beginning and throughout the differentiation scheme development. This will ensure all stakeholder needs are included early, reduce design iteration cycles and reduce the risks of missing the inclusion of important details that may otherwise not be included.

Again, the box containing the device(s) is considered to be a part of the user interface and should be evaluated in formative usability evaluations to assess how well it supports end users in determining the difference versus competitive products and, if needed, products within a product portfolio and/or dose concentrations. Finally, the commercial differentiation scheme must be included in the final HF Validation Study.

### 5.2.1 *Differentiation of demonstration or training devices*

When developing a demonstration or training device is important for users to clearly see and identify the difference between an actual device and a demonstration or training device. This is particularly critical if the device and drug content is intended to be used for emergency medical treatments, e.g., the delivery of epinephrine. If a demonstration/training device is created then it too should be included when developing a differentiation scheme. The risks associated with the importance of a user choosing the correct device, actual versus demonstration, may determine it is best to conduct formative usability evaluations with realistic use scenarios to ensure the demonstration/training device can be clearly seen and identified. As such, it may also be determined to include the demonstration/training device in the Human Factors Validation Study.

## 6. Design of instructions for use

IFUs that use illustrations and graphics rather than text to convey information may make user guides more accessible (Hettinger et al., 2014). This includes the following details:

1. Clarity of explanation
   - The task can easily be completed without errors (task scenarios)
   - The text could be understood by a user with limited reading ability
2. Continuity and consistency of instructions
   - Task step were logically ordered (task scenarios)
   - Terminology was consistently used throughout the manual
3. Visual recognition
   - The illustrations accurately represented the task (task scenarios)
   - The manual provided illustrations when they were needed
4. Aesthetic and minimalist design
   - Font size and appearance is readable and large enough for the average user

Fig. 8.4 below detail an example of instructions for use of a drug delivery syringe system. It includes details such as all components of the Sypfiny®: the barrel, plunger, dosing clip and multiparticulate drug. The steps of use are broken into sections. In this example, text supports illustrations and arrows direct the user to key actions. Use tips are also provided under areas where use challenges were identified through formative usability testing.

The layout of instructions for use will vary for combination drugs and could include: a fold out card to be included in the packaging, a booklet and/or printed use details on the outer packaging. Selecting a layout for use instructions requires a clear understanding of the user, use context as well as the drug and related pharmacokenetics. All instructions, regardless of layout, should be included as early as possible in usability tests.

See Chapter 9 for more information.

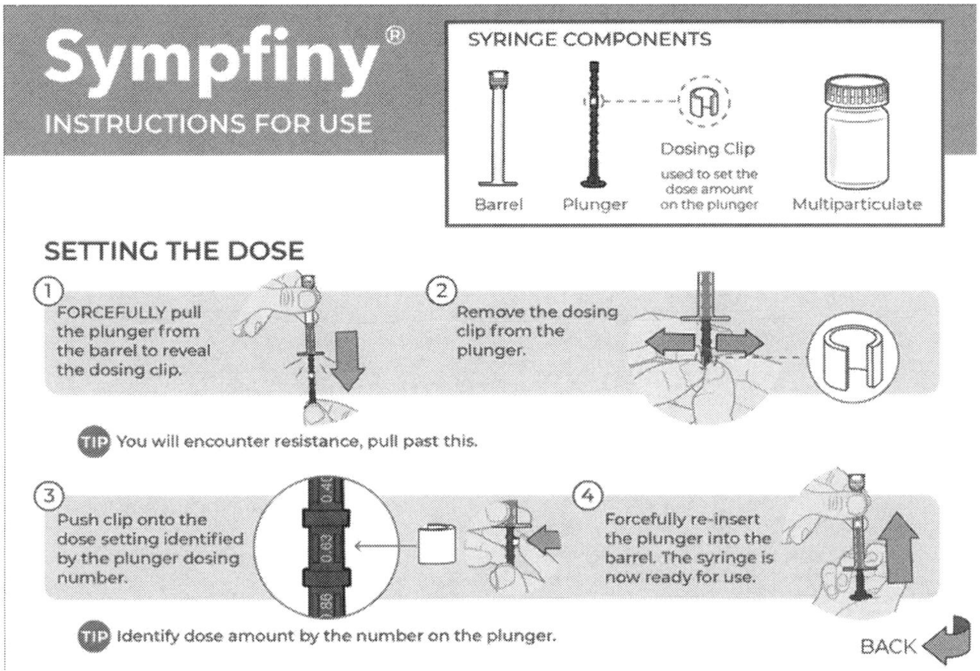


FIG. 8.4 Example of use instructions incorporating illustrations, clear steps of use supported with key text. *Image courtesy of HS Design.*

# 7. Special considerations for human factors testing of combination products

## 7.1 Pre-clinical versus clinical studies

A pre-clinical study is strongly recommended before any clinical trials are started with humans. The goal of a pre-clinical study for drug development is to identify the adequacy and ability of the device user interface to support safe and correct use. For a combination product, this includes the identification of potential use errors through human factors methodologies, including formative studies.

As it is common for drug delivery devices to have very large populations of users and may require several design iterations and many formative evaluations. Table 8.1 below provides an example of a formative usability testing plan for a device intended for general population use. A total of 12 formative studies are described including the features tested and purpose of the study. In this table, an iterative design approach is undertaken moving from low fidelity concept storyboards to fully functional designs. The purpose of the study is described in order to provide examples that a series of formative studies might be the best approach for assessing the usability of the user interface, use case scenarios as well as independent features or design attribute.

TABLE 8.1 Example table of formative testing (12 in total) the features tested and study purpose.

| | Study | Features tested | Purpose |
|---|---|---|---|
| Formative (exploratory) Testing I | 1 | All design features (physical)- using storyboards, foam models | Explore potential design, determine design direction |
| | 2 | Simulated injections, IFU, dose-dispensing, buttons, GUI | Explore details in user interface with more fidelity |
| | 3 | Simulated injection directed by GUI with IFU and training | Define needs of the GUI, IFU and training throughout dosing process |
| | 4 | Simulated injection, dose cartridge exchange | Assessing details concerning device longevity |
| Formative (fully-functional prototypes) Testing II | 5 | Options in injection means, interactive GUI, warning messages | Increasing fidelity in design, identify use errors, knowledge comprehension of warnings |
| | 6 | Simulated injections, dosage regimens, USB port use | Identify use errors regarding dosage regimens and data communication means |
| | 7 | Simulated injections in use case scenario, charger, injection tracking | Identify use errors in charging the design and tracking |
| | 8 | GUI alarms, GUI regimen and injection tracking, use case simulated injections | Identify use errors and/or points of confusion regarding alarms, dosing and tracking |
| | 9 | Injection tracking and physician communications | Identify use errors related with injection tracking and communications |
| | 10 | Simulated injection, packaging, dosing regimen, IFU | Assessing usability on near final design |
| | 11 | Set-up instructions, storage case, on-box use instructions | Assessing effectiveness of set-up instructions and usability of storage case |
| | 12 | Entire use scenario | Assuring preparation for HFE validation study |

Whereas a formative evaluation has the ability to be somewhat fluid in study moderation, a clinical study follows a specific plan and includes strict adherence to the methodology described in the study protocol. Equal to a clinical study protocol adherence is the human factors validation study. See Chapter 15, Section 1.1 on combining a Human Factors Validation Study with a clinical study for more information on the similarities/dissimilarities.

## 7.2 Do not rely on training

As described in ISO 14971, design is the most effective means of reducing risk. For the design of combination products, this translates to the importance of good design without reliance on training as the largest risk mitigation as the manufacturer would be required to demonstrate assurances that training would be carried out in an effective manner upon regulatory agency review. When training is required it is considered part of the user interface and it would be anticipated that a manufacturer would have appropriate administrative control as well as provide evidence that the training is an effective risk mitigation.

On the contrary, if providing training, it is expected to occur and occur consistently. In designing a training program it is important to consider the following:

- What will the training encompass?
- How will the manufacturer ensure that every user receives the training?
- Who will be responsible for conducting the training?
- How will the training be performed e.g., in person, video, online?
- How will consistency in the training be ensured?
- How often will training occur?
- What training materials will be provided?

The training decay time in human factors studies must be appropriate and reflect [clinical] reality. That means a manufacturer will encounter users who will be distracted in training or simply are un-interested. Additionally, for combination products, it is best practice to include an arm of a human factors validation study wherein no training is provided prior to device use. The aim of this practice is to assess the effectiveness of information materials aimed at supported safe device use.

As a result, the design of the user interface should not rely on training.

See Chapter 15, Section 4.3 for more information regarding training and Chapter 16, Section 1.4 for further information.

## 7.3 Literacy versus health literacy

Literacy is defined, most simply, as the ability to read and write. Yet, it also includes the ability to use language, numbers, images, computers and other basic means to understand, communicate, gain useful knowledge, solve mathematical problems within a specific culture. A person's ability to identify, understand interpret, create communicate and compute using written materials in varying contexts, is fundamental in all medical device design. However, literacy in general is not the same as health literacy.

According to the US Department of Health and Human Services (DHHS), Office of Disease Prevention and Health Promotion, health literacy is the degree to individuals have the capacity to obtain, process and understand basic health information and services in order to make appropriate health decisions (DHHS, 2018). The impact of health literacy is a person's ability to navigate the healthcare system, share personal health information, engage in self-care and/or chronic disease management as well as comprehend mathematical concepts (probability and risk). Low health literacy has been linked to poor health outcomes such as higher rates of hospitalization and less frequent use of preventative services (DHHS, 2018).

Improving the usability of health information in consideration of health literacy, especially in the design of user interface in combination products is especially important and can be achieved by:

1. Ask "is the information appropriate for the user?" E.g. in the product labeling, IFU or digital support material, do the materials reflect the age, social/cultural diversity, language and literacy skills of the intended users?
2. Assess the users' understanding before, during and after the introduction of information when asking knowledge comprehension tasks.
3. Ask "is the information easy to use?" E.g. by design, are the number of messages limited, does the material use plain language and focus on action?

In the design of support materials, instructions should be supplemented with visuals in order to make written communications look easy to read. Additionally, use of headings or bullets to break up the text with plenty of white space around the margins. For more information of IFU design, See Chapter 9.

## 8. Summary

This chapter discusses designing useable combination products. Patient centricity is the overall theme wherein the user is supported throughout all phases of dosing with detailed consideration toward the context of use. This approach emphasizes that the users of a combination product may have unique sets of capabilities and limitations which must be considered when designing a combination device. Factors such as the environment of use, the delivery of instructional materials, demonstration/training devices and device differentiation and labeling become of primary importance when considering a risk-based design approach. Finally, as the delivery of healthcare continues to evolve at a rapid pace, so does the development of drug delivery systems. Applying human factors to the design of combination products is not only a regulating agency requirement; it is also a humanitarian need.

## 9. Further reading


- U.S. Food & Drug Administration, Office of the Commissioner. (March 19, 2018). About Combination Products.
- U.S. Food & Drug Administration, Office of International Programs. (April, 2017). General Principles EMA-FDA Parallel Scientific Advice (Human Medicinal Products).
- U.S. Food & Drug Administration, Center for Devices and Radiological Health. (April, 2001). Guidance on Medical Device Patient Labeling.
- U.S. Food & Drug Administration, Center for Devices and Radiological Health. (February, 2016). Draft Guidance for Human Factors Studies and Related Clinical Study Considerations in Combination Product Design and Development

- ANSI/AAMI HE75:2009/(R)2013 Human factors engineering – Design of medical devices
- AAMI TIR59: 2017 Integrating human factors into design controls

## Acknowledgments

Thank you to Dr. Molly Story for her words of wisdom and review of this chapter. Thank you to Martin Rausch and Florian Schauderna for their thoughtfulness and insights which helped to enhance this chapter. Thank you to Mike Quinn and the design team at HS Design for their guidance, comments, and graphics. And, finally, thank you to the hundreds of usability study participants, in both formative and Human Factors Validation studies, for providing their perspectives which has led to the collective wisdom shared within … this chapter is as much because of each of you as it is for each of you!

## References

Center for Devices and Radiological Health. (April 19, 2001). *"Guidance on medical device patient labeling." U.S. Food and drug administration premarket information - device design and documentation processes, center for devices and radiological health.* Retrived from www.fda.gov/medicaldevices/deviceregulationandguidance/humanfactors/ucm119190.htm#guidancelabel.

Department of Health and Human Services (USA). (2018). *Who is the quick guide for? How to use the quick guide.* Retrieved from https://health.gov/communication/literacy/quickguide/Quickguide.pdf.

Edwards, E. S., Edwards, E. T., Simons, F. E. R., & North, R. (2014). Drug-device combination products in the twenty-first century: Epinephrine auto-injector development using human factors engineering. *Expert Opinion on Drug Delivery, 12*(5), 751–762. https://doi.org/10.1517/17425247.2015.987660.

FDA. (2011). Communicating Risks and Benefits: An Evidence-Based User's Guide. Retrieved from https://www.fda.gov/files/about fda/published/Communicating-Risk-and-Benefits—An-Evidence-Based-User%27s-Guide-%28Printer-Friendly%29.pdf.

FDA. (2016). Applying human factors and usability engineering to medical devices guidance for industry and food and drug administration staff preface public comment. Retrieved from http://www.regulations.gov.

Haigney, S. (2018). Regulatory developments in combination products. *Pharmaceutical Technology, 42*(12), 39.

Hettinger, A. Z., Lewis, V. R., Hernandez, A., Abts, N., Caplan, S., & Larsen, E. (2014). When human factors and design unite: Using visual language and usability testing to improve instructions for a home-use medication infusion pump. *Proceedings of the International Symposium on Human Factors and Ergonomics in Health Care, 3*(1), 254–260. https://doi.org/10.1177/2327857914031041.

IQVIA Institute. (2018). *Medicine Use and Spending in the U.S. A Review of 2017 and Outlook to 2022 - IQVIA.* Retrieved from https://www.iqvia.com/institute/reports/medicine-use-and-spending-in-the-us-review-of-2017-outlook-to-2022.

Isherwood, S. J., Mcdougall, S. J., & Curry, M. B. (2007). Icon identification in context: The changing role of icon characteristics with user experience. *Human Factors: The Journal of the Human Factors and Ergonomics Society, 49*(3), 465–476. https://doi.org/10.1518/001872007x200102.

Landers, S., Madigan, E., Leff, B., Rosati, R. J., McCann, B. A., Hornbake, R., & Breese, E. (2016). The future of home health care: A strategic framework for optimizing value. *Home Health Care Management and Practice, 28*(4), 262–278.

Masterson, F. (2018). Factors that facilitate regulatory approval for drug-device combination products in the European Union and United States of America: A mixed method study of industry views. *Therapeutic Innovation and Regulatory Science, 52*(4), 489–498. https://doi.org/10.1177/2168479017735142.

McConnell, D., & Ulrich, S. (2017). *Designing drug delivery devices for usability.* Retrieved from March 12, 2019 https://www.meddeviceonline.com/doc/designing-drug-delivery-devices-for-usability-0001.

National Institute of Health, N. (2019). Drug delivery systems | national Institute of biomedical imaging and bioengineering. Retrieved from March 12, 2019. https://www.nibib.nih.gov/science-education/science-topics/drug-delivery-systems.

Oray, S., Mountian, I., Stumpp, O., Domańska, B., Pichon, C., & Poon, S. (2017). Using patient feedback to optimize the design of a Certolizumab Pegol electromechanical self-injection device: Insights from human factors studies. *Advances in Therapy, 35*(1), 100–115. https://doi.org/10.1007/s12325-017-0645-1.

Stegemann, S., Ternik, R. L., Onder, G., Khan, M. A., & Van Riet-Nales, D. A. (2016). *The AAPS, 18*(5), 1047–1055. https://doi.org/10.1208/s12248-016-9938-6.

U.S. Food & Drug Administration. (March 19, 2018). *Office of the commissioner*. About Combination Products. Retrieved from https://www.fda.gov/CombinationProducts/AboutCombinationProducts/default.htm.

U.S. Food & Drug Administration, & Office of International Programs. (April 2017). *General principles EMA-FDA parallel scientific sdvice (human medicinal products)*. Retrieved from https://www.fda.gov/downloads/AboutFDA/CentersOffices/OfficeofGlobalRegulatoryOperationsandPolicy/OfficeofInternationalPrograms/UCM557100.pdf.

CHAPTER

# 9

# Applying design principles to instructional materials

*Renée Bailey*

Agilis Consulting Group, Orlando, FL, United States

OUTLINE



*Applied Human Factors in Medical Device Design*
https://doi.org/10.1016/B978-0-12-816163-0.00009-8

*When all else fails, read the instructions.* **Agnes Allen**

## 1. Introduction

Given growing regulatory emphasis on instructional materials (i.e., labeling and training) for users of medical and drug delivery devices, it is crucial for companies and professionals in the industry to have a practical and systematic approach to designing and evaluating instructional materials. Instructional materials are part of the user interface as referenced in key industry standards such as IEC 62366 (2015) and FDA guidance (CDER 2016; CDRH 2016). Instructional materials should be evaluated as part of the human factors strategy to show the materials promote safe, effective and accurate performance when used by the end user. Within the context of human factors, many teams and professionals can face challenges with how to effectively design and evaluate these materials and subsequently optimize them based on study data.

This chapter provides a practical guide to applying design principles during initial development of the instructional materials, or when identifying root causes and recommended improvements based on human factors study data.

Author's note: Performance based instructional design is a scientific field in and of itself and top professionals in this field have years of training and experience translating desired performance of tasks into relevant and effective instructional materials for end users. It is impractical to communicate the full science of this field in this chapter. Thus, this chapter focuses on fundamental instructional design principles and practical application of instructional design within the context of human factors for medical devices or drug delivery products. Ultimately it is beneficial and recommended to work with a trained and experienced instructional designer when developing instructional materials for medical and drug delivery devices.

## 2. What are instructional materials?

Instructional materials help support safe and effective use of a device, and they include materials such as:

- Instructions for Use (IFU), Quick Reference Guides (QRGs), printed or electronic manuals,
- On-board or electronic performance support systems (EPSS) or graphical user-interfaces (GUI) which provide on-screen navigation and task related instructions,
- Instructor-led or self-directed training and eLearning, and
- Device and package labels which provide instructions, warnings, cautions or contraindications.

For the purposes of this chapter, the focus is specifically on materials that are part of the user interface and are intended to support safe and effective use of the device. Marketing and promotional information, communications, and other supplemental information and materials are not considered instructional materials or part of the user interface, so they are not part of the scope of this chapter. However, it is always a good idea to assess the effectiveness of other materials (or any materials that are intended to guide or influence user performance).

## 3. Integrate instructional design with the human factors process

The process of designing and optimizing instructional materials integrates with the human factors process. Even though instructions and training use a different method of design compared to the device, instructional materials should be tested and optimized along with the rest of the device interface. Instructional designers use the same inputs as human factors (user profiles, use environment profiles, use cases, and task use-error analysis) when designing and developing instructional materials. And similar to the way engineering design or industrial design principles are applied to optimize the device-user interface, instructional design principles should be applied to optimize the instructional materials.

Integrating instructional design with the human factors process is as important as applying instructional design principles correctly. An integrated process allows you to efficiently design, develop and optimize instructions through formative human factors testing. This process helps avoid problems with the instructional materials emerging during validation studies when it is too late to iterate the instructional materials while still meeting submission deadlines.

A poorly integrated process can lead to countless formative studies, mandates to incorporate regulator feedback into instructional materials, or even repeating validation or summative studies due to data showing the instructional materials are not optimized. An ineffective process for developing instructional materials leads to significant added costs, timeline delays and delays with regulatory submissions and product launches.

Fig. 9.1 provides an example of an effective framework to integrate development of instructional materials into the human factors process. The final instructional interface can vary based on the nuances of different users, use environments, device complexity and associated use-related risks for any given device or system. However, by following the process outlined in this chapter, you will determine the instructional components and design needed for the final instructional interface of a given device or product.

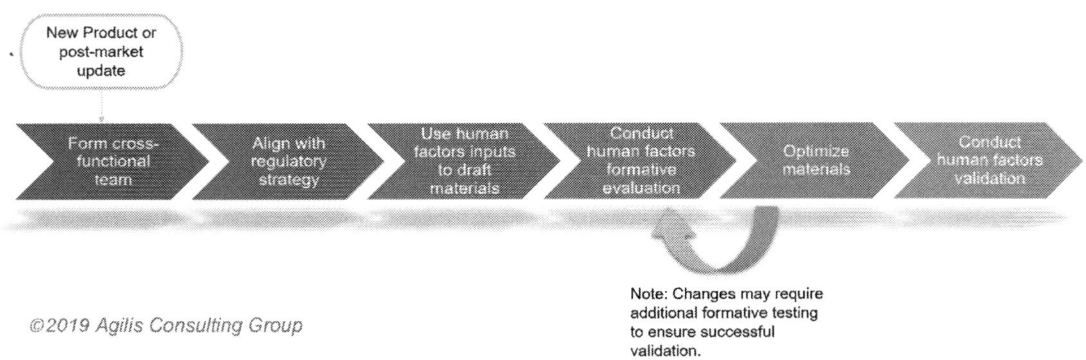


FIG. 9.1 Process for integrating instructional design with human factors.

## 4. Include instructional designers in the cross functional team

A stakeholder responsible for the design and development of instructional materials should be included early and throughout discussions about a new device or post-market update to a device. For a new product, the timing to involve instructional designers is recommended to be as early as the concept and feasibility stages when prototypes are being tested with representative users.

Cross functional teams for new products can include stakeholders from several different functional areas. Ensuring the instructional designer is included on the cross functional team early is important so they understand:

- The regulatory strategy which helps identify the requirements for the instructional materials. For example, there may be specific guidance or regulations that dictate content or placement of information in the instructional materials.
- Product design and how the device is intended to work such as features, controls, device feedback, etc. Understanding the product design helps the instructional designer draft instructional materials that communicate user-device interactions accurately.
- Risks associated with device use according to the development of the risk assessment. Instructional designers are alerted to the risks to be mitigated by the instructional materials, which helps to ensure this information is prioritized into the design of the materials.
- Human factors approach and upcoming human factors evaluations. When early formative studies are conducted with prototypes and representative users, the instructional designer can produce early drafts of instructions to include in formative testing. Early formatives allow evaluation and optimization of the instructional materials in parallel with optimization of the device design.

For post-market updates, include the instructional designer early so they understand the reasons for the updates, how the instructional materials could be impacted or how the materials need to be updated. Cross functional discussions provide the instructional designer with information they need to make decisions and updates that support the post-market update while also adhering to sound instructional design principles.

## 5. Align instructional design with the regulatory strategy

Instructional designers need to be aware of certain regulatory information early in the process. Changes to the regulatory strategy over the course of product development could impact the human factors strategy, risk assessment and subsequently the design of the instructional materials. The classification of a device, device type and the regulatory pathway for submission can link to specific regulatory requirements for labeling, which often means design requirements for the instructional materials.

For example, FDA has specific guidance documents for infusion pumps (CDRH 2014) and reprocessing (CDRH/CBER, 2015) with sections outlining guidelines and expectations of labeling (i.e., instructional materials). The guidelines provide expectations for different types of users (healthcare professionals and lay users) and use environments as well as human factors validation testing of the instructions. Other international standards that may be applicable are also referenced, such as IEC 60601-1-2 (2015) regarding electromagnetic compatibility.

The instructional designer needs to be in alignment with regulatory expectations for the specific device, class, device type and submission or post-market update strategy. By including the instructional designer early in discussions about the regulatory strategy, any additional expectations outlined by regulators can be included in the instructional design process. Additionally, if decisions are made that alter the original regulatory strategy, the instructional designer is involved and can assess any impacts to the instructional materials.

## 6. Design the instructional materials

Instructional designers are ready to start drafting instructional materials when they understand the regulatory guidelines and expectations that apply and have at least a working prototype of the device. As mentioned in the earlier framework (Fig. 9.1), the instructional materials are designed with instructional design principles, just as the device is designed with engineering design and industrial design principles.

This section describes how instructional materials are developed using human factors inputs and instructional design principles.

### 6.1 Gather industry references

Many industry resources provide guidance on developing instructional materials with regulatory submissions and usefulness for the end users in their use environments in mind. Anyone responsible for the design of instructional materials should keep abreast of various references including, but not limited to the following:

- ANSI/AAMI HE75 (2009) – Human factors engineering – Design of medical devices
- AAMI TIR49: 2013 Design of training and instructional materials for medical devices used in non-clinical environments (that also includes information useful for devices used in clinical environments)
- US FDA CDRH, Guidance on Medical Device Patient Labeling, published on April 19, 2001
- US FDA CDER, Guidance on Safety Considerations for Container Labels and Carton Labeling Design to Minimize Medication Errors, draft guidance published in April 2013
- ANSI Z535.6-2011, Product Safety Information in Product Manuals, Instructions, and Other Collateral Materials
- ANSI/AAMI HA60601-11:2015, Medical Electrical Equipment – Part 1-11: General requirements for basic safety and essential performance
- US FDA CDRH/CBER, Design Considerations for Devices Intended for Home Use, published November 24, 2014

## 6.2 Gather human factors inputs

Instructional designers use fundamental human factors inputs to make decisions about the design of the instructional materials. For example:

- **User and use environment profiles** are used to make decisions about media, writing style, formatting of instructional content, physical layout of printed materials, design of on-board instructions, etc.
- **Use cases or scenarios of use** are used to determine appropriate sequencing and scope of the instructional material for each user group and may help define distinct sections of complex sets of instructional materials.
- **Task analysis** (including PCA and risk assessment information) is used to identify skills/knowledge needed; identify tasks that are large (many steps) or complex; make decisions about media to be used; determine if training is necessary; identify details that help draft instructional content.

To understand why human factors inputs are important to instructional design, consider the following examples:

Scenario 1 – A multi-dose pen injector is used by nurses and patients to deliver the medicine. While the information both nurses and lay user patients need is the same, the nurses and lay user patients have different user characteristics. Different instructions for both user groups may be necessary. Or one set of instructions could be developed for the lay user patient assuming the nurse could follow the same set of instructions.

Scenario 2 – An implanted device includes multiple components that are used by different healthcare professionals in different hospital settings. The implant procedure involves a surgical team, post-operative care involves a critical care nurse, and a nurse educator is responsible for training the patient before the patient is discharged from the hospital. Considerations must be made to determine what is appropriate instructional material for each type of user and how those materials should be sequenced and written to support safe and effective device use.

Note: Additional information can be referenced in AAMI TIR49 (2013) Section 7.2.1.1.

## 6.3 Determine the instructional components needed

Using the human factors inputs of user profiles, use environment profiles, use cases, and task use-error analysis, the instructional designer can make preliminary decisions about the instructional components needed to support safe and effective device use.

Most devices require printed material such as instructions for use (IFU) or a user manual. This type of material is developed using the task use-error analysis as a basis to ensure each use case is covered, the instructions are written accurately based on task definitions, and risks to be mitigated in the instructional materials are addressed in the materials. But, how should decisions be made to include other types of instructional components?

Using the user profile and the task analysis, instructional designers consider some of the following elements:

- Whether use cases or tasks are completed frequently or infrequently
- Tasks which are large (with many steps) or require multiple decisions
- Tasks that are complex for the representative end user

Using this information, the instructional designer can then make preliminary decisions about the types of instructional components that may be necessary to support safe and effective device use. Some examples are provided below.

Scenario 1 – An autoinjector for a common therapy which has only one use case and a short series of steps may only require an IFU to provide users instructions.

Scenario 2 – An *in vitro* diagnostic home use kit has multiple use cases and steps for preparing to use, collecting the sample, interpreting results, labeling the samples and shipping for further analysis at a laboratory. In this case, a larger set of instructions may be needed, but training may not be feasible given it is a home use product. If human factors testing reveals users require further assistance to effectively use the kit, the instructional designer may consider developing a short video as supplemental information to the printed instructions.

Scenario 3 – A complex surgical system with many use cases and tasks, multiple user groups with different roles and responsibilities, and a high-risk profile will likely need well designed user manuals for different user groups performing tasks in different use environments. In addition, due to the complexity or criticality of performing certain tasks, a training program may also be considered. The training program may contain additional components such as eLearning, video, classroom training with hands-on practice, and post-training support or refresher training for infrequently performed tasks. Quick reference materials may also be needed to guide users with infrequent critical tasks such as emergency procedures.

Note: Additional examples and information can be found in AAMI HE75.

Keep in mind that determining which instructional components are needed is a preliminary decision that may need to be adjusted throughout the device and instructional design process. As instructional materials are included in human factors testing (and the user interface of the product may also be iterated based on formative testing), the preliminary instructional strategy may need to change. It is important to evaluate the planned instructional interface based on human factors study results throughout human factors testing and update the strategy as needed. Some examples of changing strategies are provided below.

Scenario 1 – An inhalation product has only one use case and a simple set of steps. The preliminary decision is made to only include an IFU with the product. However, after several human factors studies and revisions of the IFU, representative participants still struggle with the specific inhalation technique required to ensure the full dose of medicine is inhaled. In this case, development of additional instructional components such as a demo device and training may be considered as parts of the instructional interface since users are unable to interpret and correctly complete a critical task from the printed material alone. In this case, the strategy could be revised to include additional instructional components.

Scenario 2 – An on-body infuser has multiple use cases and critical tasks, although none of the tasks are very long or complex. It is initially thought that the end users may require training in addition to a user manual to safely and effectively use the device. However, after conducting human factors studies where participants only used the user manual, user performance showed the device could be used safely and effectively without training. In this case, the strategy could be revised to omit the instructional component of training.

Just as the device design and human factors strategy is revisited based on results of human factors formative evaluations, the instructional design strategy may also need to be revisited and adjusted.

## 6.4 Design and develop instructional materials, starting with human factors inputs to draft the primary source materials

During the instructional design process, certain materials are better to start with as they can feed into other components identified as part of the instructional interface. Early materials to draft include instructions for use, user manuals or other comprehensive materials that instruct the user through each use case and task. These materials can then be leveraged and drive what is included in other instructional components such as quick reference materials, videos, eLearning and training. Fig. 9.2 shows an example sequence of development for different types of instructional materials.

Materials such as the instructions for use or user manual are often the ***primary source materials*** that contain the full information for device use. From there, additional materials or components may be developed based on that same content. The key is to be consistent with information across the full instructional interface to avoid user confusion or use errors. Information in the primary materials should match other components such as quick reference materials or videos. Best practice is to work on materials in a sequential order, as shown in Fig. 9.2, to avoid multiple revisions on different components, or worse, having different information in different components of the instructional interface that leads to user confusion, use errors or difficulties.

Instructional designers should start designing and developing the instructional materials with the human factors task use-error analysis. The benefits of using the task use-error analysis, especially if it includes PCA analysis, include the following:

- Provides the use cases, tasks, steps and sub-steps that are needed to organize instructions effectively (assuming the use cases and tasks are defined from a **user** perspective and not a device perspective).
- Provides the criticality (based on the risk assessment) associated with each task which helps identify areas of focus in the instructions.
- Provides critical knowledge tasks the user must be able to see and understand in order to avoid or mitigate certain risks associated with using the device or product.

An important requirement of the task analysis is that it should include all the user tasks, otherwise your instructional materials risk missing those as well as critical knowledge that should be included in your human factors testing. A common shortcoming of the task analysis is that only device tasks are included, but other supplemental tasks that users must

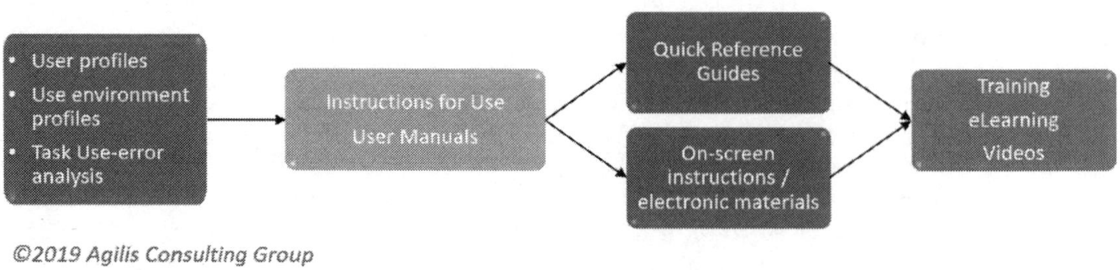


FIG. 9.2 Example sequence of instructional materials development.

complete to use the device correctly are excluded. Some examples are steps to prepare a device for use the first time, gathering supplies in preparation of device use, how to clean a device, how to transport the device, etc.

### 6.4.1 *Start with low fidelity drafts*

Instructional materials design and development is a process that can and should start early in product development and human factors testing. Evaluation of low-fidelity drafts of instructions and preliminary ideas for illustrations is beneficial and can easily be incorporated into early human factors studies. Starting early is incredibly important for effectively and efficiently iterating instructional components while building human factors data that support the design of the instructional interface.

Starting with low-fidelity drafts is efficient and cost effective because it allows the instructional designer to work in a format of instructions that is easily manipulated and revised. Using word processing software, such as Microsoft Word or Google Docs, can be very efficient for early iterations. Alternatively, iterating initial drafts of instructional materials that are in a commercial format is possible, however, this approach can be very time consuming and expensive. This can unnecessarily drain resources and increase your timeline to iterate the materials.

### 6.4.2 *Identify sections and content required*

The use cases and task use-error analysis developed by human factors can be used to help identify logical sections and sequences of instructional materials content. Consider information that may be required for end users including the following from AAMI HE75:

- Indications/Intended Use
- Device description including illustrations and identification of device parts
- Additional supplies needed to use the device
- Support information including any customer service or support phone numbers or patient hotline
- Safety information including warnings, cautions, contraindications
- Instructions for using the device which can often be divided into sections using the use cases for the device

Also consider if there are regulatory requirements for sections and sequencing such as those outlined in the FDA guidance document (2001), Guidance on Medical Device Patient Labeling.

Note: Additional information can be referenced in AAMI TIR49 (2013) – Sections 7.2.1.2 and 7.2.1.3 as well as in AAMI HE75.

### 6.4.3 *Write effective instructions*

The writing style and language used to develop effective instructions involves considering the users, use environments, and specific writing techniques. Consider the following types of users in their representative environments:

- Healthcare professional in a busy hospital environment with many distractions
- Lay user patient undergoing chemotherapy using a device in their home healthcare environment

In both cases, certain writing styles work best to communicate instructions users can quickly, easily, and accurately understand and follow.

Note: Methods for writing effective instructions are outlined in AAMI HE75 and AAMI TIR49 (2013) Section 7.2.1.5.

While AAMI HE75 provides more details than provided here, the most important aspects of writing effective instructions are considered to be:

- Write simple, task-oriented sentences with one action per step while avoiding prose and multi-sentence "steps"
- Use sequential lists to indicate a sequence of steps users need to complete
- Place warnings and cautions with the relevant step
- Use wording that has clear meaning and is not left open to interpretation

These writing techniques help avoid common issues found in human factors evaluations of instructional materials. The following are examples of some common issues and root causes seen in evaluations of instructional materials that are related to writing style.

**Example 1 – Users do not know what words or symbols mean.**

Words and symbols can be difficult for users to understand, because they often leave room for interpretation and do not provide specific or meaningful information to the user. Including wording that is more meaningful to the end users and is specific (this could be based on the PCA analysis) can clearly communicate the performance expected.

List of words and phrases users commonly misinterpret or do not understand:

- Utilize
- Initialize
- Frequently
- Large bubble
- A few times
- Modify
- < (less than symbol) or > (greater than symbol)

**Example 2 – Users do not complete all steps or skip some steps.**

When looking specifically at the instructional design and why users did not complete steps, there can be multiple reasons including (AAMI, 2017):

- The step as written takes too long (more than 20 seconds to complete).
- The step contains more than one step or action and the additional steps or actions were not seen or read.
- Writing style is prose rather than direct, active voice leading the user to ignore or skim blocks of dense text, and miss important details.
- It is unclear when a user should perform or stop performing an action or step and the user was not alerted to the cues by the instructions.
- Instructions are written in paragraph format or in a bulleted list rather than a sequential list of distinct, single actions which leads users to skim information rather than follow as instructions.

- Steps are out of sequence in the instructions.
- Additional or important information is located in another section or on a prior/next page which the user does not see.

When users miss completing a step, proper writing techniques can be followed to alleviate most issues. Evaluating the materials again in a human factors study will provide data on whether the issue has been mitigated with the update to the instructions.

### 6.4.4 Create effective illustrations and graphical elements

Drafting instructions and drafting appropriate visuals or illustrations that support the instructions is done together. As the instructional designer writes the instructions, they should also consider, "How can the instructions visually show the desired performance?" Illustrations are best used to support instructional content and tell users (AAMI, 2017):

- When to do something
- How to do something correctly
- What an outcome looks like

Illustrations also include elements of the task use-error analysis, especially if PCA analysis is included. What the user is seeing, feeling, or hearing helps inform the content and design of the illustrations.

A good way to start when developing early drafts of instructions is to take photographs. Photographs are easy and cheap to take and iterate if a concept does not work with end users. Photographs can also be turned into illustrations by a graphic designer. A graphic designer can help determine a specific styling, perspective and coloring that works best with the end users. Styling can range from very simple line drawings in black and white, to full photo-realism that shows much more realistic detail.

Although photographs can be useful during early development of instructions, illustrations are typically more effective than photographs for end users. Illustrations allow you to control how much detail is visually presented to the end user. Certain device, product or environmental aspects can be emphasized or de-emphasized in illustrations, and background clutter or distraction that is often present and difficult to control in photographs can be eliminated in illustrations.

There are also times when photographs may be more effective for end users than illustrations. An example in which photographs might be beneficial is when medical procedures are involved. Surgeons and surgical teams may find it more useful or informative to see photographs of an actual procedure using a device, rather than an illustration where the necessary realism is lost or difficult to replicate. In this situation, using photographs might be a better option than illustrations.

Note: Additional information can be referenced in the AAMI HE75 section on Instructional Materials.

#### 6.4.4.1 Elements that improve the usefulness of illustrations

Just having an image or illustration is not always enough to influence safe and effective user performance. For example, it may be unclear just from the illustration exactly what the user should do or exactly how to hold or manipulate the device. A variety of additions can be applied to illustrations to improve user interpretation of the illustration and subsequent safe and effective use of the device. Some examples of these additional elements are:

- **Captions or callout text** – provide text directly next to the image or with a line pointing directly at an aspect of the illustration. The text can help clarify things like "hold here" or "grasp firmly."
- **Placement and size relative to the text** – if all the illustrations are going to be very small, users may not be able to identify aspects of performance or the device. Try to make illustrations large enough that important details are not lost.
- **Angle and perspective** – use an angle and perspective (self-perspective or observer-perspective) that is easy to visualize, and users can quickly determine the performance depicted.
- **Hands or no-hands** – often hands and placement of fingers can influence users to perform a step exactly as depicted. There may be times when hands would obscure important elements of the device interaction or may cause user confusion. Testing illustrations with and without hands may be done during formative evaluations to see if there is any difference in user performance.
- **Cropping and enlarging areas of focus** – start with a larger illustration with all details or context of the step being performed. If a specific aspect is difficult to see or focus on, crop the illustration down and enlarge the specific area to focus on the details as needed.
- **Directional arrows** – use thick or block style directional arrows so that users can clearly determine the motion or direction required. Very thin or light-weight lines are ineffective when the illustration size is reduced for final commercial formatting.

Table 9.1 provides a list of common issues regarding illustrations that can cause confusion or use errors with participants in studies and how to correct them.

### *6.4.5 Add organizational and navigational elements*

Many components of instructional materials stand alone and must provide certain structural and navigational cues to the user. This is true for various types of media including print, electronic, on-device screens, web-based, etc. Users rarely read all printed instructional materials from beginning to end like a book, which makes it imperative to incorporate well-designed and well-placed organizational and navigational elements to ensure the overall effectiveness of the materials. If users cannot find the information, there is little chance they will read it and follow it, which could expose them (or patients) to risk of harm.

#### 6.4.5.1 Use headings, table of contents, index, and cues for page turning (as necessary) in printed booklets

For printed materials that will be in a booklet format, it is key to include meaningful headings, a table of contents, and for larger manuals, an index. Users may use these elements in various ways, but often the user looks for key words in headings, the table of contents and index to locate specific information. If the headings are not written in a meaningful way containing key words that resonate with the end users, this may cause users to not find information or to not use the instructional material at all.

TABLE 9.1 Common issues with illustrations and how to fix them.

| Issue with illustration | How to fix the issue |
|---|---|
| User is confused about multiple actions included in the illustration | Include one action per illustration and match the illustration placement with the step text |
| User cannot visualize the intended performance from a technical or engineering style illustration | Use a styling method such as simple line drawings or photo-realism illustrations avoiding dark backgrounds and ensuring very clear contrast of lines against the background |
| User is confused because colors in the illustration do not match what the user sees on the device | Use shapes or graphical treatments to call attention to device parts and keep device parts colored accurately in the illustration |
| User cannot follow hand or finger placement in the illustration | Ensure the best hand or finger placement for performing the task is included in the illustration |
| User did not notice the color graphic highlighting certain aspects of the illustration | Use color carefully and avoid using so much color it reduces its affect to call attention to important information or aspects of performance |
| Illustrations are too small and detailed for users to interpret | Crop or enlarge the most important portion of the illustration to add visual clarity |
| User did not see the illustration until after performing the step | Align illustrations with the respective text and make placement consistent, allowing users to see the illustration first and read text if and when needed |
| User could not tell what illustration was conveying, illustration was confusing | Consider alternative perspectives to clarify the illustration. Perspective may be from user point of view or observer point of view. |

Note: Some high-risk devices or systems can generate long lists of warnings and cautions that must be placed into the instructional materials. It is difficult for users to find relevant warnings and cautions in such long lists. To alleviate this issue, consider grouping warnings or cautions by specific topic areas. Use meaningful headings for each topic area and separate the long list into smaller lists under each heading. Additional information can also be located in the AAMI HE75 (2009) section on Instructional Materials.

Another consideration for printed materials is page turning. In booklets, if information is continued on the next page, an indication should be included to alert the user to turn the page for more instructions or information. If the end of a page or section gives an instruction for what the user should do next (e.g., repeat steps, go to another task, immediately perform step on the next page), the instruction needs to include what should be done and where to reference additional instructions. Often, page turns that are not identified can lead to users missing important information or instructions by not turning the page.

#### 6.4.5.2 Use clear identifiers, graphical treatments, and cues for page turns for large format printed sheets

When printed materials are large-format sheets (e.g., newspaper style), a different approach is required. First, the outside of the final folded sheet should be clearly identified as instructions. There can be multiple pieces of printed information that accompany a device and the instructions should be clearly identified to the user. Graphical treatments to help the user unfold the sheet completely may be needed. Folds can cover important information and navigational cues can help users unfold the sheet and see all the information provided. If the sheet is two-sided, a graphical element can be helpful to encourage users to turn the sheet over to find information on the other side. Large format sheets also require meaningful headings to group information and separate informational sections from instructions.

#### 6.4.5.3 Consistently organize electronic materials

Electronic materials should be organized in a similar manner to the primary source materials (e.g., instructions for use, user manuals). However, the navigation should match the standard navigation already established for the electronic delivery method. For example:

- Instructions that are presented on a web page should use standard conventions for navigating a web browser.
- On-board instructions embedded in a device should provide very clear navigation to move back and forth between screens and may leverage other standard navigation conventions.
- eLearning should provide standard navigation conventions appropriate for the user. Some lay users may have never used eLearning interfaces, so the process of navigating should be clear and may need to mirror other, familiar navigation conventions.

### *6.4.6 Apply formatting to instructional materials*

Different instructional components will have different formatting according to the media selected for delivery to the end users. For example:

- Printed instructional materials may have a variety of layouts applied from booklets, to folded leaflets to single page quick start guides or cards, etc.
- Electronic materials may be delivered in an electronic document or may be formatted for delivery over standard web browsers.
- eLearning is typically drafted using storyboards that are easily updated and manipulated. The storyboard defines all the images, graphical elements, text, navigation, etc. planned for each screen included in the eLearning. The storyboard is then used to create the electronic design.
- Classroom training materials may consist of a presentation with printed materials for the trainer and the training participant.

While minimal formatting may be applied during early formative evaluations, once you approach late stage formative testing, a formatting design close to the commercial or clinical study design should be tested. Often final formatting design and branding can influence user performance and will need to be optimized with the other aspects of the instructional interface.

Instructional materials made available in different media for the commercial or clinical study device should generally follow the acceptable guidelines for that specific media type. The AAMI HE75 (2009) section on Instructional Materials comprehensively describes several important formatting considerations for instructional materials design. The list below discusses some formatting issues that are common in human factors testing of instructional materials:

- Text formatting that makes instructional material difficult for users to read
  - Size of the text is too small (smaller than 10 pt. font size)
  - All UPPERCASE words or phrases are used throughout the text
  - **Too much bolding reducing the effect of emphasizing certain words**
  - Bolding the wrong words
  - Light, thin or compressed fonts that have poor contrast
  - White or light text on dark backgrounds which are difficult to read
- Contrast of both illustrations and text makes it difficult for users to visualize the content
  - Dark backgrounds used behind illustrations or text
  - Illustration line weights do not offer enough contrast
- Use of too much color makes it difficult for users to focus on important information or details
  - Too much color is used to emphasize information along with color illustrations, branding and graphical design elements
- Flow of information makes it difficult to easily read or navigate to different sections of instructional materials
  - Users cannot easily navigate the material or navigate to the wrong information at the wrong time
  - Users tend to have a more difficult time reading down a column and moving over to the next column and reading down again. A more natural reading pattern is across rows of information rather than down columns then across the page.
- Position of illustrations and accompanying text is not consistently aligned making it difficult for users to follow text and illustrations in the correct sequence
  - Illustrations are placed after steps or in areas where they fit on the page instead of before steps

Note: Additional information can be referenced in AAMI TIR49 (2013) Section 7.2.1.6.

#### 6.4.6.1 Additional formatting and layout considerations for printed materials

In most instances, it is best to have nearly final content and illustrations before placing printed material into a commercial or clinical study layout. This is due to the time and effort associated with iterating in a final layout design. It is often impractical for some organizations to continuously edit within the final layout and instead, a low-fidelity version is best to use until the content of the instructional materials has been optimized for end users.

Once you have a draft of text and illustrations that work well for end users, it is time to apply the commercial or clinical study design (i.e., layout and formatting) to the materials. Many factors influence the final design of the materials. Along with the full scope of the content to be included, the instructional designer also needs to know:

- **Footprint of the instructional materials.** Materials provided within the packaging of a device or product will need to fit inside the packaging dimensions (length, width and depth). If the design is a booklet, the booklet size and thickness will be limited to the packaging design and dimensions. If the design is a large-format sheet or leaflet, the final folded size of the instructional material must fit into the packaging.

- **Color vs. black and white.** Printing in color generally costs more, and there may be constraints on how much the production of instructional materials can cost. The instructional designer will need to evaluate the costs of printing early and whether color is an option or if black and white should be considered as a less expensive alternative.
- **Considerations for separate materials.** Some materials may be provided separately for a large, complex system and thought should be given to where the material will be stored and if it will also be made available electronically and displayed on a monitor or mobile device.

Some large systems with numerous use cases and many tasks may have instructions placed in a content management system. Templates and layouts may be prescribed for large booklets that are produced from content management systems. In this case, the formatting of the final product may be dictated by the templates available in the content management system. However, testing these instructional materials in their final format is still important to ensure the layout does not create difficulties or use errors.

Other cross-functional stakeholders should provide input for the printed materials including packaging and distribution as well as manufacturing. This is to ensure the automated production and packing of the instructional materials is considered before the layout design is finalized. The following may be concerns of the additional stakeholders and should be considered for the design of the printed layout:

- Limitations on sheet size and printing capabilities for large-format sheet sizes
- Limitations on color
- Automated machine folding of leaflets
- Automated picking and packing of the instructional materials on the manufacturing line

It is recommended to get input from any additional stakeholders as early as possible to agree on a format that will work for end users and can be accommodated for commercial production. These stakeholders should receive early drafts of the commercial materials to run production tests and alert the instructional designer of any modifications required.

### *6.4.7 Develop additional instructional materials or components*

As discussed and illustrated in Section 6.4 of this chapter, it is recommended to develop some instructional materials or components *after* primary source materials are developed (e.g., instructions for use, user manuals). Development of the following types of materials is usually best after the content for the primary source material is finalized.

- Quick reference materials
- On-screen or on-board instructions (EPSS)
- Training videos
- Training and eLearning

Developing materials in this recommended sequential order increases consistency and accuracy of all materials. Additionally, it reduces duplication of effort to make updates across materials, which happens often during early development and iterative human factors testing.

#### 6.4.7.1 Create effective quick reference materials

The most common issue seen in human factors testing with quick reference materials is use errors or difficulties caused by missing information. Quick reference materials can be designed to avoid this issue by.

- Including all steps and safety information for the task(s) covered
- Directing users when to reference the full instructions and where in the instructions to go
- Making it obvious that the quick reference materials only include a subset of the full instructions
- Only using quick reference materials when required:
  - In general, a preliminary assessment of the task analysis and the frequency of task completion by users will help inform the need for quick reference materials. Through the course of human factors testing, users may also express ideas about tasks for which quick reference materials would be beneficial.
  - For example, an auto-injector with one use case and a small number of steps should not require a quick reference guide. Producing a quick reference guide for marketing purposes to show the auto-injector can be used in "two simple steps" (e.g., load dose and inject) is inaccurate from a user performance perspective and an incorrect application of the quick reference guide when users should see and understand all the information related to safe and effective use of the auto-injector.
  - In another example, an implanted device that has different components used in the operating room (OR), the intensive care unit (ICU) and at home with the patient could require quick reference materials. If tasks in the ICU are completed infrequently with a separate component of the system, adding quick reference materials to instruct nurses to perform specific tasks could be required. The materials could be laminated and attached to the equipment for easy reference with explicit mention of the full user manual if additional information is needed.

Often, quick reference materials leave out steps, warnings and cautions, or information designed to mitigate use errors. Without complete information, users may not perform critical steps, may not use the device correctly and may make mistakes that could have been avoided with additional information. The design principles in this chapter all apply to quick reference material, and the quick reference material should be consistent and complete compared to the primary source materials (i.e., instructions for use, user manual).

Note: Additional information can be referenced in AAMI TIR49 (2013) Section 7.2.2 and the AAMI HE75 (2009) section on Instructional Materials.

#### 6.4.7.2 Create effective on-screen or on-board instructions (EPSS or GUI)

The on-board instructions embedded on a device display may start with storyboards developed from preliminary drafts of use cases and related tasks and steps. It is recommended the storyboarding and development of these instructions leverage the primary source instructions to ensure accuracy and completeness.

Instructional design principles can also serve as useful tools during the development of the storyboards, including how instructions, decisions, prompts, messages and warnings are presented to users in the software or graphical user-interface (GUI). Instructional designers should be part of the cross-functional design team during software development, ensuring any instructional elements of the software are optimized and promote accurate understanding of information and performance by end users.

Note: Additional information can be referenced in AAMI TIR49 (2013) Section 7.3 and the AAMI HE75 section on Instructional Materials.

#### 6.4.7.3 Create effective training videos

Developers of training videos frequently experience similar issues to quick reference materials where the training video becomes ineffective and problematic because it does not contain enough detail for users to completely understand how to use the device correctly.

There are many ways to integrate video with electronic materials posted on a website or within training programs. Videos that thoroughly depict device use could be quite long (approximately 20 min or more), even for a device with a single scenario of use. In such cases, it should be examined how to divide a long video into smaller segments that users can access and view as needed, much like sections of instructions for use or user manuals. For best results in user performance, training videos should have content that is consistent with the primary source materials (i.e., instructions for use, user manual).

Note: Additional information can be referenced in the AAMI HE75 section on Instructional Materials.

#### 6.4.7.4 Create effective training and eLearning

Development of training is another separate process within instructional design that requires special expertise to ensure training will be an effective risk mitigation as part of the user interface. It is highly recommended to work with experienced instructional designers who use a performance-based methodology to create a training program, especially for high-risk or complex devices or systems.

While preliminary training decisions are made and investigated early in the human factors program, definitive training decisions are made later in the process and according to a systematic, structured approach. Well-designed training programs are built from the task analysis and primary source instructions. Using these inputs, the instructional designer identifies the skills and knowledge users need to safely and effectively use the device, and designs training interactions to build the required user skills and knowledge. AAMI TIR49 (2013) provides explicit detail about the systematic process used to design and develop training, and this process is also referenced in AAMI HE75.

eLearning is also part of the systematic, structured training design approach outlined in AAMI TIR49 (2013) Section 5.2. eLearning may be identified as a method of training delivery during the design stages of the training program. Once the different components of the training program have been outlined, those areas to be covered in eLearning go through a storyboard process. The storyboarding process should leverage the primary instructional materials (i.e., instructions for use, user manual) but will also outline formatting, page navigation, links to additional resources and documents, planned learning interactions, graphics, illustrations, video, etc. that make up the entire eLearning portion of the training program. These elements combined are designed to ensure end users build requisite skills and knowledge as part of the eLearning component.

Table 9.2 lists issues commonly seen related to training programs and how to avoid these common mistakes.

TABLE 9.2 Common issues with training and how to avoid them.

| Issue with training | How to fix the issue |
|---|---|
| Not including instructional materials available to the user (e.g., instructions for use, quick reference guides) in the training and hands-on demonstrations with the device | Most training does not include time to train users for long-term retention and recall of knowledge or skills.<br>Best practice is to teach users how to use the materials available in the entire instructional interface to support performing tasks. This leads to users being much more likely to use the available instructional materials when they need these supports, and ultimately, they are much more likely to use the device safely and effectively. |
| Not considering the baseline skills and knowledge the user has before completing the training (or differences in baseline skills and knowledge for experienced versus naïve users) | Training should provide specific delivery methods to the trainer which accommodate different levels of user experience with the device. For example, tailor training content for experienced users to understand what is new or different about the device being trained as well as practice with the device that focuses on those differences until the user can safely and effectively use the device.<br>On the other hand, inexperienced users should be offered enough training and practice to ensure that they achieve a high-level of self-efficacy in performing tasks with the device using the provided instructional materials. |
| Not allowing users enough practice to become proficient at using the device | Users need practice to develop skills. The training should accommodate different types of users and allow them to practice using a device as many times as needed to gain proficiency at device tasks.<br>Refresher training may be needed to assess and ensure infrequent and important tasks (e.g., emergency procedures) are practiced often enough to increase retention and recall of the task performance when users need to perform these tasks in an actual use environment. |
| Not assessing skills at the conclusion of training to ensure user proficiency meeting the prescribed criteria for safe and effective performance | User performance of tasks generally needs to meet certain standards of safety and effectiveness, which relate back to the task analysis. Final assessments of certain use cases or training segments should be included in training to assess the user's abilities to perform tasks safely and effectively with the requisite skills and knowledge.<br>If users cannot perform the tasks correctly, feedback and additional practice should be provided until the assessment demonstrates the user can safely and effectively perform the task. |

## 7. Conduct formative evaluations of instructional materials

Building in the testing of instructional materials into early human factors studies helps optimize the user interface based on user performance and feedback. The goal of testing instructional materials is to optimize until:

- Instructions for use, user manuals or quick reference materials stand alone and support safe and effective device use when used
- After training, the user can perform tasks safely and effectively
  This section covers strategies for optimizing instructional materials through formative evaluations.

Note: Additional information can be referenced in the AAMI HE75 section on formative and summative testing as well as Chapters 10, 11 and 15.

## 7.1 Include instructional materials in early formative evaluations

Incorporating draft instructional materials, even low-fidelity versions, early in product development and testing is key to optimizing the materials as part of the user interface. Testing instructional materials during early formative work identifies any improvements from the user perspective such as wording or terminology that is confusing or easily misunderstood, text that is interpreted in different ways by different users, and illustration updates that would better support user performance.

There may be specific parts of the instructions that need to be isolated and tested to mitigate a use error or difficulty, such as a specific step of instructions or a particular sequence of illustrations. In this case, you may only need to test that portion of the instructions to focus on exactly what needs to be iterated or changed.

Formative evaluations can be a tool to alleviate extensive internal reviews of instructional materials. Often internal reviewers within a cross-functional team are not representative end users of the instructional materials or device. Therefore, these internal reviews may offer input that contradicts sound instructional design and may not ultimately be what representative end users need. If the cross-functional team discusses the instructional materials at length and cannot agree on certain aspects of the design, a few strategies can help:

- Ask participants to use up to 2-3 different design alternatives of the instructional materials in a formative evaluation. In this case, it is important to consider the appropriate participant sample size to allow each sub-group of users to use a different "primary version" of the materials so the sub-groups can be compared. Then the cross-functional team can make informed, meaningful and data-driven decisions based on user performance with the design alternatives.
- Have 2-3 alternate designs for a particular step or illustration that is expected to be problematic ready to present to participants in the formative evaluation. Get agreement from the cross-functional team on the primary design to be tested during the study session, and present one design to all participants. Then gather subjective feedback from participants at the end of the scenario or study session comparing the alternates to the primary design they used. This approach will help determine if participants think a design alternative would help them better understand the intended information.

## 7.2 Optimize instructional materials based on human factors data

Once human factors data are collected from a formative evaluation, it is important for the instructional designer to apply the instructional design principles in AAMI HE75, AAMI TIR49 (2013), and this chapter to identify potential changes. Often feedback is provided by study participants to change the materials in specific ways, but the changes proposed by participants may or may not be effective. Worst case scenario is that the changes may induce new or unanticipated use errors or difficulties that were not seen in previous testing. Therefore, it is imperative to conduct effective follow-up interviews with participants and good root-cause analysis to help the instructional designer make informed decisions.

**Example: optimizing in early formatives**

Instructions were designed for an infusion system used by healthcare professionals and lay user patients. The instructional materials were tested throughout the human factors program as part of multiple formative evaluations.

During the formative testing, the instructional materials were optimized as follows:

- Illustrations were revised to clarify performance by including one step per illustration.
- Steps were formatted as sequential lists rather than large blocks of text.
- Color was changed to provide better contrast between steps and safety information and illustrations.
- Sections were reorganized to flow better from the user perspective and the sequence in which users would perform tasks.

As a result of the formative testing and attempting to resolve early use errors and difficulties with the lay user task, the decision was made to improve the device design. The instructional materials were updated to reflect the new device design.

Due to including instructional materials in formative testing early, device design and instructional design was optimized in parallel with positive results.

## 7.3 Optimize after late-stage formative evaluations

Once early formative evaluations have been conducted and instructional materials have been iterated and optimized, changes in later stage formative evaluations should be implemented with extreme caution. The instructional designer and cross-functional team should carefully consider the impact of late stage changes and only make those changes directly resulting from study data. Changes that are unrelated to study data could cause use errors and difficulties not observed previously, which will cause more iterations to become necessary. The scenarios listed below highlight how making non-data driven changes after late-stage testing can be detrimental to the overall project timeline and budget.

Scenario 1 – The graphic design team was not involved in discussions of study data or changes to be made. A request was made for the graphic designers to change the line weight of a specific line in one illustration to make the device part clearer to end users. When the materials were tested again, there were new use errors attributed to the illustrations and negative subjective feedback from study participants about the illustrations. Upon further review of the instructional materials, the graphic artist had updated all the illustrations with the new line weight making them very difficult for users to understand.

Scenario 2 – Throughout early formative testing, no use errors were seen for a specific step. In late-stage formative testing, a new use error was observed performing the step and user comments suggested that an element of the illustration did not stand out and contributed to the use error. When the earlier instructional materials were compared to the newer version, it was determined that the background color had been changed causing the element to no longer stand out. The background color change was not based on a need identified by the prior formative work.

## Example: optimizing in late-stage formatives

Instructions were designed for a combination product to be used by lay user patients who have disease and age-related limitations. The predominant limitations affecting the use of the instructions were decreased vision, limited manual dexterity, lack of stamina and some possible loss of cognition.

The instructional materials were printed as large-format (e.g., newspaper style) instructions that were tested in two formative evaluations. Through the formative evaluations, the instructions were optimized with the following revisions:

- Improved illustration styling to add contrast and detail missing from original illustrations.
- Illustrations were added to more steps to make the steps of use clearer.
- Format of the printed layout was updated to make sections of the large-format sheet clearer.
- Navigational aids were added to columns on the sheet to ensure users were able to follow instructions to the next column.
- Safety information was included to avoid use errors that would lead to a missed dose or underdose.

One task in the instructions caused multiple use errors in each formative evaluation. The design included a graphical table with images and text. The design was iterated, and several alternatives were tested over the course of the formative evaluations.

Regardless of the iterations and optimization, the instructions could not be modified further to increase the correct completion of the critical task. The cause of the incorrect performance could not be corrected by the instructions, instead the device interface needed to be updated. Due to the characteristics of the end users, the instructional materials could not overcome the device design to promote correct performance of the task.

## Example: optimizing in late-stage formatives

Instructions were designed for an *in vitro* diagnostic kit to be used by lay user patients in their homes. The instructional interface included a user manual, a label sheet to identify the samples, and shipping instructions to send the samples to the laboratory.

The instructions were tested through a series of formative evaluations and optimized after each formative evaluation based on the study data. Revisions to the materials included:

- Improved illustration styling – removing dark background and using high-contrast illustrations for an older age demographic
- Included one action per illustration and one illustration per step to make the process clear for low literacy users.
- Added sequential lists for steps instead of large blocks of text.
- Revised intended use, how the kit works and other informational content to simplify prose style writing and increase comprehension for low literate users.
- Updated the sequence of sections to match regulatory requirements and to provide important information before the instructions for use.

One critical task had use errors and difficulties in each formative evaluation. The illustrations had to support user performance more than the text due to the low literate user group. Several alternate illustrations were tested until the use errors and difficulties were reduced significantly.

In the validation study, user performance was not perfect, however, the residual risk assessment showed the iteration of the instructional materials and determined no further optimization could be made to the instructional materials.

### 7.4 Optimize after validation

Just as optimizing materials in late-stage formative testing should be done judiciously, the same is true after validation testing. Changes made post-validation could lead to a supplemental validation study if the change(s) could introduce new risks or could have an impact on user performance. The impact a change could have on user performance can be difficult to predict, therefore supplemental testing may be required to gather data proving the effectiveness of the changes (especially since the final user interface is required during validation testing).

There may be times when a specific portion of the instructional materials has been especially problematic during human factors testing. The instructional content may have been iterated, multiple versions tested, and still in the validation study there are observed use errors and difficulties. In this case, consider the residual risk and the human factors data related to the instructional interface that were gathered throughout the overall human factors process. The residual risk analysis and human factors data specific to the instructional materials may show improvements in task performance, even though use errors and difficulties were not fully eliminated in the final design. If the instructional materials have been optimized as much as possible, consider building residual risk rationale using human factors data related to the instructional materials.

## 8. Summary

Instructional design is a structured, systematic process of designing and developing the instructional interface. Correct application of instructional design principles influences user behavior and promotes safe and effective use of medical devices and drug delivery products. There are many resources available to the person or team responsible for developing the instructional interface. These resources guide the application of instructional design principles for the best possible design of instructional and training materials.

To create the safest and most effective instructional interface, the instructional design process is integrated with the human factors process. Instructional materials are part of the user interface and should be designed and tested early in the overall design process, along with the device, as part of human factors studies. Based on the human factors testing data, the instructional interface can be optimized in a similar manner as the device-user interface. Iterative design and testing of instructional materials yields the best instructional interfaces that will support end users while also building the evidence needed for regulatory submissions to demonstrate the labeling and training foster safe and effective device use.

## 9. Further reading

- ANSI/AAMI HE75 Human factors engineering – Design of medical devices
- AAMI TIR49: 2013 Design of training and instructional materials for medical devices used in non-clinical environments
- US FDA, Guidance on medical device patient labeling. April 19, 2001
- US FDA CDER, Guidance on Safety Considerations for Container Labels and Carton Labeling Design to Minimize Medication Errors
- ANSI Z535.6-2011, Product safety information in product manuals, instructions, and other collateral materials
- ANSI/AAMI HA60601-11:2015, Medical Electrical Equipment – Part 1-11: General requirements for basic safety and essential performance
- US FDA CDRH, Design Considerations for Devices Intended for Home Use, Guidance for Industry and Food and Drug Administration Staff
- AAMI Workshop entitled Applying Human Factors to Improve Instructional Materials as Part of the User Interface

## Acknowledgments

Thank you to Melissa R. Lemke and Patricia (Pat) Patterson for their contributions to this chapter.

## References

*AAMI workshop entitled, applying human factors to improve instructional materials as part of the user interface, that was presented.* (March 2017).

AAMI TIR49. (2013). *Design of training and instructional materials for medical devices used in non-clinical environments.*

AAMI/ANSI HE75. (2009). *ANSI/AAMI HE 75:2009, Human factors engineering – design of medical devices* (United States), www.aami.org/publications/standards/he75.html.

ANSI/AAMI HA60601-11. (2015). *Medical electrical equipment – part 1-11: general requirements for basic safety and essential performance.*

ANSI/AAMI/IEC 62366-1. (2015). *Medical devices – part 1: application of usability engineering to medical devices.*

*FDA guidance entitled, applying human factors and usability engineering to medical devices that was published by CDRH, FDA on February 3, 2016.* (2016).

*FDA guidance entitled, Human factors studies and related clinical study considerations in combination product design and development that was published by CDER, FDA on February 2016.* (2016).

*FDA guidance entitled, infusion pumps total product life cycle guidance for industry and FDA Staff, that was published by CDRH, FDA on December 2, 2014.* (2014).

*FDA guidance entitled, reprocessing medical devices in health care settings: validation methods and labeling, that was published by CDRH/CBER, FDA on March 15, 2015.* (2015).

# PART IV

# Formative design evaluation & reporting

CHAPTER

# 10

# Heuristic analysis, cognitive walkthroughs & expert reviews

*Mary Beth Privitera*

HS Design, Gladstone, NJ, United States

OUTLINE



*Design isn't finished until someone is using it.* **Brenda Laurel**

## 1. Introduction

Heuristic Analysis, Cognitive Walkthroughs and Expert Reviews are all types of usability inspection methods using analytical techniques that do not require users per se. They can generate results in a fraction of the time and costs as empirical techniques. This chapter defines heuristic evaluations and cognitive walkthroughs, describes their differences and communicates the processes of each technique for the purposes of medical device design. Each of these techniques are considered cost-effective, flexible techniques aimed at improving

*Applied Human Factors in Medical Device Design*
https://doi.org/10.1016/B978-0-12-816163-0.00010-4

the usability of a system. These methods are valuable for the determination of usability issues and are complementary to formal usability testing with users. They are often performed by experts in usability/human factors and result in: a list of usability issues and strengths, a severity or priority ranking, and recommendations for design improvement. They can and should be used throughout the design process, often with multiple iterations as the design of the user interface increases in fidelity.

## 2. Background

Heuristic evaluation was originally proposed by Neilson and Molich in 1990 as a means of identifying usability issues found in software. The intent of this evaluation method was to evaluate a user interface by simply looking at the interface and passing judgment according to own's own opinion, in essence an expert review. Wharton, Rieman, Lewis, and Polson (1994) proposed the cognitive walkthrough method in the book "Usability Inspection Methods" as another tool to uncover problematic issues in user interface design. While these methods focus specifically on user interface design, they are also applicable to the physical design of a product, which allows them to provide an analysis of the entire user interface, both digital and physical.

For medical device design, the FDA specifically encourages the use of heuristic and expert analyses to identify use-related hazards and hazardous situations. The methods are clearly described within their guidance in Sections 6.3.2 and 6.3.3 (FDA, 2016) as an effort to identify problems and make recommendations for improving the usability of a device (FDA, 2016). Following this recommendation, each method is described in detail below.

## 3. Heuristic analysis

Heuristic evaluation focuses on improving effectiveness and efficiency with respect to user testing (Hvannberg, Law, & Lárusdóttir, 2007). A heuristic evaluation is a human factors method intended to uncover areas of improvement that improve usability. It is an evaluation of the use interaction at all touchpoints, such as input forces, fit and biomechanics. It can be used throughout the design process to identify optimum ergonomic product architecture, assess use risk and review overall usability. Conducting a heuristic evaluation involves having a small set of evaluators examine the user interface against recognized usability principles (the heuristics) (Nielsen, 1995; Nielsen & Molich, 1990).

Nielsen (1994) provides 10 usability heuristics which have been validated and built upon. They are:

1. Visibility of the system status: Always keep users informed about what is going on through appropriate feedback in a reasonable amount of time.
2. Match between system and real world: The system should speak the user's language with words, phrases and concepts familiar to the user rather than system-oriented terms.
3. User control and freedom: Users often choose functions by mistake and will need a clearly marked route to correct the issue.

4. Consistency and standards: The design should follow conventions found that comply with the appropriate standards. A standard ensures that users understand the individual interface elements in a design.
5. Error prevention: Careful design prevents a problem occurring in the first place. It is accomplished by eliminating error-prone conditions or checking for them and presents users with a confirmation option before they commit to taking an action.
6. Recognition rather than recall: Minimize the user's memory load by making objects, actions and options visible.
7. Flexibility and efficiency of use: This is the ability for a user to tailor frequent actions.
8. Aesthetic and minimalist design: Extra information competes with relevant units of information and diminishes their relative visibility.
9. Help users recognize, diagnose and recover from errors: Good error messages are polite, precise and constructive. They explicitly state something has gone wrong in language readable by users and constructively suggest a solution.
10. Help and documentation: Provide help and documentation that is easily searched, focused on the user's task and which lists steps that assist in carrying out the task.

Since it would be difficult for one person to identify all the usability problems in an interface, an evaluative team approach is warranted. This will yield overlapping results and highlight key opportunities for improved usability as each reviewer analyzes the interface independently and then compares the results. The process consists of identification of problem areas and then providing an assignment of severity. The suggested severity rankings include the following (Chen & MacRedie, 2005; Kushniruk, Monkman, Tuden, Bellwood, & Borycki, 2015; Nielsen, 1994):

- 0 = not a problem
- 1 = cosmetic problem; fix if possible
- 2 = minor; fix the problem; however, low priority
- 3 = major; important to fix
- 4 = catastrophe; important to fix before release

In order to complete a heuristic evaluation of software, the following steps should be completed:

1. Determine the number of evaluators with recommendations by Nielson of between 3 and 5 (Nielson, 1995).
2. Define heuristics and determine approach; e.g., domain specific, use of sub-heuristics.
3. Conduct evaluation.
4. Develop recommendations and document in a report.

Table 10.1 below provides a template for completing a heuristic evaluation using Nielson's methodology:

Using Nielson's work as a backbone, Zhang (2003) expanded the list of heuristics in order to modify the traditional heuristic evaluation method and make it applicable to medical devices as well as to evaluate patient safety. As a result, the list of heuristics expanded from 10 to 14; all of these were validated (Zhang, Johnson, Patel, Paige, & Kubose, 2003). A list of these heuristics and description is below (Table 10.2):

**TABLE 10.1** Heuristic evaluations using Nielson's methodology.

| Heuristic | Evaluation |
|---|---|
| 1. Visibility of system status | *Rating- brief description & justification* |
| 2. Match between system & real world | *Rating- brief description & justification* |
| 3. User control & freedom | *Rating- brief description & justification* |
| 4. Consistency & standards | *Rating- brief description & justification* |
| 5. Error prevention | *Rating- brief description & justification* |
| 6. Recognition rather than recall | *Rating- brief description & justification* |
| 7. Flexibility & efficiency of use | *Rating- brief description & justification* |
| 8. Aesthetic & minimalist design | *Rating- brief description & justification* |
| 9. Help users recognize, diagnose & recover from error | *Rating- brief description & justification* |
| 10. Help & documentation | *Rating- brief description & justification* |

**TABLE 10.2** Zhang's 14 heuristics for usability evaluation of medical devices (Zhang et al., 2003, pp. 25–26).

| Heuristic | Definition | Examples |
|---|---|---|
| 1. Consistency | Users should not have to wonder whether different words, situations or actions mean the same thing. Standards and conventions in product design should be followed. | **a.** Sequences of actions (skill acquisition).<br>**b.** Color (categorization).<br>**c.** Layout and position (spatial consistency).<br>**d.** Font, capitalization (levels of organization).<br>**e.** Terminology (delete (del) and language (words, phrases).<br>**f.** Standards (e.g., blue underlined text for unvisited hyperlinks). |
| 2. Visibility | Users should be informed about what is going on with the system through appropriate feedback and display of information. | **a.** What is the current state of the system?<br>**b.** What can be done at the current state?<br>**c.** Where can users go?<br>**d.** What change is made after an ac tion? |
| 3. Match | The image of the system perceived by users should match the model the users have about the system. | **a.** User model matches system image.<br>**b.** Actions provided by the system should match actions performed by users.<br>**c.** Objects on the system should match objects of the task. |
| 4. Minimalist | Any extraneous information is a distraction and a slow-down. | **a.** Less is more.<br>**b.** Simple is not equivalent to abstract and general.<br>**c.** Simple is efficient.<br>**d.** Progressive levels of detail. |

TABLE 10.2 Zhang's 14 heuristics for usability evaluation of medical devices (Zhang et al., 2003, pp. 25–26).—cont'd

| Heuristic | Definition | Examples |
|---|---|---|
| 5. Memory | Users should not be required to memorize a lot of information to carry out tasks. Memory load reduces users' capacity to carry out the main tasks. | **a.** Recognition versus recall (e.g., menu vs. commands).<br>**b.** Externalize information through visualization.<br>**c.** Perceptual procedures.<br>**d.** Hierarchical structure.<br>**e.** Default values.<br>**f.** Concrete examples (DD/MM/YY, e.g., 10/20/99).<br>**g.** Generic rules and actions (e.g., drag objects). |
| 6. Feedback | Users should be given prompt and informative feedback about their actions. | **a.** Information that can be directly perceived, interpreted and evaluated.<br>**b.** Levels of feedback (novice and expert).<br>**c.** Concrete and specific, not abstract and general.<br>**d.** Response time.<br>• 0.1 s for instantaneously reacting;<br>• 1.0 s for uninterrupted flow of thought;<br>• 10 s for the limit of attention. |
| 7. Flexibility & efficiency | Users always Learn, and users are always different. Give users the flexibility of creating customization and shortcuts to accelerate their performance. | **a.** Shortcuts for experienced users.<br>**b.** Shortcuts or macros for frequently used operations.<br>**c.** Skill acquisition through chunking.<br>**d.** Examples:<br>• Abbreviations, function keys, hot keys, command keys, macros, aliases, templates, type-ahead, bookmarks, hot links, history, default values, etc. |
| 8. Good error messages | The messages should be informative enough such that users can understand the nature of errors, learn from errors and recover from errors. | **a.** Phrased in clear language, avoid obscure codes. Example of obscure code: "system crashed, error code 147."<br>**b.** Precise, not vague or general. Example of general comment: "Cannot open document."<br>**c.** Constructive.<br>**d.** Polite. Examples of impolite message: "illegal user action," "job aborted," "system was crashed," "fatal error," etc. |
| 9. Prevent errors | It is always better to design interfaces that prevent errors from happening in the first place. | **a.** Interfaces that make errors impossible.<br>**b.** Avoid modes (e.g., vi, text wrap) or use informative feedback, e.g., different sounds.<br>**c.** Execution error versus evaluation error.<br>**d.** Various types of slips and mistakes. |
| 10. Clear closure | Every task has a beginning and an end. Users should be clearly notified about the completion of a task. | **a.** Clear beginning, middle and end.<br>**b.** Complete 7-stages of actions.<br>**c.** Clear feedback to indicate goals are achieved and current stacks of goals can be released. Examples of good closures include many dialogues. |
| 11. Undo: Reversible Actions | Users should be allowed to recover from errors. Reversible actions also encourage exploratory learning. | **a.** At different levels: a single action, a subtask or a complete task.<br>**b.** Multiple steps.<br>**c.** Encourage exploratory learning.<br>**d.** Prevent serious errors. |

*(Continued)*

TABLE 10.2 Zhang's 14 heuristics for usability evaluation of medical devices (Zhang et al., 2003, pp. 25–26).—cont'd

| Heuristic | Definition | Examples |
|---|---|---|
| 12. Use users' language | The language should always be presented in a form understandable by the intended users. | **a.** Use standard meanings of words.<br>**b.** Specialized language for specialized groups.<br>**c.** User-defined aliases.<br>**d.** Users' perspective. Example: "we have bought four tickets for you" (bad) versus "you bought four tickets" (good). |
| 13. Users in control | Do not give users that impression that they are controlled by the systems. | **a.** Users are initiators of actors, not responders to actions.<br>**b.** Avoid surprising actions, unexpected outcomes, tedious sequences of actions, etc. |
| 14. Help and documentation | Always provide help when needed. | **a.** Context-sensitive help.<br>**b.** Four types of help.<br>• task-oriented;<br>• alphabetically ordered;<br>• semantically organized;<br>• search<br>**c.** Help is embedded in contents. |

In some instances, the development of sub-heuristics has been effective in providing further clarity for the purpose of analysis (Chen & MacRedie, 2005; González, Masip, Granollers, & Oliva, 2009; Hermawati & Lawson, 2016). This points to the fact that in practice, the heuristics, as described may not include all the features of a design that impact usability. Nielson's heuristics are intended to be generally applicable to a wide variety of software interfaces and not intended to be comprehensive to an entire user interface nor a domain-specific set of heuristics. Kölling & McKay (2016) hypothesize that a smaller, more concise, more orthogonal (or independent) set of heuristics is easier to use than a larger one and furthers criteria for individual heuristics as:

- Being able to uniquely identify a set of actual known issues in existing systems from the target domain;
- Sufficiently orthogonal to the remaining set to avoid ambiguity in classification of identified faults.

In order to accomplish this, the entire set of heuristics was small enough to manage and supported the identification of major known problem areas of the target domain. This approach was also completed by Chen and MacRedie (2005) in order to prescribe a step-by-step or pragmatic approach for usability inspections. For facilitation, a detailed, structured checklist was developed to maximize the number of usability problems which could be identified. A comprehensive review of 70 studies related to usability for specific domains conducted by Hermawati and Lawson (2016) discusses a deficiency in validation efforts following the heuristic proposition; however, results were inconclusive due to a lack of validation quality and clarity on how to assess the effectiveness for specific domains. For future heuristic analyses in which the heuristic is domain specific, Hermawati et al. recommend:

- Robust and rigorous validation and adoption of standard measures as indicators of heuristics' effectiveness.
- Building on heuristics that already exist in a domain.
- Better definition of what constitutes expertise with respect to usability and specific domain.

In summary, a heuristic evaluation is a type of usability evaluation against a set of known usability principles. The ones presented above have been validated through research. While it is possible to develop new heuristics or sub-heuristics, care should be taken in order to validate the measure of the heuristic itself.

## 4. Cognitive walkthrough

A cognitive walkthrough is an evaluative method for the design of a user interface with attention to whether or not a novice user can easily carry out tasks within a given system ("Cognitive walkthrough | Usability.gov," 2018; Jones, 2018). It is a method in which people work through representative tasks and ask questions about the task as they go. It focuses on the tasks, the interface and the learnability. A cognitive walkthrough begins with task identification and then a performance assessment of the device's usability. It is a proven method of attaining early feedback on whether a design solution is easy for infrequent users to learn and why/why not. It is most helpful with prototypes or mock-ups before a system is implemented, when there is no access to real users. It is not intended to be a substitute for a user evaluation.

The goal of the assessment is to predict whether the user will try and achieve the correct outcome. This question examines the inherent design assumptions about the user's level of experience or knowledge. It can predict when a user's expectations of an action do not align with the actual action taken because they are using other reference points and becoming confused (e.g., a blinking light used to communicate "system charging" in one instance and in another area communicating "system working")

The process of conducting a cognitive walk-through consists of the following steps:

1. Identify specific traits or persona for infrequent users of a design solution.
2. Develop a set of tasks emphasizing the experience of a new user.
3. Designate a team member to play the role of this user having the previously determined trait(s).
4. Ask the "user" (team member playing the role) to accomplish their goal using a printed or interactive design. Ask them to verbalize what they would attempt, or do next, as they proceed.
5. If the "user" becomes lost, the evaluation team should not lead the user through the task. Instead, they should ask questions in order to gain an understanding of intent and causality.
6. Assess the "user's" ability to easily and/or quickly move through the task, paying specific attention to expected outcomes.

7. Analyze the walkthrough results, highlighting challenging areas and identifying what areas of the design require improvement.

For each action accomplished by the user, the evaluator must ask two fundamental questions, according to Spencer (2000), these are:

1. Will the user know what to do at this step?
2. If the users do the right thing, will they know that they did the right thing and that they are making progress toward their goal?

Other common questions asked during a cognitive walk-through include (Blackmon, Polson, Kitajima, & Lewis, 2002):

- Will the user try to achieve the right outcome?

Hidden or obscure controls are problematic for users. Often users can be overwhelmed with the number of options available, which can result in confusion. By reducing options, controls and/or tasks become easier to accomplish.

- Will the user notice that the correct action is available to them?

Requiring complex actions to be undertaken will be problematic for novice users. Intuitive selections are those that do not "make the user think" about what needs to be done in order to execute a task.

- Will the user associate the correct action with the expected outcome?

Feedback to the user that an action has been correctly completed or that an action is necessary demonstrates a functioning system.

- If the user performs correctly, will he/she see progress toward the intended outcome?

If feedback is missing, badly worded, easy-to-miss or ambiguous, it may result in user frustration or use error.

Table 10.3 provides a template for performing a cognitive walk-through, which enables a usability evaluation for each task.

As a result of the task orientation of cognitive walk-throughs, this method is commonly used in order to complete a full Task Analysis (see Chapter 6).

**TABLE 10.3** Cognitive walk-through template.

| Analysis questions | Task 1 | Task 2 | Task 3 | Task 4 |
|---|---|---|---|---|
| 1. Will the user try to achieve the right outcome? | *Y/N – why?* | *Y/N – why?* | *Y/N – why?* | *Y/N – why?* |
| 2. Will the user notice that the correct action is available to them? | *Y/N – why?* | *Y/N – why?* | *Y/N – why?* | *Y/N – why?* |
| 3. Will the user associate the correct action with the expected outcome? | *Y/N – why?* | *Y/N – why?* | *Y/N – why?* | *Y/N – why?* |
| 4. If the user performs the task correctly, will he/she see progress toward the intended outcome? | *Y/N – why?* | *Y/N – why?* | *Y/N – why?* | *Y/N – why?* |

# 5. Expert reviews

Expert reviews are a process in which a usability interface expert inspects a system in order to identify possible issues within the user interface. The difference between heuristic evaluations and expert reviews is often blurry within organizations; however, more generally, an expert review expands on heuristic evaluations by assessing the design against known heuristics, principles of usability according to cognitive psychology and/or human-computer interaction, the reviewer's expertise and past experience in the field. An expert review may also be a review conducted by a representative user, often internal to the organization, however may not be a human factors expert.

The results of the review are typically presented in a written form and contain detailed information with recommendations. These can serve as justification for design changes throughout the design process. As such the reviewer should have deep knowledge of usability best practices and a large amount of past experience, but not have been involved in creating the design to be reviewed in order to prevent bias.

The process of conducting an expert review consists of the following steps:

1. Determine the areas of product design to be analyzed.
2. Review existing research, literature, device use.
3. Identify all touchpoints.
4. Identify user workflow.
5. Analyze UI for adherence to user disciplinary tenets.
6. Analyze UI for adherence to published ergonomic standards e.g., ANSI/AAMI HE 75.
7. Analyze entire UI for use error risk
8. Determine strategies for improved UI/usability.
9. Determine design improvements for UI/usability.
10. Generate report.

## 5.1 Syringe example

Below is an example of an expert review of a conceptual design with the goal being to optimize the physical interaction of the device. The example presented is a combination product which mixes two medications prior to delivery. In this example, the areas of the physical interface are explored (Fig. 10.1) and the design is compared with ANSI/AAMI HE 75 standards (Figs. 10.2 and 10.3).

In Fig. 10.1, the device interaction within the hand is fully explored and illustrated. Usability weaknesses are called out in an issue list.

Fig. 10.2 highlights applicable gripping considerations as described in ANSI/AAMI HE 75. It also provides a recommendation for design improvement.

Fig. 10.3 describes usability considerations in order to provide the user adequate feedback that an intended goal has been accomplished as described in ANSI/AAMI HE 75.

By providing a review of all use interactions in the context of usability standards, the product design can be improved in subsequent iterations.

## CONCEPT Syringe

*Fit and Gripping Surfaces*

VIEW 1 VIEW 2 VIEW 3 VIEW 4 VIEW 5

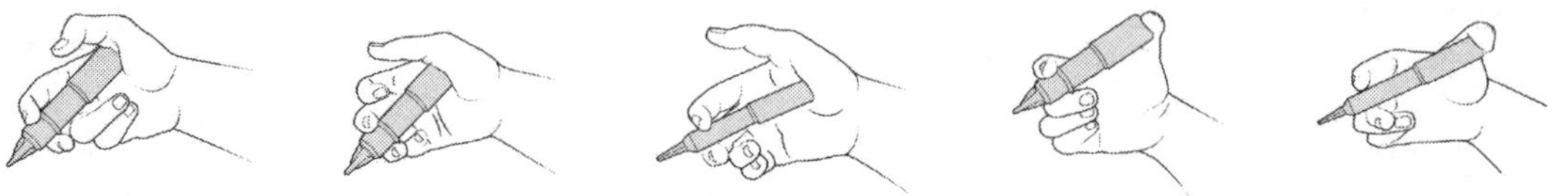

**ISSUES:**

1. No indication to Pull Tip: on/of may be confusing
2. No gripping surface to enable Pulling Tip
3. No clear indication how to hold product overall
   - ***View 3*** enables considerable force; users will likely dispense more than requirement and may empty syringe
4. Index-middle finger interaction can be problematic depending on grip style
   - ***View 4*** fingers at maximum spread provides users less control
   - ***View 5*** has the best fit between index-middle fingers with minimal spread between them
5. Dispensing is completed with a continuous flow; users will rely on visual feedback only during dispensing

hsd HS Design, Inc. Expert Review : Human Factors Analysis | 1

FIG. 10.1 Expert review of syringe.

## HUMAN FACTORS

*General Considerations Dispenser*

**GRIP:**

- **Overall Grip Considerations include size, diameter and shape.** ***Reference 16.2.4 and 16.2.4.2 in HE 75***

  ***See guidelines:***
  - ***16.19 Tool size accommodates differences in handedness, strength and work capacity***
  - ***16.21 Grip diameter enables fingers to wrap 270 degrees around the gripping surface***
  - ***16.39 For fine manipulation tasks, handle diameter should be between 0.5 and 1.6 cm***
  - ***16.44 Pommels or flanges should be used to prevent the hand from slipping.***
  - ***16.45 A cylindrical handle should be used for fine manipulation that requires gripping by fingertips***
  - ***16.62 Minimize weight to decrease tremor***

**Recommendation: Where possible use cylindrical shape (.5-1.6cm diam.), flanges for grip cues/slipping and encourage fingertip grip**

hsd HS Design, Inc. Expert Review : Human Factors Analysis | 2

FIG. 10.2 Recommendations for specific design pulled from ANSI/AAMI HE 75.

HUMAN FACTORS

*General Considerations for Dispenser*

**FEEDBACK:**

- When using a hand tool, users should receive tactile, visual and/or auditory feedback regarding the status of the tool and the task. *Reference 16.2.7.5 in HE 75*
  *See guidelines:*
  - *16.134 Provide tactile feedback*
  - *16.136 Keep handle clear of markings where they can be obscured by the user's hand*
  - *16.139 Discrete control positions should be indicated by tactile and auditory feedback.*

- Feedback should be provided to user that a motion has been accepted by the device. *Reference 7.1.2.2 in HE 75*
  *See guidelines:*
  - *7.10 Immediate feedback*
  - *7.11 Clear feedback*
  - *7.12 Redundant feedback is advantageous (multi-modal concurrent feedback will make it more likely users will be able to use a control accurately and reliably.*
  - *7.14 Auditory feedback is most effective in environments with low ambient noise levels*
  - *7.15 Confirmation of critical actions*

hsd HS Design, Inc. Expert Review : Human Factors Analysis | 3

FIG. 10.3 Specific feedback recommendations from ANSI/AAMI HE 75.

## 6. Comparability of these methods

Heuristic analysis, cognitive walk-throughs and expert reviews are similar in that they are flexible, inexpensive approaches to identify usability issues. They rely on an individual evaluation completed by usability experts, most often done as part of an analytical team and completed at any time in the device development process. In research conducted by Kushniruk et al. (2015), both heuristic analysis and cognitive walk-throughs were combined with the intent to leverage the strengths of both techniques. The commonalities between these inspection methods includes: having multiple expert reviewers yields the best results; the ability to uncover many usability problems; conducting the test in-house without users but following a Users' point of view; and the ability to be more thorough than User Testing as there are typically no constraints of the number of tasks evaluated therefore the reviewers are able to seek all the nooks and crannies of a user interface.

According to a study by Khajouei, Esfahani, and Jahani (2017), heuristic analysis and cognitive walk-throughs do not differ significantly in regards to the number of usability issues identified; however, differences were described regarding the severity and coverage of some usability attributes.

In summary, heuristic analysis is intended to broadly reveal usability issues; cognitive walk-throughs have the same intent; however, are aligned with the specific tasks involved throughout use. Expert reviews can be a combination of both (heuristic analysis and cognitive walkthroughs) and compare specific usability aspects or tasks to known industry standards (e.g., an expert review may include comparison of the force required to manipulate a control against known ergonomic standards).

Each method is appropriate for medical device design and may be most powerful as a combined approach with the intent to provide identification of usability issues on the user interface. Once an issue is detected, it is most helpful to use published standards such as ANSI/AAMI HE 75 or like standards to provide justification for design improvements.

## 7. Assessing risk and identifying design opportunities using heuristic evaluation, cognitive walk-throughs or expert reviews

Risk assessment can and likely should happen throughout the design process because using Heuristic Analysis, Cognitive Walk-throughs or Expert Reviews as a backbone to identify usability issues can assist in defining risks found in particular tasks or situations that are the result of product design or workflow. These risks can then be further mitigated by developing comprehensive design opportunities. Fig. 10.4 below highlights a detailed analysis using visual language to describe a step with sub-steps within a workflow based on previous user research (contextual inquiry) and a proposed design. Each step is described and visualized.

From this figure, the potential risk areas can be identified and risk mitigations proposed alongside defined opportunities for improved design. In completing this type of analysis, the resulting product design can have improved usability.

## 8. Assessing competitive ergonomics

Using an OTC device as an example, the photos below (Figs. 10.5 and 10.6) illustrates evaluation techniques needed to determine competitiveness and relative usability between two devices with the same intended functionality or clinical utility. For this example, 2 TENS unit/pain-relieving muscle stimulating devices (FDA approved) were purchased and analyzed for usability. This was completed solely to provide a brief example for the purposes of this book and is not intended to provide recommendations for a next-generation device.

According to the FDA, a user interface includes all points of interaction between the product and user(s), including elements such as displays, controls, packaging, product labels and instructions for use, etc. The user interface descriptions always include graphical representations of the UI, descriptions of the UI, device labeling and an overview of operational sequence of device and expected user interactions with the UI. As such, the entire use experience was considered for this evaluation.

Using the validated usability scale 0 (no problems)—4 (Catastrophic problems), each system was evaluated and compared (Table 10.4).

From this evaluation, the Healthmate Forever® scored higher in overall usability, whereas the Mini Massager proved problematic for user control, consistency and standards and error prevention, as well as flexibility and efficiency. If the intent were to provide recommendations for improved design, this evaluation would be used to prompt further analysis into the root cause of the usability issue(s) and then suggest improvements.

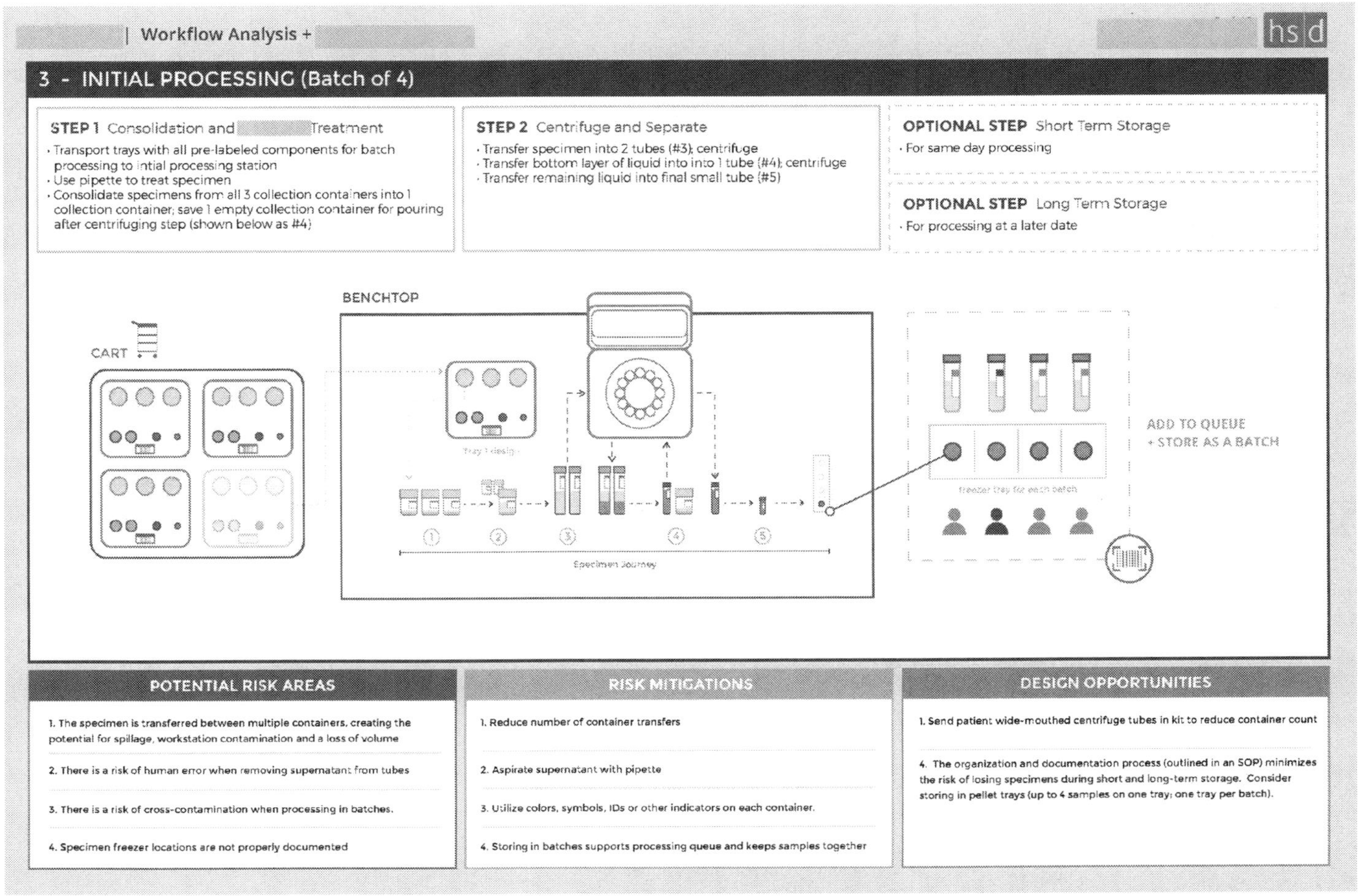


FIG. 10.4 Workflow analysis of lab equipment with potential risk areas, risk mitigations and design opportunities identified.

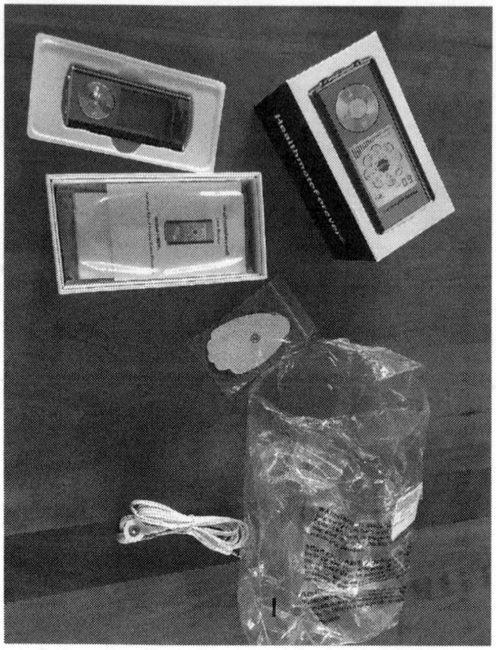


FIG. 10.5 Device 1: Healthmate forever (R) pain relief TENS & powered muscle stimulant.

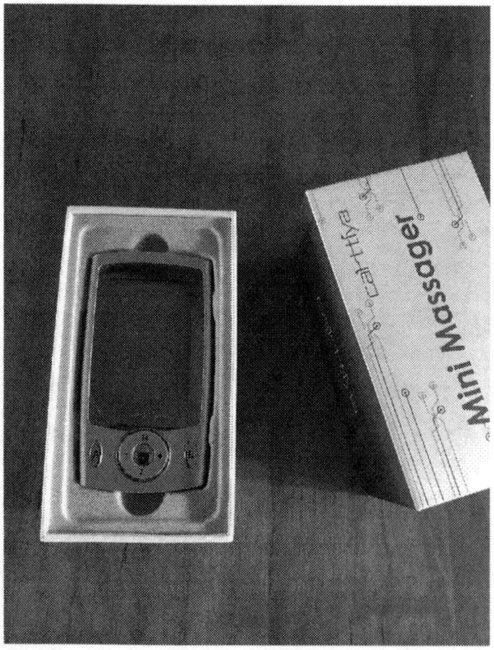


FIG. 10.6 Device 2: Mini massager tens unit by Cartiya.

**TABLE 10.4** Usability comparison of devices using Heuristic review.

| Heuristic | Healthmate forever® | Mini massager by Cartiya |
|---|---|---|
| Visibility of system status | 1- Status is clearly displayed on the interface. Both electrodes do the same task | 3- it displays both modes and allows for variability between the electrodes; however, discrimination is difficult |
| Match between system & real world | 1- Easily understood icons | 1- Easily understood icons |
| User control & freedom | 2- Changing modes is difficult | 3- Changing modes is difficult |
| Consistency & standards | 0- Uses common icons | 2- Uses some common icons |
| Error prevention | 0- easily change modes | 2- knowing which electrode is controlled by A or B is a challenge |
| Recognition rather than recall | 0 | 2- When you change modes, need to remember to change intensity |
| Flexibility & efficiency of use | 0 | 3- Double modes is confusing |
| Aesthetic & minimalist design | 1- the pictures are small in the IFU and on screen | 3- Too busy |
| Help users recognize, diagnose & recover from error | 2- Modes on screen do not have clear delineation | 3- Challenges between which electrode is being controlled by UI |
| Help & documentation | 4- Too small yet simply written | 2- Nice guide for electrode placement per ailment |

## 9. Summary

Heuristic Analysis, Cognitive Walk-throughs and Expert Reviews are common practice in medical device design and are widely used when there have not been any previous human factors work. When properly documented, they are methods which can be utilized as a Formative Usability Study, as the intent of each method described in this chapter is to identify usability issues and improve the product design.

## Acknowledgments

Special thank you to the HS Design and Human Factors Team for their work utilizing these methods and demonstrating improvements in device usability through these efforts. Special thanks go to Elissa Yancey for editing.

## References

Blackmon, M. H., Polson, P. G., Kitajima, M., & Lewis, C. (2002). Cognitive walkthrough for the web. In *Conference on human factors in computing systems - proceedings U6*. Ctx_ver=Z39.88-2004&ctx_enc=info%3Aofi%2Fenc%3AUTF-8&rfr_id=info%3Asid%2Fsummon.Serialssolutions.Com&rft_val_fmt=info%3Aofi%2Ffmt%3Akev%3Amtx%3-Abook&rft.Genre=proceeding&rft.Title=Confe. Retrieved from http://uc.summon.serialssolutions.com/2.0.0/link/0/eLvHCXMwtV1LS8NAEB4ivQgefFStjxJBvEikbpPaHLxYLEIrCFY8hjy2Gs2jkITaf-_sbjYv7cGDlyXswm6YGWa_HWa-AdglVz2t4RPq3XsUGV8r5_5VsTiHqmWFsn9QbrEpTuA3qhhHVDKODfz761VTFuyxqL-IyI_zFjq8no_1bOCk24KZ_FITpQF8n6RqL6Mij2hpB5-y.

Chen, S. Y., & MacRedie, R. D. (2005). The assessment of usability of electronic shopping: A heuristic evaluation. *International Journal of Information Management*. https://doi.org/10.1016/j.ijinfomgt.2005.08.008.

Cognitive walkthrough | Usability.gov. (2018). Retrieved from October 21, 2018. https://www.usability.gov/what-and-why/glossary/cognitive-walkthrough.html.

FDA. (2016). *Applying human factors and usability engineering to medical devices.*

González, M., Masip, L., Granollers, A., & Oliva, M. (2009). Quantitative analysis in a heuristic evaluation experiment. *Advances in Engineering Software, 40*(12), 1271–1278. https://doi.org/10.1016/j.advengsoft.2009.01.027.

Hermawati, S., & Lawson, G. (2016). Establishing usability heuristics for heuristics evaluation in a specific domain: Is there a consensus? *Applied Ergonomics*. https://doi.org/10.1016/j.apergo.2015.11.016.

Hvannberg, E. T., Law, E. L. C., & Lárusdóttir, M. K. (2007). Heuristic evaluation: Comparing ways of finding and reporting usability problems. *Interacting with Computers*. https://doi.org/10.1016/j.intcom.2006.10.001.

Jones, N. (2018). How to conduct a cognitive walkthrough | interaction design foundation. Retrieved from October 21, 2018. https://www.interaction-design.org/literature/article/how-to-conduct-a-cognitive-walkthrough.

Khajouei, R., Esfahani, M. Z., & Jahani, Y. (2017). Cincinnati health sciences library user on. *Journal of the American Medical Informatics Association, 24*(e1), 2017, e55–e60. https://doi.org/10.1093/jamia/ocw100

Kölling, M., & McKay, F. (2016). Heuristic evaluation for novice programming systems. *ACM Transactions on Computing Education*. https://doi.org/10.1145/2872521.

Kushniruk, A. W., Monkman, H., Tuden, D., Bellwood, P., & Borycki, E. M. (2015). *Integrating heuristic evaluation with cognitive walkthrough: Development of a hybrid usability inspection method*. https://doi.org/10.3233/978-1-61499-488-6-221.

Nielsen, J. (1994). Heuristic evaluation. *Usability Inspection Methods*, 25–62. https://doi.org/10.1089/tmj.2010.0114.

Nielsen, J. (1995). *How to conduct a heuristic evaluation*. Nielson Norman Group Norman.

Nielsen, J., & Molich, R. (1990). Heuristic evaluation of user interfaces. In *Proceedings of the SIGCHI conference on Human factors in computing systems* (pp. 249–256). https://doi.org/10.1145/97243.97281.

Nielson, J. (1995). *Heuristic evaluation: How-to: Article by Jakob Nielsen*. Retrieved from October 21, 2018. https://www.nngroup.com/articles/how-to-conduct-a-heuristic-evaluation/.

Spencer, R. (2000). The streamlined cognitive walkthrough method, working around social constraints encountered in a software development company. In *Proceedings of the SIGCHI conference on human factors in computing systems - CHI '00*. https://doi.org/10.1145/332040.332456.

Wharton, C., Rieman, J., Lewis, C., & Polson, P. (1994). In J. Nielsen, & R. L. Mack (Eds.), *Usability inspection methods* (pp. 105–140). New York, NY, USA: John Wiley & Sons, Inc.. Retrieved from http://dl.acm.org/citation.cfm?id=189200.189214.

Zhang, J., Johnson, T. R., Patel, V. L., Paige, D. L., & Kubose, T. (2003). Using usability heuristics to evaluate patient safety of medical devices. *Journal of Biomedical Informatics, 36*(1–2), 23–30. https://doi.org/10.1016/S1532-0464(03)00060-1.

CHAPTER

# 11

# Simulated use formatives

*Deborah Billings Broky*[a], *Tressa J. Daniels*[b], *Melissa R. Lemke*[a]

[a]Agilis Consulting Group, LLC., Cave Creek, AZ, United States; [b]UXD & Human Factors, BD Medical, San Diego, CA, United States

OUTLINE



## 1. Introduction

Human factors testing for medical devices has one overarching goal: To optimize user interface design and demonstrate to the regulatory agency that the device is safe, effective and usable by end-users as a requirement prior to market release. Formative evaluation

*Applied Human Factors in Medical Device Design*
https://doi.org/10.1016/B978-0-12-816163-0.00011-6

studies feature usability testing with simulations using prototypes in order to explore overall usability objectives (AAMI, 2009). These evaluations are intended to identify user interface design strengths, weaknesses and unanticipated use errors (IEC, 2015). Simulated use evaluations ("formatives") conducted prior to final validation testing add great value to a human factors submission. They enhance the final validation report by providing details regarding the incorporation of human factors throughout device development in order to optimize the user interface. Further, formatives help sponsors better define the uses and use environments in terms of user interface and users. Ultimately, formatives inform the use-related risk analysis for the device. Results from formatives can also be used to identify critical tasks, to update task categorizations and risk analyses based on newly uncovered or unanticipated risks, and to provide meaningful justification for including or excluding certain design elements evaluated in the final validation. Formative data can be quite valuable within a submission as regulating agencies may ask sponsors to provide formative evaluation data that helped sponsors determine the designation of tasks or interface design requirements.

It is a fact that a medical device design can have some use-related risks yet achieve regulatory clearance. Granted, in these cases, the design development history presents compelling evidence that the sponsor has considered end users in the design process and shown the use-related safety and effectiveness of the design has been optimized such that no further design mitigations would reduce or eliminate the residual risk. Simulated use formatives play a major role in assuring the usability of a device and ultimately preparing a device for human factors validation testing. For example, well-documented formative findings could potentially:

- Confirm that design elements of the user interface (including implemented risk mitigations) prevent use errors from occurring;
- Demonstrate that labeling materials promote safe and effective use of the device;
- Establish that user training provided and managed by the sponsor as part of the user interface (UI) is an effective mitigation; and/or
- Show that the validation methodology is unbiased and appropriate, use scenarios are realistic and that the correct data can be collected (i.e., conduct a formative using validation methodology to provide pilot study data).

According to ANSI/AAMI HE75 (2009), formatives provide a means to:

- Identify and prioritize tasks according to potential harm as derived from the inherent analytical techniques;
- Guide the development of use scenarios;
- Identify unanticipated use errors;
- Identify critical tasks early in the design process;
- Guide the modification of device user interface elements; and
- Clarify the dynamics of user-device interaction within the use scenarios.

While some sponsors only see additional costs to formatives as burdens with the potential to extend project timelines and require development schedules, the benefits of implementing these formatives prior to validation testing often outweigh the costs. Performing formatives early in the design cycle can identify design issues early in device development when design changes are much easier and less costly to implement rather than at the end of the development cycle. When formatives are conducted, sponsors are much more likely to optimize user interface design (including Instructions for Use, IFU) and have a successful validation without any last-minute surprises. It's recommended to not rush to human factors validation testing without considering how formative evaluations may benefit your design and submission.

## 2. What are formative evaluations?

Existing standards and guidance documents define "formative evaluations" similarly (see Table 11.1). They offer a way to identify usability problems with device design during the development cycle. Formatives do not have strict assessment criteria or endpoints like validation studies or clinical testing; rather, well-planned formatives can provide a variety of data about the design of your device user interface in order to inform design changes and optimize device design. Formatives allow sponsors to base design decisions or changes on concrete user performance data and subjective feedback instead of making decisions based on user preference or arbitrary modifications that the regulatory agency may not trust as valid.

A formative evaluation may be conducted in many ways, including: heuristic analysis, cognitive walkthroughs, expert review (see Chapter 10) or by using a **simulated use study methodology.** This chapter focuses on simulated use methodology and provides detailed explanations.

**TABLE 11.1** Definitions.

| Guidance | Definition of "formative evaluation" |
|---|---|
| FDA CDRH (2016) | *"Process of assessing, at one or more stages during the device development process, a user interface or user interactions with the user interface to identify the interface's strengths and weaknesses and to identify potential use errors that would or could result in harm to the patient or user."* |
| HE75 | *Referred to as "formative usability testing" and "quick and dirty" preliminary testing. "Usability testing that is performed early with simulations and the earliest working prototypes and that explores whether usability objectives are attainable, but without strict acceptance criteria."* |
| IEC62366 Part 1 | *"User interface evaluation conducted with the intent to explore user interface design, strengths, weaknesses and unanticipated use errors."* |

## 3. Conducting simulated use studies for formative evaluations

A simulated use study involves testing with representative users completing realistic tasks with a device user interface in a simulated and/or representative use environment. Objective (i.e., performance) and subjective (i.e., information from the user's point of view) data are obtained, analyzed and then used to inform design decisions. Simulated use studies should be considered because they can **minimize risk** of a device performing poorly going into a human factors validation test.

Simulated use studies do not have to be time-consuming and costly. There are ways to streamline the approaches to maximize the benefit of the data collected while minimizing the burden to both timeline and budget. These are discussed in further detail below.

As mentioned previously, a simulated use study is one of several options for a formative evaluation. While it is best practice to conduct a simulated use study as a formative evaluation, it is not required per se in the standards and guidance. However, a simulated use study is required for all human factors validation studies (see Chapter 15) with strict adherence to methodology proposed in IEC 62366 and FDA Human Factors Guidance. Although a simulated use formative and a validation study both use a simulated use framework, there are major differences between a formative and validation study that should be considered during planning stages, as described in Table 11.2.

**TABLE 11.2** Simulated use studies: different considerations for formatives versus validation studies.

| Simulated use study considerations | Formative evaluation | Human factors validation study |
|---|---|---|
| Methodology | Range of "rapid insight usability testing" to a more formalized methodology can be effective. The study protocol can change/be updated as necessary relative to progress in the device development process. | Formalized methodology that aligns with agency (FDA) guidance is required. Strict adherence to study protocol is expected. |
| When to test | As early or late in a device life cycle as needed to answer specific questions about the user interface design. | After device user interface (e.g., device design, labeling, packaging, training, etc.) has been finalized. |
| Test frequency | As many times as necessary to **optimize** a device user interface, which includes device, labeling, packaging and any other materials that the user will interact with. | Typically, one time to demonstrate that the device user interface is safe and effective; additional times if further device modifications are needed to address residual risk or user interface updates are being implemented. |
| Test materials | With early prototypes, later device designs or the final device design (low or high fidelity), with the option to test alternative designs. | With final design. |
| User-device interactions to test | Small or large set of specific tasks, incomplete workflow (e.g., only test tasks associated with cleaning the device) or all tasks involved in use scenarios (complete workflow). | All tasks involved in use scenarios (complete workflow). |
| Outcomes | Decide what modifications to make to the user interface based on performance data and design recommendations. | Decide if residual risk is acceptable so that the device is safe and effective, and if the device user interface has been optimized as much as practicable. |
| Data analysis | Quantitative data can include time on tasks, success rates and other data tied to successful completion of tasks. | Quantitative data include success rate and data tied to successful task completion (e.g., timing a participant to ensure that he holds an injection for the required 10 s). |
| | Qualitative data can include comments and feedback from participants, which can be elicited throughout the session and at the time a use error or difficulty is observed, if desired. | Qualitative data can include comments and feedback from participants during the debriefing interview. |
| | Root cause analysis identifies causes for use errors and difficulties and can lead to recommendations for design modifications. | Root cause analysis identifies causes for use errors and difficulties and can lead to recommendations for design modifications. |

### 3.1 Simulated use study purpose

Simulated use studies focus on safe and effective user-device interactions rather than user preference data. As such, they utilize representative users, use environments and use scenarios with the intention that users interact with the device in manners similar to real-life or representative use/s. A simulated use approach for formatives can incorporate a wide range of objectives throughout the development process. These are described in Table 11.3.

### 3.2 Formative study timing

Simulated use formatives can be appropriate during early stage device development or during later stage development; the benefits are ultimately dependent on the formative objective(s). Wiklund, Kendler, and Strochlic (2016) describes impacts to device development, estimated timelines for formative evaluations and information about decisions that could increase or decrease monetary costs for conducting formatives. Table 11.4 below provides a summary of the costs and benefits of conducting formatives at different times during the development process.

In summary, simulated use studies that are conducted early in the device development cycle can benefit a sponsor by allowing product optimization based on user performance data. However, if a sponsor only conducts a simulated use study towards the end of the device development cycle (or chooses not to conduct a simulated use study at all prior to validation), the sponsor could likely observe unanticipated use errors and use-related problems during the human factors validation testing that may require re-design of the UI, supplemental validation testing of the modified UI, or repeated validation testing altogether. Conducting a simulated use study with an early concept or prototype can expose potential use-related problems that present significant risk of harm to users. Catching these use-related problems early in the design process allows the sponsor an opportunity to revise the design to mitigate risk and provides a strong foundation for use-focused risk management.

## 4. Planning a simulated use study

When designing a simulated use formative, there are five study components to be considered. These include the following:

1. **Task selection:** What tasks should be included?
2. **Testing methods:** Which methods will be most effective in capturing the data required in order to inform design? What are the specific strategies that support the formative study objectives? What types of test environment and materials should be used?
3. **Participants:** What are the inclusion criteria (e.g., adults 18 years and older) and the exclusion criteria (e.g., must not have participated in any other studies for Device X) for participants? How many participants should be included? And why?
4. **Data collection and analysis:** What data should be collected, and how will they be analyzed?
5. **Writing the report and implementing design modifications:** How formal or informal will the report be? What should be included, at a bare minimum? What recommendations will be included that will enhance device usability? How will implemented recommendations be documented and tracked through the development process?

**TABLE 11.3** Objectives of simulated use formatives.

| Objective/focus of formative | When should it occur? | Description | Example |
|---|---|---|---|
| 1. Inform device design and help assess new design iterations | Any time design is iterated or a safety related design decision is being made (pre- or post-market) | Inform device design throughout the device life-cycle (early to late), or in post-market evaluations. Evaluate risk mitigations and modifications to device design that were implemented. | [Device X] injector was iterated to include audible feedback when injection is complete. Formative evaluation can determine if this mitigation is effective in preventing users from prematurely removing the injector before administering a full dose. |
| 2. Test specific or hypothesized use-related problems | Early in the design process is better | Focus on potential problems associated with certain aspects of the user-device interaction versus the entire process(es). Identify potential use-related risks or further explore risks identified in other analyses. | Pen injectors have known issues with users not holding the device long enough for a complete injection. Formative evaluation can determine if this is also an issue with [Device X] injector. |
| 3. Verify assumptions early in the design process | Early in the design process is better | Verify or iterate assumptions of users, environments, tasks, simulated use scenarios and potential use errors, as well as help identify critical tasks. | [Device X] is a diabetes therapy, and the designers assumed users would understand terminology like units of insulin, volume, needle gauge, etc. A formative can either verify these assumptions or debunk them to help with the design of effective packaging, labeling and training. |
| 4. Determine effectiveness of labeling and/or training | Earlier in the design process is better to test initial version of labeling and training. | IFU focused formative: Show that participants can safely and effectively use the device when the IFU is explicitly used. | Users are directed to use [Device X] labeling to complete simulated use scenarios to determine if labeling is an effective mitigation to prevent use errors from occurring. This is a clean, unbiased way to evaluate labeling. |
| | After labeling/training have been finalized, can be used to ensure these UI elements are optimized and effective prior to the validation study. | Training formative: Show that representative training is effective. Determine representative learning decay and when no training may be appropriate, even if training is guaranteed (e.g., an infrequently used emergency device). | |
| 5. Evaluate validation study methodology (Pilot Study) | Late, once validation methodology is determined and user interface is near final. | Conduct pilot testing designed to "practice" study methodology and ensure Moderator guide, use scenarios and knowledge task questions are clearly communicated and generate necessary data. Helps to determine readiness to proceed to the validation study in order to avoid costly situation where a validation turns into a formative. | A formative can determine if simulated use scenarios for [Device X] are appropriate, and if representative users understand Moderator instructions (in order to avoid introducing any potential study artifacts into a validation). |

TABLE 11.4 Timing of simulated use formatives in the development process.

| Timing | Benefits | Costs |
|---|---|---|
| Early in design life cycle (i.e., conducted with low fidelity prototypes) | **1.** Mitigate risk early in design before changes become more costly.<br>**2.** Identify use-related risk and unanticipated use errors.<br>**3.** Address small problems before they become bigger.<br>**4.** Reduce probability of use-related problems post market.<br>**5.** Provide evidence that UI design decisions (e.g., symbols, text, images or colors in an IFU) were based on user performance data. This allows a sponsor to justify design decisions to the agency.<br>**6.** May reduce overall device timeline to market because it can identify design problems early and reduce the amount of re-work that is needed late in the development process. | **1.** May increase project timeline up front, but allows for planning and more assurance going into the validation.<br>**2.** Requires project budget in the near-term, depending on the fidelity/complexity of the formative methodology. |
| Late in the design life cycle | **1.** Will identify most use-related problems that could lead to an unsuccessful validation.<br>**2.** IFU testing may provide support for risk mitigation.<br>**3.** Pilot study can identify any issues with validation methodology a priori so that they can be fixed. | **1.** Conducted with near-finalized device; device design that cannot be easily changed. Manufacturing lines may need to be changed based on late stage user interface changes.<br>**2.** Can slightly increase a project timeline.<br>**3.** Can increase project budget. |
| No simulated use formative evaluations planned | **1.** Initially, very expeditious project timeline.<br>**2.** No extra resources need to be allocated. | **1.** Potential for **surprises** in the validation (e.g., unanticipated problems) and questions from the agency.<br>**2.** Risk mitigations may be **expensive** and **time-consuming** to address. |

Each of these components are described in additional detail below. While this list is not exhaustive, it represents high-level questions that should always be considered when planning a simulated use formative evaluation.

## 4.1 Participant task selection

Participant task selection refers to the task that a user will be asked to complete during a simulated use study. Often times, the question of "Do I need to test everything?" comes up. The simple answer is no. The first inclination in the design process may be to test everything because perfecting the design is the ultimate goal. However, it is important to remember that having users perform all tasks in one simulated use study can take a great deal of time and may result in fatigue and even frustration. In order to get the most out of allocated budget, consider prioritizing task selection criteria on the following:

**a.** User tasks that are relevant to safety or could potentially lead to harm;

**b.** User tasks that may be unusually complex and/or could require the user to look for help (e.g., in an IFU or from a nearby colleague);
**c.** User tasks that may occur rarely or infrequently but are critical to safe use (e.g., critical alarms);
**d.** User tasks that are relevant to the ease of use, appeal or competitiveness of the device; and
**e.** Areas of the user interface that the designer wants specific feedback on.

Focusing task selection on the items above likely will fill the time available for a testing session. If the tasks selected increase the session time to an unreasonable length, consider testing portions of a workflow rather than an entire workflow from start to finish. This allows the opportunity to gather feedback concerning the highest priority portion of the workflow while keeping the session length reasonable. For example, in a task such as changing user settings on a programmable infusion pump: if it is known that a user has no issues locating and accessing user settings, then try skipping over that and going to the next setting screen or menu where the user can interact with the specific setting which is unknown and warrants evaluation.

Note that if the purpose of the simulated use formative is to test a validation methodology (pilot study), it should include all tasks that are to be included in the human factors validation testing protocol. This will better predict overall readiness to proceed to the validation study.

If time permits, it is generally best practice to collect data on all tasks regardless of task categorization (whether they are critical or non-critical) as formatives can often lead to updates in assumptions or re-categorization of tasks. It is easy to dismiss data that is determined to be meaningless if a task is deemed unimportant or non-critical, but it is challenging or, in cases without video recording, impossible to go back and collect some types of data. This is especially true for participant interview data, which can help identify root causes of observed use errors or difficulties.

## 4.2 Development of simulated use testing methods

Creating the formative testing protocol and Moderator guide enable the Moderator to collect the right data at the right time. When followed, this assures thoroughness, repeatability and comprehensiveness. Examples and discussion are presented below.

### *4.2.1 Protocol development*

Developing, documenting and approving a testing protocol prior to the formative evaluation are important tasks. These steps formalize the testing so that it can be replicated, if required. The testing protocol can then be referenced in the sponsor's Design History File (DHF), which in turn provides traceability and evidence of human factors work throughout the design life cycle. While a formative evaluation does not require an extremely detailed and formalized protocol like a validation test; the formative protocol should still be well-developed and documented. If using a "rapid insight usability" formative approach, the need for traceability and documentation remains to ensure that results can be reported in the Human Factors Engineering/Usability Engineering (HFE/UE) or validation study report (and included in the regulatory submission).

#### 4.2.1.1 Contents of a protocol

The following elements should be included in designing a formative research protocol:

*Research question(s) and objectives*

- Clearly articulate the research question/s so that data collection is aimed to provide evidence. This forms the framework for the study. For example, a formative study may include the following question: Which packaging design is easiest to open, allows users to quickly find the instructions and is most intuitive for guiding users to safely store the device components?
- Identify all specific objectives for the study to ensure that data collection meets the goals. The objectives may change during a formative evaluation, especially if new issues are identified during testing. If objectives change, document those changes carefully. An example of a study objective is: Determine if representative users can understand and follow the device IFU without causing errors or failures that could result in harm to users or patients. This example is focused on the IFU and should specifically target collecting participant performance data associated with IFU usage along with comprehension tasks.

*Identify participants, test materials, test environment and training*

- Document the number and type of participants, along with any specific inclusion and exclusion requirements plus justification for decisions. For example: 5–8 injection experienced and 5–8 injection naïve participants for an autoinjector formative.
- Estimate session length, keeping in mind the potential of fatigue by participants or test personnel. It is best to maintain test duration between 1 and 2 hours, although duration is always dependent on the activities completed as well as the additional time for interview questions. For example, a 1.5 hour study session may include the following activities: study introduction (10 min), completion of 3 use scenarios (45 min), completion of 8 knowledge task questions (15 min) and Moderator interview questions/probing to collect participant subjective data (20 min).
- Outline key characteristics of the actual use environment and how to simulate important characteristics within the environment. In a simulated use study, there is the opportunity to test "worst case" conditions such as low lighting, high noise, wet gloves, etc. This is important as any of these environmental conditions could impact safe and effective use of the device. A technician who reprocesses medical instruments and surgical equipment in a sterile processing department is expected to wear personal protective equipment (PPE) when performing reprocessing activities. Therefore, a simulated use environment should include access to PPE to reflect conditions under which users complete reprocessing activities in the actual use environment (e.g., wearing wet gloves, full body fluid resistant gown, face mask or goggles, etc.). Wearing PPE may impact how well users can see and manipulate the instruments during reprocessing, so it is very important to account for this equipment in the simulated use environment so that results from the study can be generalized to the actual use environment.

- Identify any training that the participants will receive as part of the study, including any training materials and periods of training decay between training and testing. This may involve a statement about how the training provided in the simulated use formative corresponds to real-world training for participants. For example, if cardiologists are your intended users and all cardiologists receive a sponsor-led 2-hour in-service training before using the device independently, consider including this in-service training as part of the formative evaluation. Because a cardiologist could potentially receive in-service training and then use the device in their clinic on the same day, it may be appropriate to include a minimal training decay period between the in-service training and the testing. An alternative approach is to conduct the simulated use study without training to demonstrate that intended users are capable of using the device with no training, which perhaps might be a worst-case scenario in real life.
- Develop a study documentation plan that includes the versions of the UI being evaluated (e.g., IFU version, software version). Sometimes a high-level fidelity UI is used, and other times low fidelity prototypes are used. There is no hard and fast rule for the level of fidelity needed, however it should be sufficient in fidelity to enable the user to interact with the device in a representative way without introducing study artifacts which may impact user performance either positively or negatively. Using low fidelity prototypes has the advantage of counteracting participant bias and proclivity to only respond with positive feedback. Low fidelity suggests to the participant that the design is incomplete and negative participant subjective feedback about the UI is acceptable, whereas higher fidelity often gives the perception of a polished and complete device. For some lay users, the more refined the prototype, the more likely it is for participants to accept the design as complete without providing critical comments. Table 11.5 below provides examples of the different levels of prototype fidelity that could be used in a formative evaluation, including paper prototypes, grayscale wireframes and interactive workflows.

***Identify tasks to be tested and determine how to collect the data***

- Determine which tasks should be included in the formative and clearly indicate means of data collection. As previously mentioned, not all tasks associated with a device use have to be evaluated in a simulated use formative. For example, if no packaging is available at the time of testing, the formative evaluation tasks may exclude tasks associated with the packaging. If a new risk mitigation has been implemented, focus might be solely on tasks that involve that mitigation.
- Identify the best structure regarding the use scenario/s to incorporate tasks as well as how this may impact your data collection and analysis. This includes considering factors such as task sequencing, timing and repetition (e.g., opportunity for learning effects in actual use). For example, a formative could include repeated use scenarios for a pen device that is used to administer medication once or twice daily. This methodology may show any learning effects that a user may experience, given that the pen is intended to be used on a frequent basis.

**TABLE 11.5** Examples of prototype fidelity for formative evaluations.

| Examples of acceptable prototype fidelity for formative evaluations | Illustration of prototype | Best uses |
|---|---|---|
| **Low fidelity prototypes:** Low cost materials are used to mock up a user interface. For example, paper sketches drawn by hand to represent a digital display. Other materials such as cardboard, foamcore or clay can be used to generate user interface prototypes. | 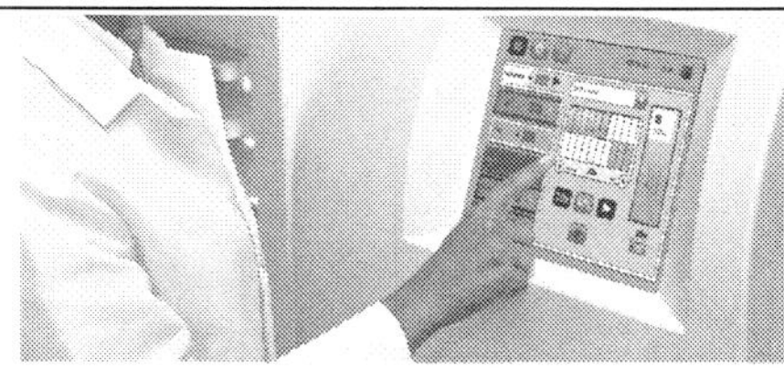 Fig. 11.1 Simulated use study with foamcore prototype and paper user interface; *image provided by HS Design.* 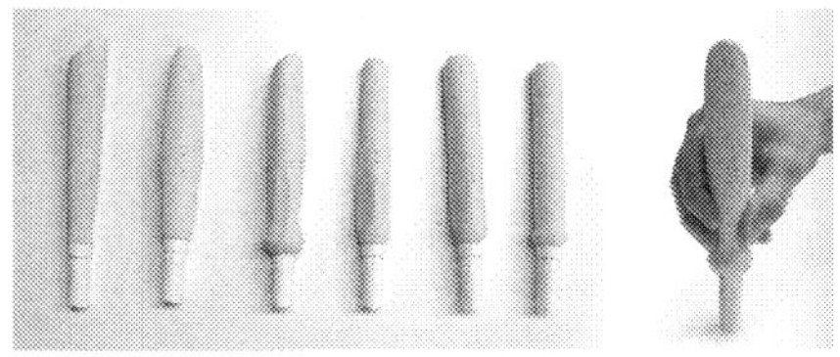 Fig. 11.2 Foam models; *photo courtesy of HS Design.* | A paper protoype of a software UI or a low fidelity prototype of hardware is best used to determine user needs. |
| **Static grayscale wireframe and 3D models:** Upgraded version of the paper prototype in digital form. Human factors engineers often display these in PowerPoint and in some cases print them just as you would a paper prototype. For hardware, 3D printed models can be used to generate prototypes of user interfaces such as hand tools. | 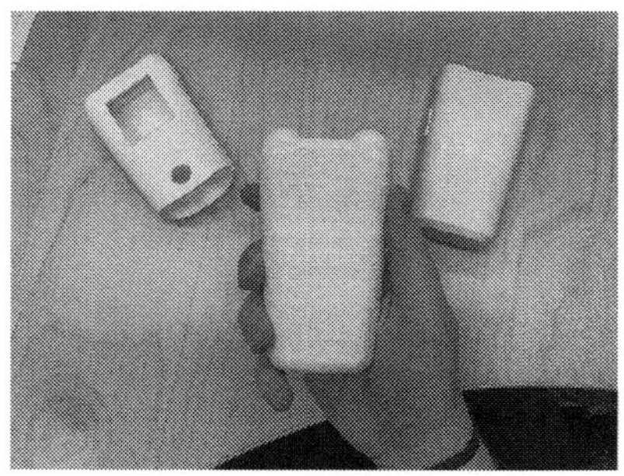 Fig. 11.3 High fidelity 3D *printed* prototypes; *photo courtesy of HS Design.* | Greyscale wireframes are used to evaluate the workflow without visual clutter of a full interface. 3D printed models are more time and cost effective to develop when a design is still being iterated. The 3D model can be weighted to test the ergonomics, and levers and knobs can be interactive as well. |

*(Continued)*

TABLE 11.5 Examples of prototype fidelity for formative evaluations.—cont'd

| Examples of acceptable prototype fidelity for formative evaluations | Illustration of prototype | Best uses |
| --- | --- | --- |
| **Interactive workflows or interactive prototypes:** Much more representative of what the final device might look like. Often color is applied, buttons are clickable, fields are editable, parts are movable, and the interaction provides a realistic experience of device use to the user. High fidelity hardware prototypes may be the initial appearance models of your device, meaning that the prototype looks like and feels like the final product and can even have functioning moving hardware components as needed for testing purposes. |   Fig. 11.4 Interactive software prototype; *photo courtesy of HS Design.* 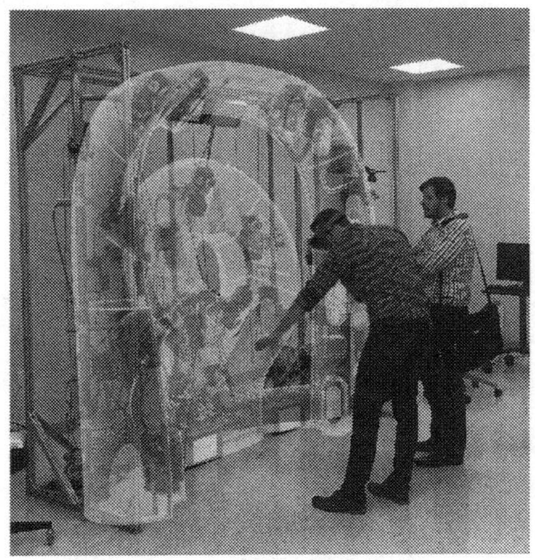 Fig. 11.5 Virtual reality prototype using Hololens to provide scale; *photo courtesy of HS Design.* | Higher fidelity and interactive prototypes provide the optimum usability evaluation. |

- Operationalize data collection. For example, what is the most appropriate way to assess whether a user has administered a complete dose of therapy? Timing may be appropriate if time is tied to a complete dose (e.g., 5 s to inject a dose from an autoinjector after dose initiation), or perhaps identifying if any medication is observed when a user removes an autoinjector from simulated skin (e.g., wet injection). A combination may be more appropriate in some cases, but it is important to identify how to collect data a priori so that data are collected in an unbiased way.
- Determine how observational (objective) and subjective data will be collected for each task, and outline specific strategies for doing so. Qualitative feedback (i.e., subjective feedback from participants) is promoted by FDA and is important to document (FDA, 2016). Participants oftentimes share thoughts in the form of subjective data that cannot be observed. Ensure that the study methodology outlines how both objective and subjective data will be documented and analyzed for each task. Subjective data are especially important for root cause analysis, and FDA encourages the inclusion of verbatim participant quotations in analyses and reports. This may include video recording from multiple perspectives in order to adequately observe device use and enable a review of comments provided throughout the simulated use formative study for root cause analysis. Alternative to video recording may be simply audio recordings or diligent notetakers.

#### 4.2.1.2 Formality of study protocol

A simulated use formative protocol can be less detailed than a validation study protocol; however, there are advantages to providing sufficient details. Consider following FDA Human Factors Guidance (FDA, 2016: Appendix A) that outlines different sections of a HFE/UE report as the framework for formative testing protocols. While this may seem to require "extra" work, in the long run, it helps ensure that human factors documentation for the device is consistently formatted, traceable and presented in a way that can easily be integrated into the HFE/UE report, including details that agency reviewers are familiar with.

An adequate protocol ensures consistent methodologies are applied across participants, which ultimately enhances conclusions as a result of evidential data. It will also allow easier comparisons of methods and performance outcomes across formative evaluations and the validation study should it become necessary to show how human factors data informed design decisions (i.e., design traceability). For example, observing performance improvement in a validation study because a design mitigation was implemented as a result of a formative evaluation further indicates the mitigation was effective. Details such as number and type of participants, scenario scripts and order of tasks evaluated provide the documentation trail that will allow you to make important comparisons, including within the final HFE/UE report.

Detailed documentation can also help provide rationale for including or excluding user groups and/or certain tasks, as well as provide an argument for the inclusion of design elements to improve performance when otherwise you may receive questions or push back from regulating agencies.

**Using formatives as evidence of good design requires good documentation**

A sponsor with a combination device anticipated FDA concerns about using a stop sign symbol in the device IFU that could potentially lead to patient confusion. The sponsor conducted a formative evaluation focused on user understanding of the stop sign within the context of operational use of the device. One of the study objectives was to determine: Did representative users correctly interpret the "Stop Sign" symbol in the IFU? Using a non-biased approach, findings showed that users overall recognized the symbol in the way it was intended to be interpreted; i.e., users knew to stop and make sure they followed the action outlined in the IFU text where the stop sign symbol appeared.

The sponsor was able to provide FDA with evidence to argue that the inclusion of the stop sign symbol in the IFU helped mitigate risk to the users (patients) by reiterating the importance of stopping at a critical point in the device use process. **The response was accepted by FDA, which highlights the importance of documenting design changes and study findings, even in formatives**.

### *4.2.2 Development of Moderator and Notetaker Guide*

The Moderator and Notetaker Guide is one of the more important pieces needed to run a consistent, replicable formative evaluation. It ensures that the Moderator and Notetaker (if available) can collect all necessary data in a systematic and similar way for all participants. Although a Moderator-Notetaker team is ideal for collecting simulated use data, a Notetaker may not always be available for a formative evaluation. A detailed and efficient guide is especially important when only one person will be moderating as well as collecting data.

A Moderator and Notetaker Guide can be in electronic format or paper-based, depending on the needs of the Moderator and the Notetaker (if available). Paper-based may be more appropriate if the formative requires a lot of movement around the testing environment, but will then require data entry to compile and analyze all data electronically. An electronic format is more appropriate when all user-device interactions occur in one place (e.g., at a table). This type of guide functions as a more efficient data entry tool and allows the Moderator (and Notetaker) to enter data directly into the form for data analysis following the completion of the formative. See Table 11.6 for an example of a Moderator and Notetaker Guide.

The Moderator and Notetaker Guide should include at a minimum:

- Moderator script to ensure consistency in instructions across participants;
- All tasks outlined, with success criteria and scoring method to record observation data; and
- Space to document Moderator (and Notetaker) observation notes, subjective comments from participants during the tasks and probing/interview data on difficulties and use errors.

TABLE 11.6 Example of a Moderator and Notetaker Guide for a combination device.
WF-1: Program a syringe infusion pump
Our first software task will start with a syringe infusion for Midazolam as listed on the order card I have handed you. Let's imagine you have already manually primed the syringe for 0.5 mL, please show how you would turn on the pump and program the infusion for your patient in the NICU.

| Steps | Scoring | Notes |
|---|---|---|
| *1. Press power button* | *C CI NC X* | |
| *2. Patient Setup* | *C CI NC X* | |
| *3. Press 'Current Patient'* | *C CI NC X* | |
| *4. Select 'Neonatal Care' for Care Type* | *C CI NC X* | |
| *5. Place badge near badge scan icon to unlock* | *C CI NC X* | |
| *6. Syringe Programming* | *C CI NC X* | |
| *7. Enter dosage* | *C CI NC X* | |
| *8. Press Start* | *C CI NC X* | |
| **Would you describe that task as easy or difficult? Can you tell me about that?** | | |
| **FOLLOW-UP QUESTIONS** | | |
| **1. What does the term "Patient Setup" mean to you?** | | |
| **2. When do you think that you would use this "Patient Setup" function?** | | |
| **3. What do you think about the location of the "care area" options?** | | |
| **4. What do you think about the name of the button as "Current Patient"?** | | |
| **5. [*If don't like name ask*] What would you name the button instead?** | | |
| **6. Did you have any difficulty on this task? [*If so*] Can you tell me about that?** | | |

*C*, completed successfully; *CI*, completed with issues; *NC*, did not complete; *X*, not assessed.

### *4.2.3 Strategies for conducting simulated use studies*

Many different strategies can be employed to conduct simulated use studies, but the decision should be made based on the objectives of your formative. Table 11.7 describes 9 strategies that can be used to support your study objectives.

## 4.3 Participants

Identifying the right type and number of participants for your simulated use formative is important because it impacts the usefulness and application of the data. A total of 5–8 participants per distinct user group is generally recommended in literature. According to Faulkner (2003), a sample of 5 users can detect a minimum of 55% and an average of 86% of problems with a user interface. ANSI/AAMI HE75 (2009) provides further rationale for a sample size of 5–8 per user group based on the law of diminishing returns. That is after 5 participants, subsequent participants will be less likely to identify any new UI issues, and any gains in new meaningful information will be limited.

**TABLE 11.7** Specific strategies for simulated use formatives.

| Strategy | Description | When to use |
|---|---|---|
| 1. Rapid insight usability test | Early, quick formative usability test conducted using prototypes or partial designs to test certain aspects of the UI. Formatives can be short in duration with around 5 participants, and participants do not necessarily need to be representative users (although they should not be people on the design team or perhaps even within the sponsoring company in general). | Used to quickly evaluate certain design elements early in the design process and to evaluate low-hanging fruit (e.g., Do IFU graphics depict what is needed, do the images work with the text provided so that they are easily interpreted, is any new terminology being included?). For example, this strategy can be used to evaluate new device terminology and whether lay users (and perhaps users with low literacy, depending on intended users) are able to understand certain terminology specific to your device. Words that product development teams view as simple or that have inherent meaning can be quite confusing for lay users. |
| 2. Partial task evaluation | Explores only part of the operational context of use for a device instead of all user-device interaction tasks. This can minimize study duration and focus on the data that are truly important to ongoing design efforts. | Can be used when problems have been identified (or hypothesized) with only certain aspects of the UI. For example, if the device developers of an infusion pump suspect that the software menu options for delivering a bolus may be confusing to users, a part-task evaluation can focus only on the tasks relating to programming a bolus.<br>Can also be used when only certain parts of the UI are ready to be evaluated. For example, if not all software development has been completed. |
| 3. Rapid iterative testing and evaluation (RITE)/ prototyping | RITE methodology takes advantage of sponsor capabilities to make changes to the UI "on the fly" (during or immediately following test sessions) in order to test and iterate potential design mitigations for observed problems that occur in the same study. In other words, if participants are observed to have problems, data analysis and root cause are discussed and design changes may be made in between participants or before the end of the session. Then, these new changes are presented to the participant in the session and/or tested with subsequent participants in the same study. | Used when sponsors want to evaluate design changes (usually labeling changes) on the fly to mitigate problems observed in the study. Labeling revisions are often the focus for this strategy, as design can be changed relatively easily. Labeling changes can be low fidelity (e.g., pasting different image over the existing image in the IFU) or high fidelity (e.g., creating a completely new IFU), depending on resources and time. |
| 4. Comparison of alternatives | Presents the participants with two or more alternative UI designs (often labeling designs) and asks the participant to compare the alternatives and provide subjective feedback. This usually occurs after the participant has completed use scenarios with one of the alternatives. This is not the same as preference testing and should be based on objective performance data to support the subjective user feedback. | Used when a sponsor wants to evaluate several design alternatives, which may include different images, layouts, etc. This strategy will help to determine if one concept/design is better than another from the user performance perspective. During early formatives, a sponsor may choose to evaluate a competitor's device as an alternative to identify best features of the existing device (see AAMI HE75 for additional information). |

**TABLE 11.7** Specific strategies for simulated use formatives.—cont'd

| Strategy | Description | When to use |
|---|---|---|
| 5. Eliciting subjective feedback | Different strategies to elicit subjective feedback can be used but should be outlined in your testing protocol.<br>Talk aloud methods are when Moderators ask participants to verbally talk through the use scenario and tasks they are performing to better understand the cognitive processes used to interact with the UI. However, talk aloud methods are not generally representative of actual use and may be awkward for participants.<br>Moderator probing to elicit feedback "along the way" can generate valuable feedback if participants do not independently talk aloud.<br>Uninterrupted task performance is when the Moderator does not intervene as participants complete a task, which gives more realistic data on user-device interactions. In this method, probing usually occurs after the use scenario has been completed. | Method for eliciting subjective feedback should be based on the objectives of your formative. For example:<br>• Use talk aloud methods when conducting exploratory evaluations, typically early in the design process.<br>• Use probing "along the way" to elicit feedback when observations are noted to help better understand participant thoughts and actions.<br>• Use uninterrupted task performance as you near validation testing and want a more accurate gauge of user performance. |
| 6. Explicit direction to use IFU versus representative use | Moderator either *explicitly* directs the participant to use the labeling or presents the labeling but does not instruct the participant to use the labeling, which is representative use because a Moderator would not be directing users in the actual use environment. | Method for directing participants to use the IFU should be based on the objectives of your formative. For example:<br>• Direct users to use the IFU when you want to evaluate the effectiveness of the labeling.<br>• Allow users to choose to use the labeling (or not) if you are conducting a pilot test prior to a validation or want to observe representative use where users are not instructed to use any labeling in actual use environments. Some researchers may choose to test the device first without specific instruction to use the IFU, followed by asking the participant to repeat the use scenario(s) while using the IFU. This has the potential to bias participant interaction with the IFU, and may also be inconvenient due to time constraints. A cleaner method would be to conduct a separate, performance-based IFU formative evaluation to demonstrate labeling effectiveness. |
| 7. Introduce problems/device error state or alarms | Introduces problems, alarms or error states on the device user interface to observe how participants respond. For example, can users perceive, interpret and successfully act on (resolve) an alarm without causing harm? Use scenarios should be developed to expose participants to device states (i.e., problems and alerts) that could be otherwise difficult to observe in a simulated study with normal use conditions. | Use when testing critical alarms or alerts that are critical to the safe and effective operation of your device. Note that alarms and alerts that occur infrequently and may not occur during regular use but that are critical to device safety are important to FDA. For example, a critical low battery indicator for an assistive cardiac care device may occur fairly frequently, and an imminent shutdown warning may occur rarely. However, correct responses to both of these alarms are critical to delivering life-saving care and should both be evaluated as part of the UI. |

(*Continued*)

**TABLE 11.7** Specific strategies for simulated use formatives.—cont'd

| Strategy | Description | When to use |
|---|---|---|
| 8. Testing both labeling and device | Test performance without Moderator intervention or direction to use the IFU, then ask participants to use labeling for second time through use scenario. When not directed to use the IFU, many participants choose not to use any labeling. This strategy allows sponsors to evaluate the IFU after collecting representative performance data. This approach can bias how participants interact with the device because they have already used it once—the Moderator may not get a clear picture of the effectiveness of labeling, so this is not always an appropriate method. | Can be used when the sponsor wants to test representative use of the device UI (i.e., uninterrupted task performance, not directing the user to use the IFU) but also wants to evaluate the effectiveness of the IFU in the same study.<br>An alternate may be to ask participants to use labeling first if the objective of the formative is to evaluate the labeling. |
| 9. Evaluating different aspects of training | Includes representative training for participants in the formative evaluation using appropriate training materials, training criteria and adequate training/learning decay. | If the sponsor will be providing and managing user training as part of the UI, you can evaluate your training program through a formative evaluation to show effectiveness of training as a risk mitigation.<br>Alternately, sometimes a sponsor will opt to include training and no training arms of the study to demonstrate that training helps mitigate additional risk, but that users are still able to safely and effectively use the device even if training is not provided (as a worst-case scenario). |

Participants who are representative intended users should be recruited for simulated use formatives, as the overarching purpose is to observe representative users interact with the device in a simulated and representative use environment. However, using a broader and less specific participant population may be appropriate during early formatives when "rapid insight" testing, or when recruitment of specific participants who are representative of intended users, may be more difficult or time-consuming (e.g., patients who are diagnosed with a rare medical condition). Regardless, care should be taken not to include stakeholders as participants in simulated formatives (e.g., designers, R&D employees, project managers, regulatory, etc.) or to generalize findings to representative users that may not be applicable.

Consider also recruiting user populations with certain characteristics and limitations that could impact device use and may be associated with higher risk for use-related problems (e.g., elderly patients, colorblind, left-handed, patients with comorbidities, etc.). This allows designers to catch any potentially problematic design elements for these users before final design decisions are made.

## 4.4 Collecting & analyzing data

Formative data collection focuses on both objective (quantitative) performance and subjective (qualitative) feedback. During sessions, a Moderator will take notes in the Moderator and

Notetaker Guide of their observations of potential issues participants have while interacting with a user interface, while a Notetaker (if available) can assist the Moderator by recording all observations of issues as well as successes. The Notetaker will also record subjective feedback provided by participants during the more conceptual discussion sections of the study and will have the opportunity to inform the Moderator of any missed use events that can be subsequently investigated.

There is value in collecting both quantitative (objective) and qualitative (subjective) data for formatives. One builds on the other and can give a bigger picture of the possible issues associated with a device. For example, a participant may tell the Moderator that they really like a specific workflow (qualitative data), yet the Moderator may observe that the participant does not use the device feature correctly during performance of that workflow (quantitative data). The Moderator could then redirect the participant back to that feature and probe for more information on the user's ideas, opinions and beliefs of how it works. Often times the Moderator can show the participant how to use the feature correctly and gain additional insights.

Objective data or quantitative data usually come in two forms: (1) time on task and (2) success rates and error rates. Time on task should be included in testing when there are tasks that have time components associated with successful completion (e.g., injection hold time) as well as tasks that would require quick response times in an emergency situation (e.g., response time for an alarm). It is also helpful to include time on task measurements with the device and with the competitor's device. For simulated use studies, time on task can also be informative for non-emergency tasks to help make the device design more efficient (e.g., time spent navigating a menu structure for a programmable infusion pump). Success rates and error rates on the other hand should always be measured. Recording observed errors helps focus the designers on areas of the user interface that need improvement and offer the opportunity to probe the participant on what would have helped them complete the task without difficulty.

Subjective data or qualitative data can be gathered in two ways: (1) collecting and following up on participant feedback and (2) probing on observed body language that conveys confusion or frustration. Typically, qualitative data include user comments about the device made while participants are interacting with the device or afterward as the Moderator probes on observed issues as well as potential unobserved issues. However, body language or other non-verbal signs should not be ignored. Moderators often observe "confused looks" which should prompt Moderators to ask what the participant is thinking. Other non-verbal cues include folding arms, long sighs and long pauses between actions. All of these cues can mean that there is a struggle with the design of your device that should be explored within the context of the session.

Data collection should be recorded if possible using video cameras placed around the room to capture interaction between participant and device as well as any comments regarding use experience with the device. The research team may also take still photos during the sessions. However, keep in mind that all participants must give consent to be recorded in any manner, which is typically outlined in the informed consent document that all participants must sign before participating in the study. Audio and video capture enables the Moderator to review performance and subjective data if necessary after the formative has been completed. Video data also can provide evidence of study observations for regulatory reviewers and/or members of the design team who may want additional context about use problems, if requested. The data collection and coding for a formative usability test is usually designed to identify potential issues and understand the root causes of those potential issues.

Observed design issues should be captured and documented for further investigation of the root causes.

Interviewing, which probes into root causes, establishes participant awareness of potential issues. Thus, they can explain their expectations of what causes may have contributed to the error occurring and provide opinions on what they would do in real life upon encountering the issue. If the research team observes issues during the formative, it is helpful to compare the subjective feedback from participants against the task analysis and PCA analysis (see Chapter 6) to help determine root causes of issues.

### Tips for moderating simulated use formatives

- Take your time; give the participant time to think. Silence is not bad.
- Do not phrase questions that sound accusatory and make the participant feel that it is their fault for use errors (e.g., Ask "How did you know to do that?" instead of "Why did you do that?").
- Do not ask leading questions (e.g., "When you did x, y, and z, did you use the IFU to help you correct your mistake?").
- Address the participant as the expert in the room so they feel empowered to share their thoughts and opinions.
- If the participant appears to be getting upset or frustrated, remind the participant that you value their opinion about the device, they are not alone in experiencing difficulty and this type of experience is exactly why the sponsor is bringing in users to improve the device.
- If participant is unaware that an issue occurred, an alternative to explicitly pointing out the issue (and potentially upsetting the participant) may be to request the participant talk through certain steps first. In many cases, the participant self-identifies the issue when using this approach.
- Body language speaks volumes! Be aware of what you are communicating to the participant. For example, use friendly and non-threatening body language (e.g., do not lean back and cross your arms when talking with participants). Fig. 11.6 depicts active listening.
  - Sit or stand comfortably.
  - Make eye contact.
  - Acknowledge the user.
  - Show understanding.
  - Listen actively.

**FIG. 11.6** Interviewer who is actively listening to participant. *Photo provided by HS Design.*

The Moderators and Notetakers can use the following coding examples to categorize their task observations in real-time during study sessions so that data are organized appropriately:

- **Completed (C):** Successful completion of a task without any observed or reported use errors or close calls.
- **Completed with Issues (CI)**
  - **Close Call**: Completion of a task with an observed or reported use error that participant self-corrects or where a participant almost experiences a use error but recovers and completes the task without a use error before causing harm.
  - **Difficulty:** The task was completed by the user, but difficulties were observed or expressed in the process. The user was able to complete the task. Difficulties might be revealed by multiple attempts to perform the task, anecdotal comments about the task's difficulty, or taking longer than expected to perform a task.
- **Did Not Complete (NC):** A task that was not completed, completed incorrectly or required assistance from the Moderator.
- **Not Assessed (X):** A task that was not assessed due to insufficient time in the study or a prerequisite task not being completed. Tasks that are not assessed should be documented in the report but removed from the data set for that particular task when calculating success rates and error rates for each task. For example, if 1 of 8 participants did not have time to complete the final knowledge task question, the performance data tables in the report should indicate that 7 of 8 participants responded to the question. The success rate should then be calculated out of 7 participants instead of 8.
- **Other observations to document:**
  - **Use Error:** User action or lack of action that was different from that expected by the manufacturer and caused a result that (1) was different from the result expected by the user, (2) was not caused solely by device failure and (3) did or could result in harm.
  - **UI Elements that Support Participant Performance:** The task was completed by the user with assistance from IFU, online help, quick reference guide, user documentation and/or simulated supervisor or helpline. When the user chooses to use an additional UI element like the ones described above to assist in task completion, the Moderator should document what part of the UI the user is using, how they are using it and whether the user was able to successfully complete the task with the support of the additional UI element. These data should be reported in conjunction with the scoring of "Completion" and "Completion with Issues," depending on the observed performance outcome.
  - **Study Artifact:** Performance that was caused by an aspect of the study environment or study materials, rather than the participant actions or decisions. Artifacts will be removed from the data set (i.e., reduced N for that specific task) when calculating success rates for each task. Examples of study artifacts include: technical issues with the device (e.g., an implantable pump programmer encounters a bug, and menu selection does not display properly), use scenario instructions are not clinically

relevant for the participants (e.g., a doctor states that she would never do tasks a, b, and c because her nurse routinely performs those tasks for the doctor), participant misunderstanding of Moderator instructions (e.g., the participant misunderstands a knowledge task question), and aspects of the use scenarios that may be perceived as artificial tasks from the participant's perspective (e.g., differentiation between similar devices).

Once the study is complete, data analysis begins. Results are compiled into a summary report that includes participant demographic information, as well as the empirical and subjective findings for each task. The specific use issues that occurred for each task are identified, along with the root cause of each issue. Additionally, the major themes of subjective feedback provided by participants regarding each task are identified.

### 4.5 Documenting the report & recommendations

As noted previously, documentation provides evidence of applied HF throughout the device development and includes all formative studies. The simulated use formative report should note exactly what was done (or reference the testing protocol), what data were collected and analyzed, what the findings were, and what specific mitigations were recommended or implemented as a result. It should link specific design recommendations to data (objective and subjective). For example: A user is supposed to enter 2 into a keypad, but incorrectly enters 23 and comments that the spacing between the keys is too narrow. A recommended design change linked to this task may be to increase the key size and spacing on the keypad, and literature can be referenced to identify key size measurements and layout recommendations that may be appropriate for the user population (see HE75 for specific guidelines on control spacing). If a design modification is warranted based on performance data, propose detailed solutions as appropriate. Document any UI iterations based on study results in the report for later reference, including as the HFE/UE report is drafted and the preliminary analyses are summarized. It is also helpful to describe the user characteristics and demographics data for formative participants, along with detailed root cause analysis (Chapter 13) for all difficulties and use errors.

The bottom line is that while the formative report does not need to be a formalized document, **best practices suggest that a strong report should document everything to provide transparency for the regulatory reviewer as well as a documentation trail for the design team who may need to look back across the history of the design iterations.**

## 5. Developing recommendations for improved design

Performance data from formatives often lead to recommendations for design modifications that are hypothesized to improve participant performance and reduce or eliminate risk. A residual risk analysis can be completed at the end of a later stage formative evaluation and should assist the sponsor in determining which, if any, modifications should be made to the device UI.

### Considerations for design modifications

1. **Based on residual risk, should design modifications be implemented?** Determine if one or more device design mitigations are necessary to reduce or eliminate residual risk. In other words, did the sponsor conduct a residual risk analysis on formative findings and determine that the device has not been optimized? Is there a way to further reduce the risk associated with using the device? When residual risk is found to be unacceptable, mitigations should be implemented and validated (even if that means a delay in production or device approval). Risk management options include the following, which are ordered for preference and effectiveness according to ANSI/AAMI/ISO 14971: (1) modify design of device itself; (2) incorporate specific safety measures in the device itself; and (3) include or modify labeling and/or training.
2. **Is there traceability?** Are design modifications traced to findings from formative evaluations? It is good practice to link design changes with data that suggest the modification is necessary and to avoid arbitrary changes that are not linked to user performance data.
3. **Are you prepared for a successful validation study?** Have the modifications been tested to verify that the risks were successfully mitigated and new risk was not introduced? Some sponsors choose to test modifications prior to a HF validation study, which minimizes the risk of an unsuccessful validation. Other sponsors choose to proceed straight to the validation study in hopes that the modifications will be shown to be successful in the validation study.

## 6. Summary

Conducting simulated use formatives is good human factors practice and should be an important consideration during device development. With a relatively small sample size and a wide variety of methods and testing strategies, formatives can inform the design process and help to reduce the overall amount of risk to end users and patients. These evaluations are useful tools for sponsors throughout the development process, and early formatives are highly recommended. Careful documentation of formative evaluations, design features directly linked to performance data and critical task categorization rooted in preliminary analyses findings can help lead to a HF validation study in which the use related safety and effectiveness of your device is successfully optimized and validated as required by FDA. The included references at the end of this chapter provide further information about simulated use formatives.

## Acknowledgments

Thanks to HS Design for providing photos for this chapter. Special thank you to Elissa Yancey for editing.

## References

AAMI/ANSI HE75. (2009). *Human factors engineering — design of medical devices.*

ANSI/AAMI/IEC 62366-1. (2015). *Medical devices — part 1: Application of usability engineering to medical devices.*

ANSI/AAMI/ISO 14971:2007/(R) (2010). *Medical devices — application of risk management to medical devices.*

Faulkner, L. (2003). Beyond the five-user assumption: Benefits of increased sample sizes in usability testing. *Behavior Research Methods, Instruments, and Computers, 35*(3), 379—383.

FDA CDRH. (2016, February 3). *Applying human factors and usability engineering to medical devices.*

Wiklund, M. E., Kendler, J., & Strochlic, A. Y. (2016). *Usability testing of medical devices* (2nd ed.). Boca Raton, FL: CRC Press.

## Additional resources

Charlton, S. G., & O'Brien, T. G. (2001). *Handbook of human factors testing and evaluation*. Taylor & Francis.

Dumas, J., & Redish, J. C. (1999). *A practical guide to usability testing*. Bristol, UK: Intellect Books.

Nielsen, J., & Mack, R. L. (1994). *Usability inspection methods*. New York, NY: John Wiley & Sons.

Rubin, J. (1994). *Handbook of usability testing: How to plan, design and conduct effective tests*. New York, NY: Wiley & Sons.

Sauro, J., & Lewis, J. R. (2016). *Quantifying the user experience: Practical statistics for user research* (2nd ed.). Elsevier Inc.

Weinger, M. B., Wiklund, M. E., & Gardner-Bonneau, D. J. (2010). *Handbook of human factors in medical device design*. Boca Raton, FL: CRC Press.

Wiklund, M., Birmingham, L., & Larsen, S. A. (2018). *Writing human factors plans & reports for medical technology development*. Arlington, VA: AAMI.

Wiklund, M. E., & Wilcox, S. B. (2005). *Designing usability into medical products*. Boca Raton, FL: CRC Press.

# PART V

# Safety related risk

CHAPTER

# 12

# Use-focused risk analysis

*Sophia Kalita, Melissa R. Lemke*

Agilis Consulting Group, LLC., Cave Creek, AZ, United States

OUTLINE



*Risk is like fire: if controlled it will help you; if uncontrolled it will rise up and destroy you.* **Theodore Roosevelt.**

## 1. Introduction

The use of medical devices inherently includes a certain amount of risk. This is true of devices that are considered to be simple as well as devices that are more complex. For example, blood pressure monitors developed for home use may appear simple and as having minimal risk. However, the size and placement of the cuff can impact the accuracy of blood pressure measurements, and potentially increase the risk to patients if treatment decisions are based on incorrect measurements. On the other hand, a device such as an X-ray machine is complex and may have obvious risks. However, the amount of risk for an X-ray machine is beyond those that are obvious. In "Set Phasers on Stun" (1998), Steve Casey discusses a situation where a technician was adjusting the settings of an X-ray machine while a patient

*Applied Human Factors in Medical Device Design*
https://doi.org/10.1016/B978-0-12-816163-0.00012-8

was inside the machine. The patient was exposed to a high level of radiation and died as a result of a use-related error. These examples highlight that, for both simple and complex devices, manufacturers need to analyze the risks beyond those that are inherent, including the need to carefully analyze use-related risks.

Risk analysis is the process of identifying potential hazards and assessing risks associated with the hazards. Risk analysis is one element and the foundation of the overall risk management process (see Fig. 12.1). According to ANSI/AAMI/ISO 14971 (2016), manufacturers are required to follow the risk management process and systematically implement policies, procedures, and practices to analyze, evaluate, control, and monitor risk. It is an iterative process; thus each of the elements should also be iterative. Further, the specific element(s) within the risk management process should be applied in the appropriate phases of the design and development process. Specific to risk analysis, it should be initiated early in the design process (i.e., device concept, device prototype) and revised continuously throughout design development as well as after the device is launched in the market.

There are multiple lenses through which the manufacturer can assess risk. To analyze the complete risk profile of a device, all lenses should be considered.

- Design-focused risk analysis determines the potential risk related to the design or functionality of the device.
- Process-focused risk analysis determines the potential risk related to the manufacturing or assembly of the device.
- Use-focused risk analysis determines the potential risks related to how the device is used by intended users in the intended use environments.

To illustrate each lens, consider a bandage and the hazard of inadequate adherence to the skin. For the specific hazard, the material properties of the adhesive would be identified in a design-focused risk analysis. In a process-focused risk analysis, the application of the adhesive to the bandage would be identified. In a use-focused risk analysis, use errors such as placing the bandage on wet skin would be identified.

For human factors, the use-focused risk analysis is critical to identifying and categorizing user tasks and to determine if design improvements are required to reduce or eliminate use-related risk. With emphasis on the use-focused risk analysis, this chapter discusses the risk analysis process and application of risk analysis to human factors.

### Terms and definitions

Note: All definitions of terms are adopted from ANSI/AAMI/ISO 14971 (2016), unless noted otherwise.

**Abnormal use** – conscious, intentional act or intentional omission of an act that is counter to or violates normal use and is also beyond any further reasonable means of user interface-related risk control by the manufacturer (IEC 62366-1:2015)

**Harm** – physical injury or damage to the health or people, or damage to property or environment.

**Hazard** – potential source of harm.

**Hazardous situation** – circumstance in which people, property, or the environment are exposed to one or more hazard(s)

**Misuse** – incorrect or improper use of the medical device.

**Normal use** – operation, including routine inspection and adjustments by any user, and stand-by, according to the instructions for use or in accordance with generally accepted practice for those medical devices provided without instructions for use (IEC 62366-1:2015)

**Residual risk** – risk remaining after risk control measures have been taken.

**Risk** – combination of the probability of occurrence of harm and the severity of that harm.

**Risk analysis** – systematic use of available information to identify hazards and to estimate risk.

**Risk control** – process in which decisions are made and measures are implemented by which risks are reduced to, or maintained within, specific levels.

**Risk estimation** – process used to assign values to the probability of occurrence of harm and the severity of that harm.

**Risk evaluation** – process of comparing the estimated risk against given risk criteria to determine the acceptability of the risk.

**Risk management** – systematic application of management policies, procedures, and practices to the tasks of analyzing, evaluating, controlling, and monitoring risk.

**Severity** – measure of the possible consequences of a hazard.

**Use error** – user action or lack of action that was different from that expected by the manufacturer and causes a result that (1) was different from the result expected by the user and (2) was not caused solely by device failure and (3) did or could result in harm (FDA guidance, 2016).

**Use safety** – freedom from acceptable use-related risk (FDA guidance, 2016).

## 2. Process for use-focused risk analysis

Use-focused risk analysis is a step-by-step process (see Fig. 12.2). It should start by defining the intended uses, intended users, and intended use environments. The next step should identify the use characteristics of the device and user interactions that could affect safe use. When identifying characteristics, an approach can be to ask a series of questions from the perspective of the intended users. As a guide, Annex C of ANSI/AAMI/ISO 14971 (2016) provides a list of questions to identify the device characteristics that could impact safe use. Then the known and foreseeable hazards related to normal and abnormal use of the device should be identified. FDA guidance (CDRH 2016; Section 6.2) provides sources of information for identifying known use-related hazards. It is important to consider all potential hazards and situations including those that may result from device failure at this point in the process. The combination of use-related hazards that can lead to hazardous situations should then be identified. The final step should estimate the risk of harms associated with the hazardous situations.

Following risk analysis, risk evaluation and risk controls are to be implemented, and risk analysis iteratively repeated throughout the design and development process, as part of the overall risk management process.

As an example, consider an auto-injector containing a drug to treat diabetes. The intended use is to deliver the medication to patients with diabetes, the intended users are adult patients

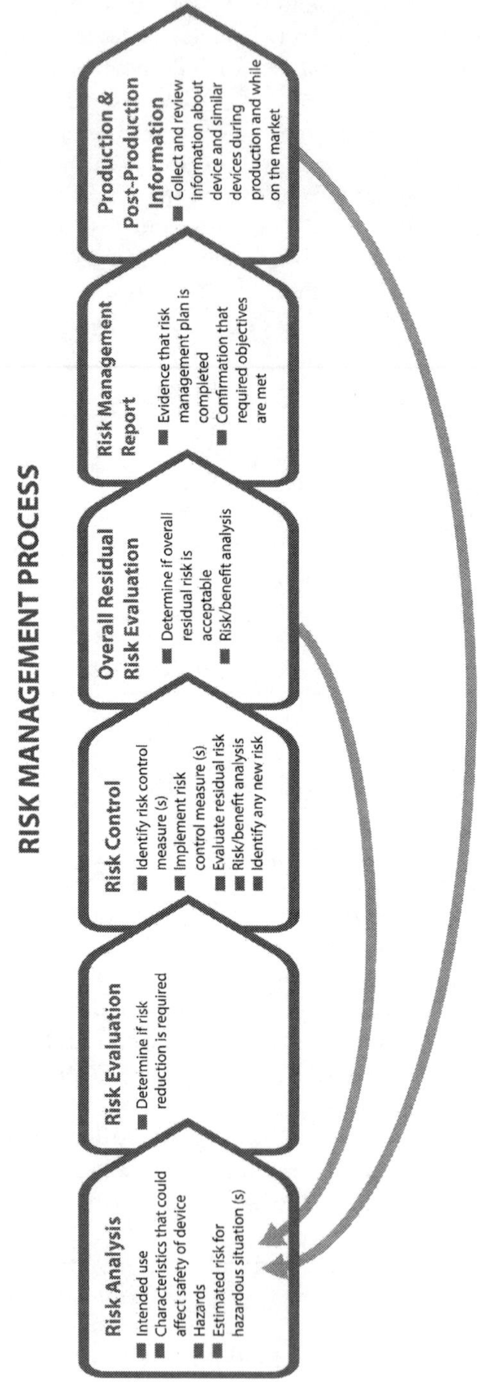


FIG. 12.1 Overview of risk management process.

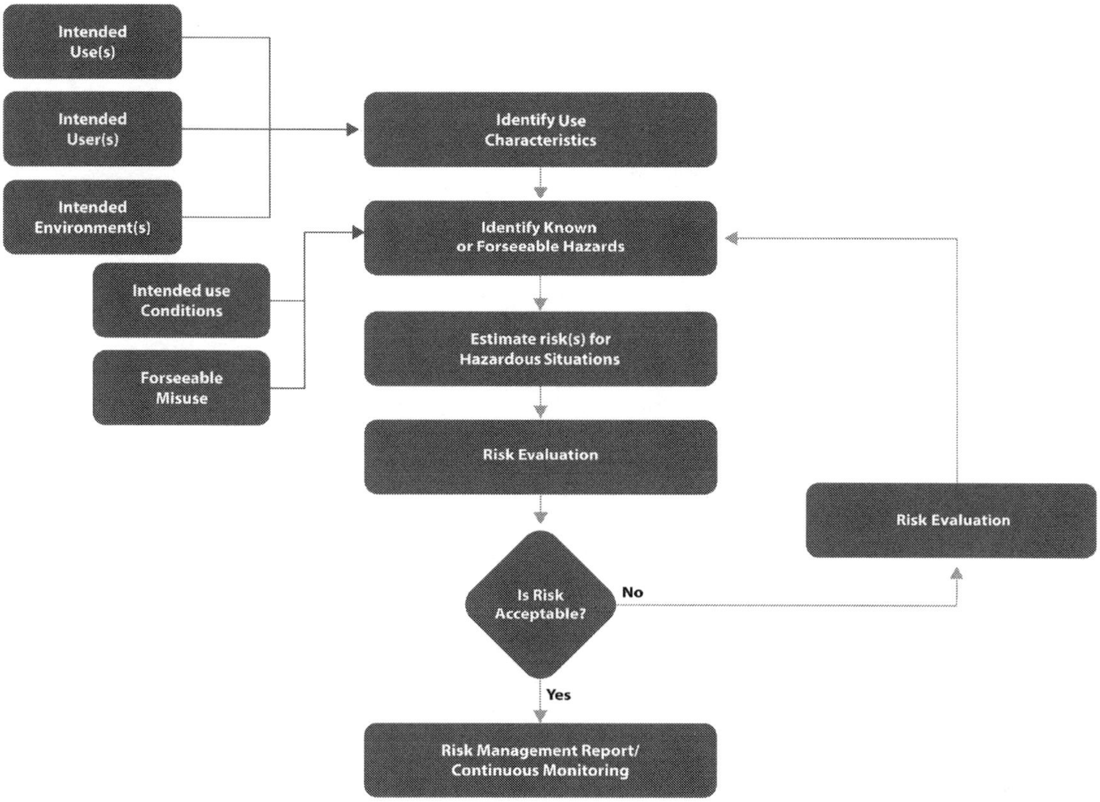


FIG. 12.2 Process for risk analysis.

with diabetes, caregivers, and healthcare professionals (HCPs). The intended use environments are non-clinical and clinical settings. For this particular example, use characteristics of the auto-injector are the drug needs to be properly mixed before use, and a proper mix is determined by viewing the drug through a window on the auto-injector. If the window is not large enough for users to view the drug, it will be difficult or impossible for users to determine whether the drug is properly mixed. Therefore, a known or foreseeable hazard is that the window is too small to indicate to intended users whether the drug is properly mixed.

As mentioned above, the final step of use-focused risk analysis should identify the combination of hazards that can lead to hazardous situations and then estimate the risk for each hazardous situation. Fig. 12.3 below is an illustration from ANSI/AAMI/ISO 14971 (2016) that depicts the components of risk and the approach for estimating risk. One component of risk is the sequence of events, which is a single event or combination of events that leads to the occurrence of a hazardous situation. A hazardous situation may result in harm to the user. The risk of the hazardous situation is then estimated by a combination of the probability of occurrence of the harm and the consequence or severity of that harm. Note that the probability of occurrence of harm involves two factors: the probability of the hazardous situation occurring and the probability of the hazardous situation leading to harm.

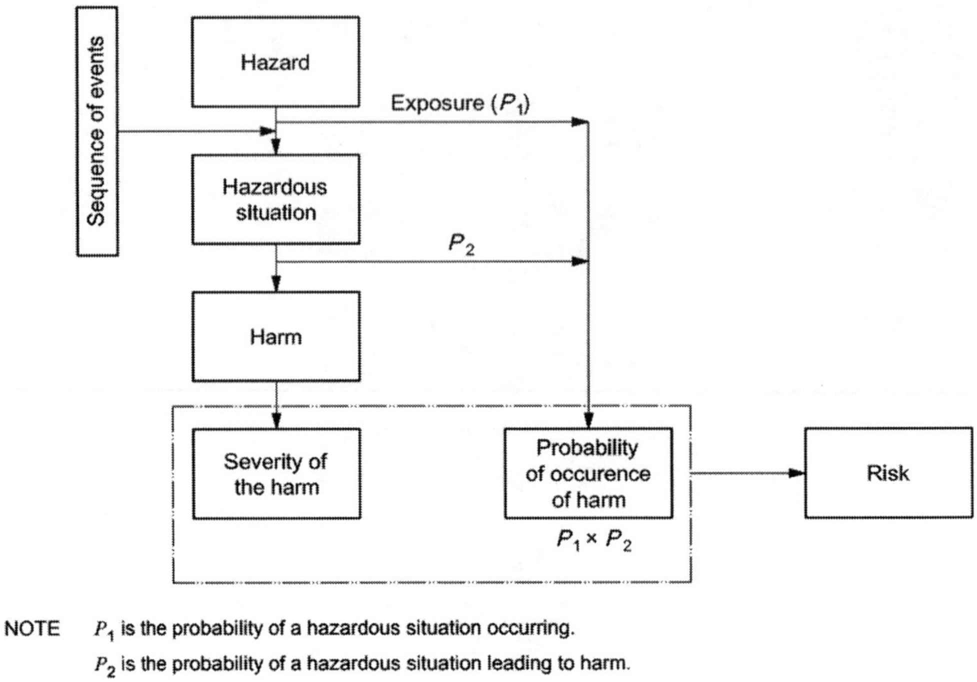


**FIG. 12.3** Components of risk and risk estimation process (ANSI/AAMI/ISO 14971, 2016).

Continuing with the example of the auto-injector and the hazard of the window being too small to indicate if the drug is properly mixed, the sequence of events that would lead the hazard to a hazardous situation is the user incorrectly determining the drug is mixed. The hazardous situation is the user administering the drug when it is not properly mixed. Injecting improperly mixed drug could lead to harms of hypoglycemia or hyperglycemia. The estimated risk for the hazardous situation would then be assessed with the severity of each harm and the probability of occurrence of each harm.

Severity of harm is determined through severity levels established by the manufacturer, which can be defined in different ways as long as they are clearly defined and applied consistently. ANSI/AAMI/ISO 14971 (2016) provides guidance on establishing severity levels and includes several examples. One common example of severity levels is provided in Table 12.1 below.

Similar to severity of harm, probability of occurrence of harm is determined through probability levels established by the manufacturer. Probability levels also are to be clearly defined and applied consistently. ANSI/AAMI/ISO 14971 (2016) provides guidance and examples for establishing probability levels, and one example is provided in Table 12.2 below.

The two factors that comprise probability of occurrence of harm requires consideration of the entire sequence of events, starting with the occurrence of the hazard through to the occurrence of the harm. Key inputs to estimate probabilities are existing information or data

TABLE 12.1 Example of severity levels adopted from ANSI/AAMI/ISO 14971 (2016).

| Category | Definition |
|---|---|
| Catastrophic | Results in patient death |
| Critical | Results in permanent impairment or life-threatening injury |
| Serious | Results in injury or impairment requiring professional medical intervention |
| Minor | Results in a temporary injury or impairment not requiring professional medical intervention |
| Negligible | Inconvenience or temporary discomfort |

TABLE 12.2 Example of probability levels adopted from ANSI/AAMI/ISO 14971 (2016).

| Category | Range |
|---|---|
| Frequent | $\geq 10^{-3}$ |
| Probable | $<10^{-3}$ and $\geq 10^{-4}$ |
| Occasional | $<10^{-4}$ and $\geq 10^{-5}$ |
| Remote | $<10^{-5}$ and $\geq 10^{-6}$ |
| Improbable | $<10^{-6}$ |

(e.g., relevant historical data, analytical or simulation techniques, experimental data, clinical literature data, production and post-production data). If probabilities cannot be estimated, the potential consequences should be identified for risk evaluation and risk control.

After estimating risk, each hazardous situation should be evaluated to determine whether risk reduction is required. Note that ANSI/AAMI/ISO 14971 (2016) leaves the specification of acceptable or unacceptable risk up to the manufacturer (i.e., the standard does not specify acceptable or unacceptable risk). If the manufacturer determines that risk reduction is required, the manufacturer should identify and implement risk control measures. In the example of the auto-injector, the risk associated with using the auto-injector to administer improperly mixed drug requires risk reduction. The risk can be reduced by implementing a risk control measure of redesigning the auto-injector (preferred and most effective risk control) or including warnings in the device labeling (e.g., packaging or instructions for use). It is important to note that multiple standards, including ANSI/AAMI/ISO 14971 (2016), recommend the following order for preferred and effective risk reduction:

1. A medical device should be designed to be inherently safe.
2. If inherent safety by design is not practicable, protective measures in the medical device are appropriate.
3. The least preferred approach for risk reduction is written warning or contraindication.

If any risk control measure is identified or implemented, the risk analysis process should be repeated to identify and evaluate residual risk. If the residual risk cannot be further reduced or eliminated, the manufacture may develop a risk/benefit analysis. This analysis may be based on available literature or data to determine that the medical benefits of the device outweigh the residual risk.

### 2.1 Risk analysis approaches: top-down versus bottom-up

Two high level approaches are generally used to develop a risk analysis: a "top-down" or a "bottom-up" approach. A "top-down" approach is an effect-cause model which identifies a single harm or undesired outcome and then works down to the causes of those outcomes. Conversely, a "bottom-up" approach is a cause-effect model which identifies possible failure modes and then works up to the associated harms.

Consider the auto-injector example mentioned earlier in this chapter. In a top-down approach, administering the incorrect dose would be first identified and then the process would be carried out to determine all the aspects of device use that could lead to that harm. In a bottom-up approach, the hazards associated with the viewing window would be brainstormed and then the process would be carried out to determine all the associated harms.

### 2.2 Risk analysis techniques

There are a variety of techniques that may be used for risk analysis. Annex G of ANSI/AAMI/ISO 14971 (2016) provides information on some of the available techniques, and Table 12.3 summarizes those techniques. Many of the techniques are interdependent and, if possible, should be used collaboratively to develop the most thorough use-related risk analysis.

It is important to note that risk analysis techniques are analytical processes. Although they have proven to be effective in the risk analysis process, it is important to confirm and augment the findings with empirical processes (i.e., processes that engage users). Since analytical processes do not include actual use by device users, empirical processes such as simulated-use testing ultimately may identify additional hazards and risks not identified during analytical risk analysis process (CDRH, 2000). At that point, human factors testing is important to obtain a robust risk profile of the device.

## 3. Application of use-focused risk analysis to human factors

### 3.1 Correlation of use-focused risk analysis to HFE/UE

Human factors considerations and approaches should be incorporated into the design, development and risk management processes of medical devices. Specifically, risk analysis should be conducted from a use perspective to identify and address characteristics that could affect safe and effective use of medical devices. When incorporating human factors into risk analysis, hazards are referred to as use-related hazards, hazardous situations are referred to as use errors, and risk evaluation is performed through human factors testing.

**TABLE 12.3** Summary of some risk analysis techniques.

| Technique | Description | Objectives |
|---|---|---|
| Preliminary Hazard Analysis (PHA) | Top-down approach performed in the early stages of a project as an initial step for a concept or when minimal information is available | • Identify potential hazards, hazardous situations, and events that may lead to harm<br>• Identify potential cause(s)<br>• Determine probability of occurrence<br>• Determine severity<br>• Identify risk control measures |
| Fault Tree Analysis (FTA) | Top-down approach that can be used in all design phases | • Define an undesired consequence or "top event"<br>• Analyze the causes or combinations of causes that lead to the top event<br>• Determine probability of occurrence<br>• Determine severity<br>• Identify risk control measures |
| Failure Mode and Effects Analysis (FMEA) | Bottom-up approach that can be used in all design phases | • Identify device tasks<br>• Identify potential failure mode(s)<br>• Identify harm(s)<br>• Determine probability of occurrence for each failure mode<br>• Identify severity of each harm<br>• Identify risk control measures |
| Hazard and Operability Study (HAZOP) | Bottom-up approach that can be used in all design phases with a mature design | • Identify operational parameters<br>• Identify potential hazard(s)<br>• Determine potential failures for each hazard<br>• Determine potential causes for each failure<br>• Identify risk control measures |
| Hazard Analysis and Critical Control Point (HACCP) | Systematic approach that can be used in the latter phase of design and development and particularly geared toward production control | • Conduct hazard analysis<br>• Determine critical control points (CCP)<br>• Establish critical limits<br>• Monitor each CCP<br>• Establish corrective actions<br>• Establish verification procedures<br>• Establish record-keeping and documentation procedures |

According to FDA guidance (CDRH, 2016), use-related hazards are typically related to one or more of the following conditions:

- Device use requires physical, perceptual, or cognitive abilities that exceed the abilities of the user,
- Device use is inconsistent with the user's expectation or intuition about the device operation,

- Device use is affected by the use environment and the impact is not recognized or understood by the user,
- Use environment impairs the user's physical, perceptual, or cognitive capability when using the device,
- Device is misused in ways that were not anticipated, or
- Device is misused in ways that were anticipated, but risk control measures were not applied.

FDA guidance (CDRH, 2000) states that because of the many different combinations of ways and conditions a user can interact with a medical device, use-related hazards can be difficult to anticipate or identify. As a guide, Annex E of ANSI/AAMI/ISO 14971 (2016) provides examples of potential sequence of events which can lead to potential use errors. Additionally, the following steps are defined in the FDA guidance (CDRH, 2016) as essential to ensure use-focused risks are addressed during the product life-cycle:

- Identify known and foreseeable use-related hazards,
- Determine use errors that could lead to the use-related hazards,
- Develop and implement measures to eliminate or reduce use-related hazards, and
- Demonstrate safe and effective use through human factors validation testing.

The international standard for application of usability engineering to medical devices, IEC 62366-1:2015, specifies ANSI/AAMI/ISO 14971 as the approach for a comprehensive risk management process, and the usability engineering process is similar to the risk management process. IEC 62366-1:2015 states that risk management is a decision-making process for determining acceptable risk whereas usability engineering is a design and development process for the user interface to reduce the possibility of use errors that could result in risks associated with usability.

It is important to note that risk estimation is not always accurate for use-related hazards. Many use-related hazards or use errors cannot be anticipated prior to simulated-use testing or actual use. Also, simulated-use testing may uncover certain use-related hazards that are a low probability during actual use (CDRH, 2000). Further, depending on the manufacturer's approach for estimating risk, a use-related hazard with high severity may be identified as a low risk because the probability of occurrence is low. Therefore, FDA guidance (CDRH, 2016) states that severity of harm is more meaningful than the probability of occurrence for determining risk reduction and categorizing critical tasks, especially since users tend to be less able to appropriately react to use-related hazards or use errors that have a low probability of occurrence.

## 3.2 Tracing use-focused risk analysis to task analysis

FDA guidance (CDRH, 2016) indicates the task analysis is used to systematically break down the user interactions with a device into discrete sequences of tasks. The discrete tasks are then used to help identify user interface components involved, potential use errors, and potential harms for all use errors. FDA guidance (CDER, 2016) indicates the use-focused risk analysis should be used to help identify critical tasks that should be evaluated in a HF study as well as to help determine the specific tasks and use scenarios to include in testing (see Chapter 6).

TABLE 12.4 Example of severity levels and corresponding task categorization.

| Category | Definition | Example task categorization |
|---|---|---|
| Catastrophic | Results in patient death | Critical |
| Critical | Results in permanent impairment or life-threatening injury | Critical |
| Serious | Results in injury or impairment requiring professional medical intervention | Critical |
| Minor | Results in a temporary injury or impairment not requiring professional medical intervention | Essential |
| Negligible | Inconvenience or temporary discomfort | Essential |

Both the risk and task analyses play an important role in HFE/UE. Although risk and task analyses are separate analyses, they can be interdependent. If the risk analysis is developed first, it can be referenced to ensure all tasks and use errors are included in the task analysis. Conversely, if the task analysis is performed prior to the risk analysis, the task analysis can be leveraged to ensure all user-device interactions are included in the risk analysis. Thus, the risk and task analyses can be leveraged against each other to identify gaps in either analysis. They also should be iterated together to ensure information is consistent between the analyses.

The risk analysis also informs the task analysis and vice versa. A key component of the task analysis is the identification of critical tasks for human factors testing. FDA guidance (CDRH, 2016) defines a critical task as a user task which, if performed incorrectly or not performed at all, would or could cause serious harm to the patient or user, where harm is defined to include compromised medical care. Note that the critical task definition in the FDA draft guidance (CDER, 2016) excludes "serious" and only specifies harm. The severity of harm determined in the risk analysis is a key input for categorizing tasks in the task analysis. If a use-related hazard has a severity level that results in serious harm to the user per the CDRH guidance (2016) or harm per the CDER draft guidance (2016), the corresponding task in the task analysis will be designated as critical. The severity level example provided in Table 12.1 above has been modified to provide an example of how severity level and task categorization correspond to each other (see Table 12.4).

Much like risk analysis itself is a cyclical/iterative process, the interaction between risk analysis and task analysis is also a cyclical/iterative process as the user interface changes throughout the product life cycle with the end result being a task analysis document that captures and categorizes all potential use-related hazards and risks appropriately.

## 4. Summary

Inputs for a use-focused risk analysis require a thorough understanding of how the device could be used and misused by intended users. It is also important to consider the device could be used in situations other than those intended or anticipated by the manufacturer. If similar devices are on the market, available and relevant information from those devices can be leveraged to determine known use-related hazards. For foreseeable hazards, simulated-use testing with prototypes and intended users can provide insight.

Once the known and foreseeable hazards and subsequent hazardous situations are identified, it may be difficult to identify or anticipate the probability of occurrence of harm. This can be the case early in the design process as well as throughout the design and development process. In such cases, the consequence or severity of the harm is the critical element for manufacturers to focus on when applying risk controls for use-related hazards in the stages of the risk management process to reduce or eliminate the risk. It is then important to confirm and augment the findings in the use-focused risk analysis with human factors testing to obtain a robust risk profile of the device.

## Acknowledgments

Thank you to Deb Billings Broky for contributions to the Introduction, Terms and Definitions, and Risk Analysis Approaches. Thank you also to Leah Taylor for contributions to the Introduction and toward correlating use-focused risk and task analyses.

## References

ANSI/AAMI/ISO 14971:2007/(R). (2016). *Medical devices— Application of risk management to medical devices.*

Casey, S. M. (1998). *Set Phasers on Stun: And other true tales of design, technology, and human error.* Santa Barbara: Aegean.

IEC 62366-1. (2015). *Medical Device — Part 1: Application of usability engineering to medical devices.*

U.S. Department of Health and Human Services Food and Drug Administration Center for Devices and Radiological Health. (2000). *Medical device use-safety: Incorporating human factors engineering into risk management.*

U.S. Department of Health and Human Services Food and Drug Administration Center for Devices and Radiological Health. (2016). *Applying human factors and usability engineering to medical devices guidance for industry and food and drug administration staff.*

U.S. Department of Health and Human Services Food and Drug Administration Center for Drug Evaluation and Research Center for Biologics Evaluation and Research. (2016). *Human factors studies and related clinical study consideration in combination product design and development: Draft guidance for industry and FDA Staff.*

CHAPTER

# 13

# Root cause analysis

*Tim Reeves, Christina Mendat, Eric Shaver*
**Human Factors MD, LLC., Charlotte, NC, United States**

OUTLINE



*Rather than being the main instigators of an accident, operators tend to be inheritors of system defects created by poor design …* ***James Reason***


https://doi.org/10.1016/B978-0-12-816163-0.00013-X

## 1. Introduction

A hospital patient received a 10-fold overdose of fentanyl when a nurse mis-programmed an infusion. The medication order called for an infusion rate of 4.5 mL/h; the nurse entered 44.5 mL/h[1]. What caused this error and how can we prevent the error from happening again? Our immediate response might be to lay blame at the foot of the user. Perhaps the nurse was tired? Or busy multi-tasking? Or in a hurry to end her shift? After all, not only was she careless in entering the rate, but she failed to check her work. To prevent the error from happening again, we might choose to reprimand or suspend the nurse, or require that she submit to remedial training. But these "mitigations" would not have been effective. This example was one of dozens of reported events involving key bounce – a problem linked to the design of the infusion pump keypad. Key bounce (or double bounce) occurred when the number on the pump was pressed once by the user but registered twice by the device (Fig. 13.1).

The key bounce example offers some important lessons. First, often the immediate response to errors is to blame the user, especially when errors seem easily avoidable in hindsight. Second, the obvious cause of an error may mask an underlying problem, such

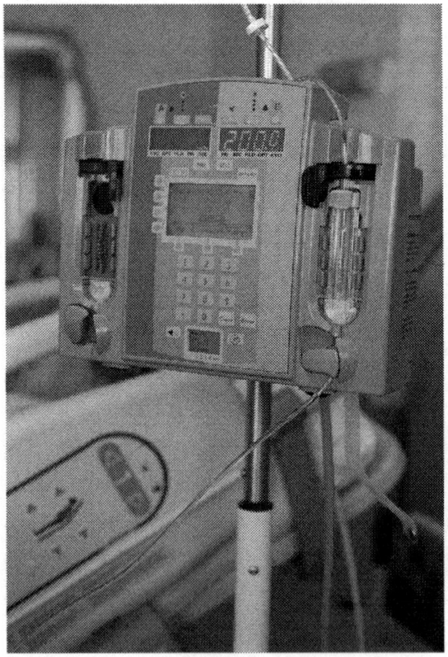

**FIG. 13.1** Example infusion pump with membrane interface.

[1]FDA Manufacturer and User Facility Device Experience Database (MAUDE) MDR report key 1201808, reported 22 September 2008.

as flaws with the design of the device. Third, designers cannot hope to correct errors and prevent them from happening in future if the wrong causes are identified. And fourth, identifying the **real** causes of error requires an approach that investigates beyond immediate and obvious suspects.

To make medical devices safer and to optimize their designs, manufacturers need to understand when users are likely to have difficulties using devices and the factors that contribute to or cause errors. This isn't just good practice. Regulators require that manufacturers conduct root cause analyses of use issues that arise during validation testing in order to determine the residual risk and priorities for implementing additional risk management measures (Food and Drug Administration, 2016). This is not solely in retrospect, after errors happen. For example, to succeed in developing safe products, manufacturers need to adopt a prospective approach to identifying potential errors and their causes early in the design process. Evaluative techniques—such as task analysis (Chapter 6), use failure mode and effects analysis (Chapter 12), fault tree analysis (Chapter 12), and expert and heuristic reviews (Chapter 10) — are examples of prospective tools for predicting when errors might happen. Later in the development process, when testing design prototypes in formative studies (Chapter 11) or commercially representative products in validation tests (Chapter 15), manufacturers adopt a retrospective approach to identify the factors that caused or contributed to errors that did happen. Additionally, following release of a product, manufacturers review complaints and other post-market data to see if use errors motivated the complaints, and if so, what can be done to address those errors.

Root Cause Analysis is a collective term describing a range of tools and techniques for discovering the causes of errors. These approaches have been used effectively in other industries, primarily as a means of understanding how accidents happen. In healthcare, the Joint Commission (Parker, 2015) as well as organizations such as the Institute for Safe Medication Practices (ISMP, n.d.) recommend Root Cause Analysis for investigation of patient safety events, including those involving medication errors. But as discussed above, medical device manufacturers need to understand the causes of errors in order to design safer devices, and Root Cause Analysis is the primary method for doing so.

This chapter outlines a framework for understanding the causes of errors people make when using a medical device. It begins by exploring differing views of error, including a discussion of differing taxonomies that have been proposed for characterizing the causes of use errors. It describes the objective of Root Cause Analysis and reviews methods for conducting Root Cause Analysis that are found helpful in practice. It describes how these methods can be applied to understanding the causes of errors with medical devices. Finally, this chapter discusses pitfalls in conducting Root Cause Analysis with examples of good and bad descriptions of root causes.

## 2. Why does use error happen?

### 2.1 The system view of use error

The above example of blaming a nurse (user) for an infusion error represents what Sydney Dekker has labeled the **Old View** of human error (Dekker, 2014). The Old View of human error assumes systems as inherently safe. Errors arise from aberrant human behavior. People can be forgetful, inattentive, poorly motivated and careless and when they are, bad things happen. From this perspective, our nurse was inattentive, careless and, as a result, negligent.

Proponents of the **New View** assume systems are not inherently safe. Systems must be designed to be safe. The New View recognizes that human factors are at play – that humans have limitations – and that systems must be designed to accommodate human factors to ensure safety. The New View accepts that people are "locally rational" – that people do what makes sense to them at the time, given their goals, attentional focus and knowledge (Woods & Cook, 2000). When errors arise, we need to understand what made sense to the person, at the time, even though it may have contributed to the error.

**New versus old views of human error**

The New View holds that human errors are a consequence and not a cause; a symptom of an underlying problem with a system's design.

The New View of human error informs our use of the term "use error" rather than "human," "operator" or "user" error. Coined by William Hayman (1995, pp. 1–2), use error provides a more descriptive term capturing what happened — there was a problem with use of the device — that is neutral about the problem's cause.

Systems in medicine can be defined broadly or more narrowly. A healthcare "system" includes patients, insurers and providers. A hospital system includes management, policies, staff and equipment. In the context of understanding the causes of use errors with medical devices, it's helpful to define a system as a device, its users and use environments.

## 2.2 Understanding the causes of use errors

There are several areas to consider when understanding the causes of use error, including error taxonomies and a mismatch between the user and the device design. Each of these are discussed below.

### 2.2.1 *Use error taxonomies*

Several frameworks for classifying errors have been developed. Error categories or taxonomies can shed light on the etiology, or different causes of use errors. These are useful when assessing causality.

### 2.2.2 *Errors of commission and omission*

Perhaps the simplest taxonomy differentiates between errors of commission and errors of omission. An error of commission arises when a user does something he or she should not have done, such as choosing the wrong medication or entering an incorrect flow rate on an oxygen valve. Errors of omission arise when a user fails to do something he or she should have done, such as not holding an auto-injector in place long enough to deliver all of a medication. Unfortunately, this simple commission/omission taxonomy does little to further an analysis of root causes: the taxonomy is helpful in describing what happened, but is silent on the question of why it happened.

### 2.2.3 Mistakes, slips, and lapses

A more helpful route to the "why" of use errors is to ask whether the outcome of the action (or non-action) was intended by the user (Norman, 2013; Reason, 1990). If an action was intended, but the action is inappropriate, the use error was a *mistake*. Mistakes are deliberate but incorrect actions (or non-actions).

If the action was unintended and inappropriate, the use error is a *slip*. The action was unintended but "slipped" out. A common mechanism for a slip is *capture*, in which a more frequently performed behavior "captures" or "takes over" a similar, but less familiar one. For example, a *capture error* is made when a nurse is programming a new infusion pump and, because the exact sequence of key presses is similar but not identical to the pump he is most familiar with, he enters the familiar sequence and incorrectly programs the pump. *Description errors* are slips that occur when the objects of different actions are close together or visually similar, as when the wrong control on an EKG is adjusted because it's close to other controls in look and feel – that is, a user may "describe" the control as the same.

*Lapses* refer to errors that involve the unintended omission of an action that was appropriate but not performed (typically due to forgetting). *Loss-of-activation errors* are lapses where the user's goal is forgotten in the middle of a sequence of actions (e.g. a radiologist forgetting what he is looking for after retrieving and displaying a comparison study) or a user omits a step in a routine sequence (e.g. failing to remove the inner cover of a pen needle before attempting to inject).

The Mistakes, Slips and Lapses taxonomy can be helpful because different kinds of errors require different kinds of mitigations (Wickens, Lee, Liu, & Gordon-Becker, 2003). Many mistakes reflect a lack of knowledge that is best addressed by providing more information to the user via improved user interface displays, improved labeling, and training. Slips often result from user interface options (e.g. controls) that are too similar in appearance or too close physically. Lapses can be addressed through practice and training or by providing user interface prompts or visual and auditory signals that serve as retrieval cues and reminders.

### 2.2.4 Perception, cognition and action errors

In interacting with devices, users first perceive information (perception), process that information (cognition) and act on that information (action) (Card, Newell, & Moran, 1983). *Perception errors* arise when users have difficulties perceiving – when they are unable to see, hear or feel information provided by a device; e.g., difficulties reading text on a display or failing to hear audio information. Alternatively, users may perceive information accurately but make errors in processing and assimilating that information. *Cognition errors* include forgetting, misinterpreting, miscalculating, overgeneralizing, drawing incorrect inferences and making incorrect decisions (e.g. forgetting to remove the safety release on an auto-injector). Further users may perceive and process information correctly but have difficulties in the execution of actions. *Action errors* include difficulties or failures when lifting, pinching, twisting, turning, pushing or otherwise acting on a device.

Categorizing errors as errors of perception, cognition or action can aid in narrowing the search for the causes of use errors. This is the method described in Chapter 6. However, use errors are products of a system that includes the device, the users and the use environment/s. Analysts need to explore whether the design of the device led or contributed to the user's perceptual, cognitive or action errors. Knowing that a user misread the value displayed on a screen (a perception error) helps in focusing the investigation. Nonetheless, it begs the question: Why was the value misread? Was the font large enough, given the viewing distance? Was the contrast high enough, given ambient lighting? Was glare from the display a factor?

### 2.2.5 Use errors reflect a mismatch between users and device design

The most useful approach to thinking about the causes of use errors is to adopt a system view which holds that use errors arise because of a mismatch between the components of the system (user, device, environment) or more specifically between the design of the device and its users. This means either:

**a.** A mismatch between the requirements imposed by the design of the device and the capabilities of the user. In other words, between what users *must* due to use the device safely and effectively, given the design of the device, and what users *can* do; or

**b.** A mismatch between how the device *actually* works and how users *expect* the device to work (the users' mental model of device operation).

Use errors can arise when the use requirements imposed by the design of the device (what users must do) exceed a user's perceptual, cognitive or physical (action) capabilities (what users can do). This is an example of a mismatch between the requirements for use and the user. In turn, use errors can arise when the user, based on their prior experience with the device or similar devices (or lack of experience), expects the device to work differently than it does. This is an example of a mismatch between what a user expects and how a device operates.

The third component of the system, the use environment, can contribute to use errors by either altering a user's capabilities (what a user can do) or by altering how the device behaves in ways that may not be obvious to the user. Examples of *environmental factors influencing the user* include high levels of ambient noise limiting a user's ability to distinguish sounds played at lower volumes; fatigue influencing a user's judgment and decision-making; and nitrile gloves limiting a user's ability to grip and act on a device. In each of these examples, the device design doesn't accommodate differences in user capabilities across variations in use environments.

A simple oxygen gauge offers an example of *environmental factors influencing a device*. Some oxygen gauges may become inaccurate as ambient temperatures drop and the pressure inside the tank decreases. As a result, a paramedic on a 911 call during a cold January evening in North Dakota may assume a tank contains more oxygen that it does if she is not aware of the effect of temperature on the gauge's performance.

The goal of Root Cause Analysis of a use error is to locate the source of the mismatch between the user and the device that led to the error. Knowing whether a user's difficulty was perception, cognition or action related can help direct the investigation into which aspect of the design is at odds with the user. As examples, an error reading an SPO2 level points to potential issues in the size, placement or presentation style of the displayed value; forgetting to remove the safety on an auto-injector implicates a lack of visual cues or reminders; difficulties attaching a connector suggest that there may be issues with the size, shape or torque required to make a secure connection.

Errors in cognition may point to inconsistencies in the placement or behavior of controls and display elements; missing, ambiguous or misleading feedback from the device; missing or false design affordances (Norman, 1990); missing or inconspicuous placement of information; a lack of visual prompts or reminders; the use of unfamiliar terms or symbols; poor error and troubleshooting messages; and a lack of visibility of the device's state. All of these design issues can contribute to a mismatch between how a user thinks a device operates and how it actual does operate.

To address a mismatch between the design of the device and its users, designers have two choices: they may either change the device (how it operates and/or interacts with users) or change the user (through labeling or training; or limiting use to those with the requisite qualifications such as limiting use to healthcare professionals only).

## 3. Root cause analysis methods

Earlier, Root Cause Analysis was defined as a collective term for a range of tools and techniques in order to discover the causes of use errors. In this section, three additional methods are discussed which have been found to be helpful. The methods include: Sakichi Toyoda's Five Whys; Kaye, North, and Peterson's (2003) UPCARE Model and Wiklund, Dwyer, and Davis's (2015) Structured Approach.

### 3.1 The five whys

Developed by Toyota founder Sakichi Toyota, the Five Whys is an iterative approach intended to extend the analysis of root causes beyond the obvious or top-of-mind candidates (Ohno, 1988). The strength of the Five Whys' method is its simplicity. As an example, to carry out a Five Whys' analysis, analysts begin with an outcome—in this case, a diabetic patient experienced hypoglycemia—and simply start by asking, "Why?"

The patient experienced hypoglycemia.

1. **Why?** Because of an over delivery of insulin.
2. **Why?** Because the patient received an unwanted bolus of insulin, while replacing the insulin cartridge (see Fig. 13.2).

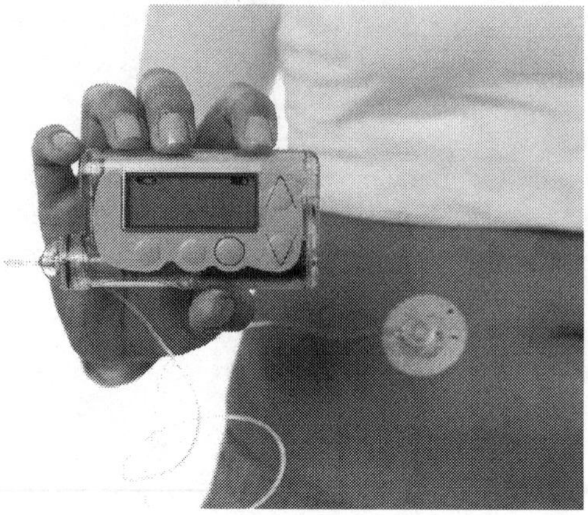

FIG. 13.2 Insulin pumps push insulin from a refillable cartridge in the pump through tubing into a diabetic patient's body. To avoid unwanted insulin delivery, users must remove the tubing from their body prior to refilling the cartridge.

3. **Why?** Because the patient left the tubing attached to their body while replacing the insulin cartridge and insulin in the tubing was accidently pushed out and into her body.
4. **Why?** Because the patient forgot to detach the tubing from her body before replacing the cartridge.
5. **Why?** Because the pump user interface does not require the user to remove the tubing before replacing the cartridge and there is nothing in the user interface of the pump to remind users to do so.

The answer to the first why is: The patient's sugar levels went dangerously low because of an overdose of insulin. But it is incomplete, which leads to further questioning. By asking five whys (more or less), the analyst can eventually dig down to reach the root cause. These questions should be directed at understanding what aspects of the user interface design of the device, if any, contributed to the use error.

The simplicity of the Five Whys is also its Achilles heel. The method assumes there is a single causal factor or linear chain of events. However, errors often have multiple causes or several contributing tangential factors rather than a single causal chain leading to the error. Nevertheless, the spirit of the Five Whys embodies an important move beyond the obvious or immediate cause, so that by asking probing questions, a root cause can be identified.

## 3.2 The UPCARE model

We've outlined how users perceive, process and act on information when interacting with a medical device along with how difficulties or failures in perception, cognition or action can contribute to use errors in the discussion of the Perception, Cognition and Action Error taxonomy. The Perception, Cognition and Action model of human performance (Card et al., 1983) provides the theoretical basis for Kaye, North and Peterson's UPCARE model. This is a tool designed to help analysts describe and analyze medical device use problems.

The PCA in the acronym U**PCA**RE represents perception, cognition and action issues. The U stands for problems arising from Unmet user needs (user requirements that were not considered or satisfied by the user interface design) and the R stands for Results. The results are the impact of the problems on users and patients. In this model, the E represents human-factors-oriented Evaluation methods to further analyze the problem.

The UPCARE model is a useful tool for exploring potential gaps between what a device does versus what it needs to do, as well as what the device requires for safe use versus what users are capable of while in use environments. Kaye et al. (2003) generated the model following interviews with healthcare professionals (mostly nurses) and others about the nature of the problems these users most often encountered when interacting with medical devices. Their insight was to note that many of the use issues that users reported could be linked to either unmet needs (what the user needed and expected, but the device didn't provide) or perception, cognition or action issues that stemmed from limitations in the device's user interface design.

Unmet user needs may include aspects of the setup, configuration, use, monitoring and repair of devices, during which the devices fail to provide the necessary information, functions or the support needed by users to ensure safe and effective use. Kaye et al.'s interview participants reported that some devices failed to provide a clear, unambiguous indication of whether the device was working properly or whether troubleshooting actions were effective. Other devices failed to notify the user of a critical change in the patient's condition or failed to provide support in troubleshooting problems with device operation. Still other devices lacked needed features, like an emergency stop or the ability to undo changes to device settings or therapy parameters. Finally, some devices left users confused about how to interpret the clinical meaning of the device's output (such as a diagnostic test result).

Perception, cognition and action issues reported by interview participants reflected problems participants had interacting with devices. These included issues perceiving information (perception), processing and assimilating information (cognition) or carrying out physical actions in response to information (action).

Perception issues included: issues seeing displays because the displays were not bright enough or prone to glare and issues hearing alarms because alarm volumes were too low or their frequency too high.

Cognitive issues included: confusing menu structures, overuse of abbreviations and non-standard names, overly dense displays, confusing feedback, excessive alarms and inadequate instructions and training.

Action issues included: challenges assembling and connecting components, keystroke errors and the degradation of devices following repeated use.

Based on professional experience, the UPCARE model provides a useful framework for directing the Root Cause Analysis. Knowing that a use error is perception, cognition or action-related helps focus the investigation into which aspects of the device user interface may be causing or contributing to the error.

### 3.3 Structured approach

In the 2015 book by Wiklund, Dwyer, and Davis titled Medical Device Use Error: Root Cause Analysis, the authors describe a structured, seven-step framework for conducting Root Cause analysis with medical devices. The framework assumes that user interface design flaws are a key contributor to use errors, and that the goal of Root Cause Analysis is to identify these flaws. The framework consists of these seven steps:

1. **Define the use error.** Analysts should be clear on what use error occurred and on the sequence of events that led up to the error. In describing the error and event sequence, it's important to reference all available data, including observations of video, session notes and interview transcripts.
2. **Identify provisional root causes.** Provisional root causes are the top-of-mind or obvious root causes. It is helpful to document these, as long as the analysis attempts to challenge these preliminary hypotheses and consider the possible role of other factors.
3. **Analyze anecdotal evidence.** A potentially useful source of information includes an account by the user who committed an error. Participants in usability studies should be interviewed to understand what they recall thinking or doing at the time the use error arose (see Box: Interviewing Study Participants About Observed Use Errors). A user's recreation of the use error may shed light on the underlying cause.
4. **Inspect the device for user interface design flaws.** As outlined above, use errors are symptomatic of a mismatch between the user's expectations, capabilities or predispositions and the design of the device. To understand the reason for the gap, analysts can compare the design of the device to established principles of good design, including design heuristics (Zhang, Patel, Johnson, Chung, & Turley, 2005), best practices and design guidelines (HE:75, 2009).
5. **Consider other contributing factors.** Analysts should consider how other factors may contribute to or suborn use issues. Contributing factors include *user characteristics* such as impairments, habits and prior experiences; *environmental factors* include lighting and ambient noise levels; and possible *test artifacts*, such as unclear instruction to the study participant.
6. **Develop a final hypothesis.** While the preceding analysis may have identified several root causes, it's important to "weed out" those that inappropriately blame the user and are supported by the evidence gathered in earlier steps in the analysis. Analysts should use their professional judgment to arrive at the best possible explanation for the use error.
7. **Report the results.** Reports may be terse or expansive depending upon audience. The example (See Box) describes one style of reporting root causes. Regulators may prefer a tabular format.[2]

[2]In their 2016 Guidance, the FDA describes a tabular presentation of use errors (task failures) that summarizes possible root causes and related candidate risk controls. While the table provides a quick summary, analysts may want to include a more detailed review of the Root Cause Analyses for their product elsewhere in their Human Factors Engineering Report.

## Reporting on the results of root cause analysis

When reporting root causes, be sure to provide a succinct description of the use error and the potential for harm as background for the reader. The example below (Table 13.1) illustrates a terse but complete description of the Root Cause Analysis of a use error that arose when connecting lines during a validation study.

**TABLE 13.1** Reporting root causes example.

| | |
|---|---|
| Use error | Over-twisting line connector |
| Critical task | CT 1 – Connect components |
| Risk identifier | R3 (User over-tightens) |
| Occurrences | P4, P7 |
| Description | Both participant 4 and 7 over-tightened the connector, breaking it. |
| Participant comments | P4: "It didn't feel secure, so I kept twisting."<br>P7: "I guess I expected to feel a click or snap or something. I didn't hear anything, so I kept twisting." |
| Potential harm | Delay in therapy while connector replaced. |
| Root cause | Absence of feedback — the connector does not indicate when the connection has been established.<br>Weak materials — the connector materials are not sufficiently strong to accommodate likely twisting forces. |

## 3.4 Combining methods

An effective approach to Root Cause Analysis merges aspects of the Five Whys, the UPCARE model and Wiklund et al.'s (2015) Structured Approach. The Structured Approach offers a systematic, repeatable formula for completing Root Cause Analysis. The UPCARE model's delineation of problems by unmet need, perception, cognition or action errors helps focus the analysis in steps 2 through 5 of the Structured Approach. The Five Whys emphasize the need to look beyond the immediate, obvious assessment of root cause by posing follow-up questions directed at understanding which aspects of the user interface design of the device contributed to the observed use error.

In identifying potential root causes (Steps 2 through 5 of Wiklund et al.'s Structured Approach), analysts can explore possible root causes by asking questions aimed at uncovering aspects of the design of the device user interface that may be contributing to the use issue. A non-exhaustive listing exemplifying the kinds of questions that can focus the analysis can be found below. The questions are sorted by perception, cognition and action-related issues.

**Questions to Pose in Identifying Which Aspects of the User Interface Design May Be Contributing to Use Problems:**

**Perception Errors**

- Does the intensity and pitch of auditory signals allow them to be easily heard across all use environments?
- Is the size, contrast and color of labels and displays visible?
- Is all relevant information clearly labeled, visually salient and easy to find?
- Does the display minimize in bright lighting conditions?
- Is the volume of the auditory signal sufficiently loud given the use environment?
- Is the frequency of the auditory signal well within the user's hearing range?
- Is the auditory feedback "unique" relative to other sounds in the use environment?
- Are important visual indications conspicuous? Are they in the optimal location? Are they an appropriate size?
- Is the information readable given the location/angle of the reader?
- Do all interactions result in immediate visual, auditory or tactile feedback?
- Are warnings conspicuous?
- Are instructions easy to locate inside the product's packaging?
- Are alerts and alarms both visual and auditory?

**Cognition Errors**

- Does the device user interface (including the labeling) use familiar terms, abbreviations and/or symbols that are consistent and unambiguous?
- Does the device limit the steps users are required to remember? Does the user interface provide helpful visual, auditory, or tactile prompts or reminders?
- Are operational steps logical, related and memorable?
- Are controls labeled? Are labels positioned consistently relative to the controls?
- Are units of measure referenced and displayed consistently?
- Are controls organized (segmented and grouped) in a way that makes identification easy and error-free?
- Is the design consistent with established conventions?
- Is necessary information organized by a visual hierarchy that guides information finding?
- Does the organization of data and controls on the display have an underlying logic or follow a meaningful workflow?
- Are graphics and illustrations used to reinforce or enhance communication?
- Does the device operate in the same way as other devices the user is familiar with?
- Does the device prompt the user to confirm critical data entry or actions?
- Do warnings describe the potential consequences of the use error?
- Does the device operation mirror the users' typical workflow?

**Action Errors**

- Are controls sufficiently spaced?
- Are controls easily accessible given the user's positioning and posture?
- Is the mapping between the movement of controls (up, down, left, right) consistent with the device response?

- Are controls appropriately sized and shaped?
- Are handles appropriately sized and shaped?
- Are safeguards in place to limit inadvertent activation of controls?
- Have pinch points been eliminated?
- Are the force requirements to manipulate controls within the range of expected users?
- Are connectors unique and identifiable?

## 4. Common pitfalls in root cause analysis

Identifying the root causes of use errors can be challenging. Despite practicing Root Cause Analysis for years, one can expect to encounter situations where the underlying causes of use errors are difficult to identify. However, there are common pitfalls in RCA you can avoid. The list of pitfalls include:

### 4.1 Not conducting an analysis at all

It has been noted anecdotally that regulators lament that some sponsors simply fail to provide an analysis of root cause for issues identified during validation testing. Instead, these sponsors reported the use errors that arose, assessed the risk to be acceptable and claimed that their device could be used safely and effectively without further mitigations; all without exploring the causes of the use errors.

It's important to recognize that manufacturers are now required to reduce the risk associated with use-related hazards to as low as possible. It simply isn't enough to claim that a risk is acceptable without an analysis of whether the risk can be reduced further —an analysis that is impossible to do without exploring the underlying causes of the use issue.

### 4.2 Not incorporating the user's perspective

When conducting Root Cause Analyses, it can be helpful to know what the user was attending to at the time of the use error. Knowing where a user's attention was directed and what he or she recalls thinking about when the error was made *may* help narrow the search for underlying causes. This view is shared by regulators as well. The international usability engineering standard, ANSI/AAMI/IEC 62366-1:2015, encourages analysts to determine the root cause of use issues based on user performance and subjective comments from users. The FDA Guidance on human factors goes further, recommending that analysts always discuss use errors with test participants to understand what happened from the participant's perspective. From the FDA's vantage point, "*the test participant's perspective on use errors can provide helpful insights and reasons for the use error and sometimes includes suggestions for design improvements.*" (FDA, 2016, p. 37).

Unfortunately, the FDA's position is somewhat at odds with research from social and cognitive psychology, which offers many reasons to be suspect about the explanations people offer for their own behaviors (Nisbett & Wilson, 1977). Much of cognition happens below the level of conscious awareness leaving people unaware of the factors affecting their behavior. However, manufacturers may be penalized by regulators if they fail to consider the user's perspective on the causes of use errors. As a result, we suggest analysts explore use issue/s with test participants, but focus the analysis of participants' comments on what they recall in relation to the use error (the content of their thoughts) and treat with skepticism participants' explanation for why they made an error.

**Interviewing study participants about observed use errors**

When exploring a use error with the participant in a usability study, interviewers should focus their questions on what the test participant recalls attending to and thinking about when the use error arose. Our list of best practices for root cause-oriented interviews include:

1. Establish whether the participant recognizes that they made an error rather than direct the participant to the error. One non-leading approach is to ask the participant "How was that?" and "If you did the scenario again, is there anything you would do differently?"
2. Establish whether the participant understands how the task should be performed. "Walk me through the steps required to complete the task."
3. Establish whether the user understands the risk associated with the use error. "So if that happened in real life, what could happen? Is there anything you would be concerned about?"
4. Explore what the participant recalls attending to at the time. What do they recall perceiving, thinking about and doing?

## 4.3 Restating the use error

Avoid root causes that simply restate a use error:

- "The test participant had a wet injection because he didn't hold the injection device in place long enough to deliver all of the medication."
- "The user did not receive a full dose of insulin because she entered the wrong bolus on her pump."

For the wet injection scenario, not holding the device in place is the use error and not a root cause of the error. The same is true for the mis-programmed bolus scenario. Both "root causes" simply describe the use error and offer nothing about its cause.

## 4.4 Blaming the user

In the discussion of Why Does Use Error Happen, the Old and New Views of human error were compared. Recall that the Old View holds that errors arise from aberrant human behavior: that users can be forgetful, inattentive, poorly motivated and careless, and when they are, use errors happen. It is especially tempting to blame a use error on a test participant if all other participants in the study performed the same task(s) without an issue. Surely, if all other participants got it right, the cause of the error was this participant? No.

If manufacturers embrace the New View of human error and assume that all errors, however infrequent, may be symptoms of an underlying design issue, our Root Cause Analyses should be directed toward identifying whether there is a design deficiency, or some kind of mismatch between the user and the design of the device.

In actual use, people may not attend to all of the things they should. They may be distracted; they may be nervous and feel rushed; they may overgeneralize from past experience; they may forget things; they may misinterpret things; and they may not read instructions. However, these "human factors" are known at the outset of every design project. Well-designed products accommodate for these and other limitations and predispositions.

### 4.5 Blaming test artifacts

Most formative and summative (validation) tests conducted in support of the development of a medical device are simulated-use studies. Test participants are asked to simulate use of a device in an environment created to mimic its actual use. By their nature, simulated-use tests provide only an "approximation" of actual use. As a consequence, some people may perform differently in simulated-use than they would in actual use. One objective of Root Cause Analysis, when conducted to explore a use issue observed under simulated use, is to determine whether the observed issue is a legitimate use issue. Would the issue arise in actual use? Or is the issue an artifact of the simulation itself?

Test artifacts do happen. For example, in simulated-use tests of drug injection devices, it's common for some test participants not to check the product's expiry date or examine the appearance of the liquid medication prior to simulating injection into an injection pad. Some or all of these errors may reflect legitimate study artifacts, as these test participants may decide not to conduct these checks because the drug (typically a placebo) is injected into an injection pad and not a real person. Moreover, a participant may feel that a check is unnecessary: Why would the researcher have them use an expired or spoiled product?

However, problems arise when analysts dismiss legitimate errors as test artifacts. For example, participants may not check expiration dates or the appearance of a medication for other, legitimate reasons. Test participants may not understand that these checks are necessary. Before attributing an error to a study artifact, it's important to establish that the participants who made the error understand how the task should be performed and what hazards they could be exposed to if they failed to perform that task or used it incorrectly in actual use.

Problems also arise when tests have numerous study artifacts. This can be a sign that the test was poorly designed and make it difficult to generalize from simulated to actual use conditions.

## 5. Examples of root causes

Table 13.2 below offers examples of Root Cause Analysis. This illustrates both "weak" and "better" RCAs with examples drawn from issues with injection devices.

**TABLE 13.2** Root Cause Analysis examples.

| Use error | User comment | Type | Weak | Considerations | Better |
|---|---|---|---|---|---|
| User forgets to remove the inner needle cover before injecting. | I should have removed it. I just forgot. | Cognition | The user forgot to remove the inner cover. (Restates the use error rather than providing a cause.) | Are two needle covers for one needle consistent with user's expectations? Is the user being asked to remember a lengthy sequence of steps? Are the steps meaningful (and thus memorable)? | Users may expect a needle to have a single cover only. As a result, they assume the needle is ready to inject after removing the (first) outer cover. |
| User does not attach the needle securely. | I thought I heard a click. I guess I didn't twist hard enough. | Perception | The user mistakenly thought they heard a click when attaching the needle. (Blames the user.) | Does the device provide visual, auditory or tactile feedback in response to user actions? Does the device clearly indicate its state? | The device fails to provide clear feedback to indicate when the needle is securely attached. |
| The user did not prime the new pen before first use. | With my current pen, I just get a new one, dial up the dose, and inject. | Cognition | Habit. (Incomplete; Focuses on the user, not the device.) | How do other pens the user is familiar with work? Is the user's mental model of the device consistent with how the device operates? | Given prior experience with a similar device, the participant did not expect that the study pen required priming. |
| The user pulled the needle out of the skin while attempting to inject, creating a wet injection. | It was my first time. The plunger is a bit hard to push. I was trying to hold the needle steady, you know, not going in too far. I guess I didn't notice that the needle was coming out. | Action | Lack of practice. (Misses limitation in design of the device given the drugs viscosity.) | What force is required to expel the medication? Does user performance improve with repetition? | The drug's viscosity requires that some users engage the larger muscles in their forearm to push the plunger, making it more difficult to keep the syringe in a stable position. |
| User does not know the pen should be refrigerated between uses. | I wasn't sure where you should keep it. There isn't any information in the instructions about it, so I assume I can keep it in my medicine cabinet. | Cognition | The user was unable to locate the information in the instructions. (Blames the user; misses important design flaw.) | Where did the user look for the instructions? Which section? Is the information organized in a way that makes sense to the user? Does the visual presentation (salience of the information) stand out? | The user expected information on where to keep his/her pen would be placed in the section of "Storage and Disposal Information" rather than the section on Warnings and Precautions. |

## 6. Summary

To ensure the safe and effective use of medical devices, designers need to anticipate where users of their devices may have difficulties; and, should difficulties arise, they need to understand the factors that may have caused or contributed to those difficulties. Root Cause Analysis is a tool for evaluating the causes of use issues. The most effective approach to RCA is to treat use issues as symptomatic of underlying design issues that arise from a mismatch between what, given the design of the device, users are required to do versus what they can do, or from a mismatch between how users expect a device to operate and how it does operate. Rather than view user limitations as the cause of use errors, analysts should assume that use errors arise because the perceptual, cognitive and action requirements for safe use of the device exceed what some users are capable of or that the device operates in a way that is inconsistent with some users' mental models of the device. Ultimately, the root cause may be design deficiencies arising because human factors were not adequately considered during the design process.

Finally, to identify the causes of use errors, whether during development, validation, or post-market surveillance, Root Cause Analysis should follow a systematic approach. Determining whether an error is perceptual, cognitive or action-related can help focus efforts to identify those aspects of the design of the device user interface that may be problematic. A systematic approach enables analysts to get beyond the immediate or obvious causes of use issues, while avoiding some of the pitfalls common to Root Cause Analysis.

## 7. Further reading

- Dekker, S. (2014). *The field guide to understanding "human error."* Boca Raton, FL: CRC Press, Taylor Francis Group.
- Reason, J. (1990). *Human error.* Cambridge, UK: Cambridge University Press.
- Wiklund, M., Dwyer, A., & Davis, E. (2015). *Medical device use error: Root cause analysis.* CRC Press, Taylor Francis Group.

## Acknowledgments

Special thanks go to Elissa Yancey for editing.

## References

ANSI/AAMI/IEC 62366-1:2015. (2015). *Medical devices - part 1: Application of usability engineering to medical devices.* https://doi.org/10.1002/rcm.2619.

Card, S. K., Newell, A., & Moran, T. P. (1983). *The psychology of human-computer interaction.* Hillsdale, NJ: L. Erlbaum Associates, Inc.

Dekker, S. (2014). *The field guide to understanding human error.* Boca Raton, FL: CRC Press, Taylor Francis Group.

Food and Drug Administration. (2016). *Applying human factors and usability engineering to medical devices: Guidance for industry and food and drug administration staff.* U.S. Department of Health and Human Services.

Hayman, W. (1995). *The issue is 'use,' not 'user,' error.* MDDI (May 1995).

HE:75 A. (2009). *Human factors engineering — design of medical devices objectives and uses of AAMI standards and recommended practices.* Medical Instrumentation.

ISMP., Root cause analysis workbook for community/ambulatory pharmacy, 2014.

Kaye, R. D., North, B., & Peterson, M. (2003). UPCARE: An analysis, decription, and educatinal tool for medical device use problems. In *Proceedings of the eighth annual international conference of industrial engineering theory, applications and practice* (pp. 1–6).

Nisbett, R. E., & Wilson, T. D. (1977). Telling more than we can know: Verbal reports on mental processes. *Psychological Review, 84*(3), 231–259. https://doi.org/10.1037/0033-295X.84.3.231.

Norman, D. A. (2013). *The design of everyday things*. New York, USA: Basic Books, Inc.

Ohno, T. (1988). *Toyota production system*. Portland, OR: Productivity, Inc.

Parker, J. (2015). *Root cause analysis in health care: Tools and techniques*. Joint Commission Resources.

Reason, J. (1990). *Human error*. Cambridge, UK: Cambridge University Press.

Wickens, C. D., Lee, J., Liu, Y. D., & Gordon-Becker, S. (2003). *An introduction to human factors engineering*.

Wiklund, M., Dwyer, A., & Davis, E. (2015). *Medical device use error: Root cause analysis*. CRC Press, Taylor Francis Group.

Woods, D. D., & Cook, R. I. (2000). Perspectives on human error: Hindsight biases and local rationality. *Safety Science, 34*, 193–214. https://doi.org/10.1016/S0925-7535(00)00013-8.

Zhang, J., Patel, V. L., Johnson, T. R., Chung, P., & Turley, J. P. (2005). Evaluating and predicting patient safety for medical devices with integral information technology. *Advances in Patient Safety, 2*, 323–336. https://doi.org/10.1037/e448232006-001.

CHAPTER

# 14

# Known use error analysis & post market surveillance

*Mary Beth Privitera*
HS Design, Gladstone, NJ, United States

OUTLINE



*As a rule, conventions only become conventions if they work.* ***Steve Krug***

## 1. Introduction

Human Factors Guidance (FDA, 2016) indicates the usefulness of identifying known use errors prior to beginning medical device design. The practice can lead to further understanding of the range of harm associated with use error; it also enables the development team to

Applied Human Factors in Medical Device Design
https://doi.org/10.1016/B978-0-12-816163-0.00014-1

incorporate safety by design and reduce the probability of occurrence and/or harm associated with use error by analyzing events that have been previously reported, hence they are known use errors.

This chapter discusses Known Use Error, including:

- barriers to the identification of use errors
- the use of database searches
- interviews and focus groups to uncover use errors
- the relationship of human factors in post-market surveillance (with examples).

## 2. Known use error analysis

"Use error" refers to "user action or lack of action that was different from that expected by the manufacturer and caused a result that (1) was different from the result expected by the user and (2) was not caused solely by device failure and (3) did or could result in harm (FDA, 2016)." Understanding the causation of these use errors can produce feedback about the design and possibly improve overall usability of a device. That drive toward improvement and safety is the essence of Known Use Error Analysis.

### 2.1 Barriers to identifying known use error

There are several barriers to the identification of Known Use Errors. They include:

- Rarity and likelihood. Use errors with a high-risk impact might be difficult to identify as they are likely to be rare or under-reported.

Reporting use errors is voluntary from healthcare providers. As a result, they may be hesitant to describe the situations as they may feel that blame could be placed upon them or that risks of litigation are too high if they report.

- Complicated descriptions. Use errors can be difficult to describe.

Without a clear understanding of language which describes aspects of design related to overall usability, use errors become increasingly difficult to describe without assigning blame; e.g., avoidance of statements such as "the nurse mistakenly entered the wrong value."

- Unclear reporting. Information available via reporting databases may not always communicate the cause-effect relationship; i.e., root cause analysis may be obscured based on the information provided.

Databases record many important details regarding an incident (e.g., product lot number, manufacturer, general situation and outcome); however, they often do not include sufficient detail so that the cause can adequately described.

- Lack of data. Data entry for many databases do not typically include use error or usability issues as a reason when an issue is reported.

This indicates further searching within the provided narrative is required in order to identify underlying causality.

## 2.2 Database searches for known use error identification and analysis

Database searches are commonly used to determine Known Use Errors for devices which operate like a new device or at the onset of a re-design of an existing device. The process includes the following steps:

1. Plan database research: FDA summaries, journal articles, competitive articles
2. Search Databases: external and internal database searches of product event reports which identify use issues, sequence of events and associated harm (if any).
   a. MAUDE database is broad by nature. Search criteria can be limited by the scope of the product and design work that is in development. Once a search has been completed, it can be exported into Excel for further analysis.
   b. Searching for general use error terms (such as "inadvertent," "neglected," "mistake," "misunderstood," "incorrect," "revered," etc.) can assist in the identification of known use errors when searching reported narratives
3. Ask users: new or prior users may have direct experience with use errors to create a robust list of query topics
   a. Discussions may assist in narrowing the area of inquiry; be sure to communicate and define common abbreviations or terminology that could be used to explain the technology or procedure.

FDA Human Factors Guidance (2016) provides the following list of relevant internet sites which may provide information regarding known use-related problems (FDA, 2016). They include:

- FDA's Manufacturer and User Facility Device Experience (MAUDE) database
- FDA's MedSun: Medical Product Safety Network
- CDRH Medical Device Recalls
- FDA Safety Communications
- ECRI's Medical Device Safety Reports
- The Institute of Safe Medical Practices (ISMP's) Medication Safety Alert Newsletters; The Joint Commission's Sentinel Events.

Within the process, it is important to leverage what has already been documented and analyzed for similar products. This saves time and can assist in building a taxonomy for use when further searching additional databases.

## 2.3 Using interviews or focus groups to determine known use errors

Individual and group interviews generate qualitative information regarding the perceptions, opinions, beliefs and attitudes of users (FDA, 2016).

- describe their experiences with existing devices
- provide specific instances they have experienced
- provide their opinions and perspectives regarding a new design (FDA, 2016).

See Chapter 5 Contextual Inquiry for more information on best practices for interview techniques.

The moderator is responsible for setting the scope, introducing discussion topics or exercises and ensuring all participants are reasonably engaged. This requires paying careful attention to maintain the discussion aligned with study objectives as well as promoting participation from quieter participants.

Focus groups can be conducted during design conceptualization early in the design process and be used to solicit opinions and attitudes toward existing devices or about a design concept in order to explore attributes of the design which may or may not be important to the users. These groups assist in the refinement of the design so that it better meets the needs of and/or gain acceptance by the users.

Focus groups are always discussion-based and do not necessarily adhere to a question-and-answer format. The conversation may ebb and flow in natural dialogue that is assisted by tools. Participants may be asked to complete a pre- or post-focus group questionnaire or provide ratings comparing devices or conduct hands-on exercises.

Participants of focus groups may be heterogenous (e.g., care providers and patients) in order to stimulate diverse dialogue and generate new insights. They may also be homogenous in order to promote sharing of similarities, promoting common perspectives or exploring details more closely.

## 2.4 Analysis process

Once information from database searches and/or interview-focus groups is gathered, an analysis is conducted to determine:

- Events resulting in harm
  - Was the device implicated in the issue?
  - Was the issue the result of use error or potential use error?
  - Is the root cause identifiable?
- Event categorization
  - By error
  - By known root cause
  - By device design issue

By identifying those events which result in harm and providing a categorization, further understanding of device use and risk can be determined. This information is required as all known use errors and use-related problems should be considered in risk analysis for a new device and included if they apply to a new device (FDA, 2016).

### Legacy or user interfaces of unknown provenance (UOUP)

Legacy user interface, or UOUPs, are user interfaces that were designed prior to the publication of IEC 62366-1. It is a requirement that they be evaluated using IEC 62366-1:2015 Annex C. Annex C outlines the steps of a scaled usability engineering (UE) program, which include the following:

1. Determine Use Specification

This includes the device's medical indication, users, use scenarios, use environments

**Legacy or user interfaces of unknown provenance (UOUP)** *(cont'd)*

and operating principles. This information may be derived from legacy data, predicate product information, interviews with subject matter experts and prior research.

2. Review Post-Market Information

This includes gathering available data (e.g., database search or internal post-market surveillance files) in order to identify safety events and/or use errors relating to overall device usability.

3. Document Hazards and hazardous situations related to usability

This includes hazards, hazardous situations or events where a use error occurred as a result of inadequate usability. In some instances, it can be helpful to categorize complaints and provide justification that the situation is use-related or not.

4. Risk Control Verification

This includes conducting a risk assessment to ensure all identified hazards and hazardous situations related to usability are mitigated and reduced to an acceptable level. See Chapter 12.

5. Residual Risk Evaluation

Conduct a residual risk analysis that confirms the user interface is acceptable. If the user interface is found unacceptable (i.e., further changes or additional mitigations are required in order to further reduce risk), the user interface may no longer be eligible for status as a UOUP. In these instances, a full usability engineering program following IEC 62366-1 is warranted.

## 3. Post-market surveillance

Once a medical device is available for use, several requirements have been established by the US FDA to ensure ongoing post-market surveillance (PMS) of device safety and effectiveness (FDA, 2016; Ross, Blount, Ritchie, Hodshon, & Krumholz, 2015). This can also be found in the EU in order to comply with Medical Device Directives -90/385/ECC Active Implantable Medical Directives (AIMD), 93/42/ECC Medical Device Directive (MDD) and 98/79/EC In Vitro Diagnostics Device Directive (IVDD). As outlined, these require each manufacturer to institute and maintain up-to-date, systematic procedures to review experience gained from devices in the post-production phase and then the implement appropriate means to apply any necessary corrective action (Tariah & Pine, 2014). According to Advamed, PMS is defined as the active, systematic, scientifically valid collection, analysis and interpretation of data or other information about a marketed device that may reveal unforeseen adverse events, the actual rate of anticipated adverse events or other information necessary to protect public health and safety (Advamed, 2018).

PMS activities are designed to generate information regarding use of a device, to expediently identify device design issues or use problems and to accurately characterize real-world device behavior and clinical outcomes (Tariah & Pine, 2014). In a study by Zippel et al. (2017), it was found that medical device manufacturers use standardized production management or quality assurance processes that can be transferred quickly and inexpensively for PMS purposes (Zippel & Bohnet-Joschko, 2017). The main variability found in this study was to the extent PMS was executed within participating companies (Zippel & Bohnet-Joschko, 2017).

There are a number of reasons, aside from meeting a legal requirement, for a medical device manufacturer to initiate post-market studies or augment FDA requirements. These include:

- internal performance monitoring
- design and product maintenance
- evaluation of expanded indications for use (Ross et al., 2015).

In a study completed by Ross et al. (2015), it was reported that manufacturers conducting post-clinical research studies left a perception among physicians or the general public that the medical device was still experimental. The study noted that this type of research may raise risks of both liability and criticisms from competitors. In contrast, the risks of not conducting post-market research included not being able to address scientific questions that arise from clinicians and not generating data that is important to payers (Ross et al, 2015).

## 3.1 Post-market surveillance and human factors

The regulatory requirement of PMS is only increasing, as demonstrated by the announcement that the FDA is expanding its investment in the National Evaluation System for health Technology (NEST) to allow more rapid build-out of the system to assist the FDA in collecting and analyzing real-world evidence (AAMI, 2018). This includes considering human factors within a PMS system (Gross, 2017). Fig. 14.1 below illustrates the application of human factors across the entire lifecycle of a device including monitoring the design in the field.

Within the FDA, there are two principal authorities governing PMS: the Medical Device Reporting (MDR) Regulation and the PMS Statute. Both are noteworthy in terms of human

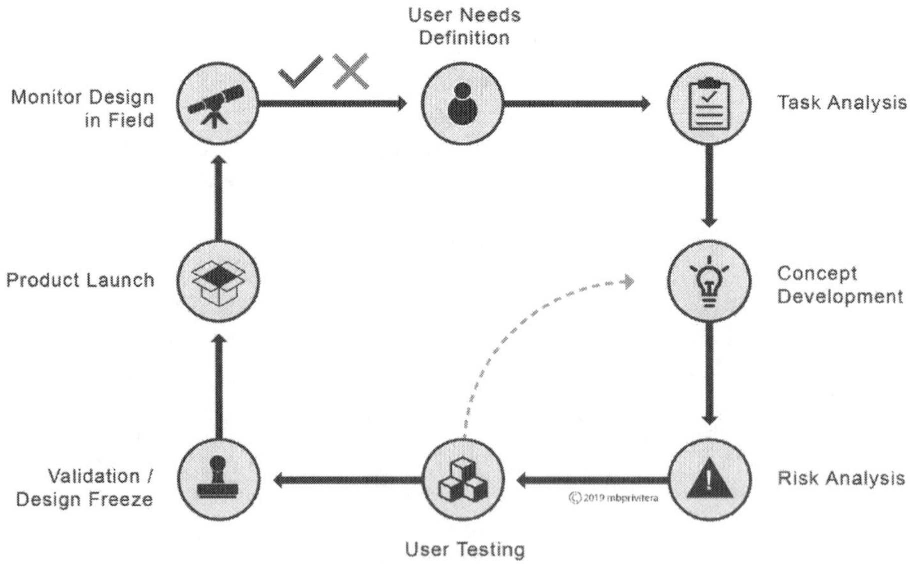


FIG. 14.1 Post market surveillance includes monitoring use error in the field.

factors principles and practices (Gross, 2017). The MDR Regulation provides passive surveillance, whereas the PMS provides for active surveillance through formal studies. The means for reporting adverse events is provided by two principal programs at FDA: a voluntary one known as MedWatch and the mandatory one, MDR. Mandatory reporting is required of user facilities, distributors and manufacturers. These entities are required to report device-related deaths and serious injuries and have time-sensitive response mandates. Submissions to FDA are required to fill out FDA form 3500 A and include terms that describe what happened, encompassing human factors-related problems as well as those less likely to be HF-related, such as electromechanical or software problems (Gross, 2017).

Examples of HF terms directly related to design problems reported using FDA form 3500 A might include "difficult to deflate" or "difficult to position." Those indirectly related might be "misreading of display" or "Mis-calibration; " terms directly related to labeling might be "mislabeled" or "incorrect instructions" (Gross, 2017). Similarly, manufacturers will be required to select evaluation terms (codes) from a coding manual addressing methods of evaluation, results and conclusions which also may include HF-related terms.

For HF, the PMS statute is likely more applicable to devices that are life-sustaining or supporting as opposed to permanently implantable devices due to the user interface prominence (Gross, 2017). As stated by Dr. Thomas Gross, Director of the Division of Post-market Surveillance, CDRH, *"human factors principles and practices have an important role to play in PMS at the FDA … human factors concepts need to be incorporated more fully in use of these tools (PMS)."* This suggests full feedback loops from device manufacturers, agency evaluators and end users in partnership as part of the adverse event discovery process (Gross, 2017).

## 3.2 Role of HF in PMS

The role of the human factors specialist in PMS is to ensure that personnel understand use error and to enhance the search/trending of use error events. Typical categories would include the use environment, user/s involved, the point of the process where the error occurred and other contributing elements, such as other devices being used at the time. This effort should tie use error risk analysis to customer complaint and inquiry categorization. It involves searching for key human factors descriptors within event narratives or arming customer support personnel with a series of questions that ascertain root cause information. The goal is to align the categories in the complaint and inquiry systems to use error/Root Cause Analysis. This may require additional complaint categories be added to PMS data collection tools (AAMI Human Engineering, 2014).

## 3.3 Challenges of PMS (in general)

There are several challenges to PMS for both clinical providers as well as device manufacturers. According to AAMI TIR 50, these include:

1. Many use errors and adverse events are not reported.
2. Many use errors and adverse events are reported with insufficient detail for analysis.
3. Many manufacturers do not have good systems in place for managing complaints and reports.

4. Many manufacturers do not have good systems in place for disseminating/communicating information to all the appropriate places and personnel within their company.
5. People reporting adverse events have a limited understanding of what constitutes a use error.
6. People are reticent to submit reports/complaints for a variety of reasons.

## 3.4 Applying HF process in PMS

Once complaints and use errors have been identified, the next step in applying human factors processes in PMS is to gather as much information regarding the situation as possible. AAMI TIR 50 Annex A provides example questions for collecting use error or close call data (AAMI Human Engineering TIR 50, 2014):

1. What happened?
2. What did the user intend to do?
3. What did the user end up doing? Were workarounds necessary? If so, what were they?
4. Where did the event happen?
5. Was the patient or operator harmed?

This may involve interviewing the people involved, referring to a task analysis (Chapter 6) in order to map complaints to tasks or using cognitive walkthroughs of the tasks (Chapter 10) to further illuminate/recreate what is actually happening.

Once the data is gathered, a root cause analysis (Chapter 13) can be performed, then assessed for the need to act. Some types of errors will reach the threshold for a Corrective and Preventive Action (CAPA) and will move into the CAPA process. Some errors may be new and may not reach the threshold for action. Nevertheless, they should be logged in to the manufacturer's internal process and monitored moving forward. Some errors may spark the need for a design change using the design change process.

### Scalpel case study

A manufacturer replaced the green color scalpel with the blue scalpel as seen in Fig. 14.2.

Both of the scalpels are FDA cleared, simple in shape and designed as a single use scalpel device. However, the user interface is

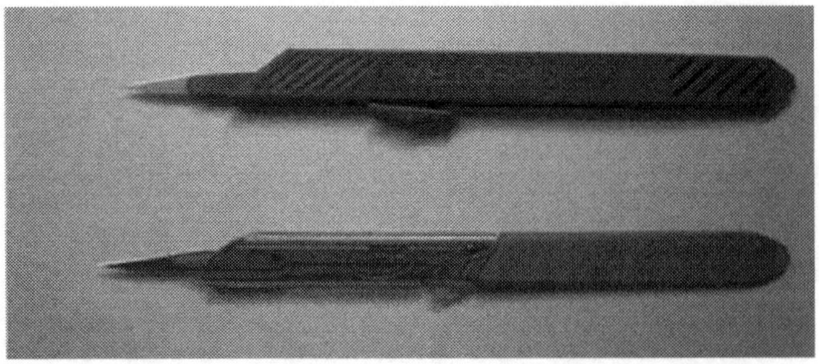

**FIG. 14.2** Original green (light black in print versions) scalpel, replaced with blue (dark black in print versions) scalpel. Both FDA-approved devices; however, the user interface varies.

**Scalpel case study** *(cont'd)*

fundamentally different. The mechanism of activation (for blade use) as well as its post-use covering are different.

Upon release, it was discovered (through PMS) that nurses believed the blade would come out of the cone-shaped end (like the green scalpel); however, when they did this, they got a surprise "scalpel stick" in the hand. As a result, customers requested the green scalpels back, and the manufacturer complied with the request.

## 3.5 Application human factors post device release (during device modifications)

Generally speaking, the human factors activities involved in device modification are similar to those when developing a new device. Ideally, the development team should be able to make use of the results of prior task analyses, usability testing, contextual inquiry, etc., during the design process. However, if such data is unavailable (for example, in the case of legacy device), there may need to conduct some of these activities from scratch.

Human factors professionals should assess the impact of changes to an existing device on those users who will migrate to the new device; i.e., the potential for negative transfer, need for retraining or other special measures, etc.

In some instances, behaviors that are poor ergonomically may exist solely due to resistance in re-training. See the Scalpel Example above. This should be evaluated in next-generation designs and assessed based on risk.

Table 14.1 below provides common examples regarding the need for human factors post-production or market release.

**TABLE 14.1** Example for post-market changes and the need for human factors engineering.

| Type of post-market change | Visible to customer? | HFE |
|---|---|---|
| Material change (Material A to Material B) | Not visible; may be visible | Usually not performed |
| Component | Visible | Depends (See box for scalpel example) |
| Dimensional changes, sub-assembly changes Ex: You add few more inches to a delivery device to make it more competitive | Depends (sometimes it is visible; sometimes its not) | Depends<br>Decision depends on whether the change impacts the user practice/clinical steps |
| Supplier change due to disruption, strategic reasons, etc. | Depends (sometimes it is visible; sometimes it's not) | Usually not performed if it is not visible. If it is visible, then it might be required. |
| Process changes | Not visible | No |

## 4. Summary

Known Use Error Analysis and Post-Market Surveillance activities are required by regulating agencies. Each method builds upon the other—strong post-market surveillance builds the database necessary for known use error analysis. This builds a strong knowledge base of use-related behaviors and conditions which currently exist in the market. In analyzing use errors known in the field regarding a specific device, competitive, or comparative device, the design of a novel device can be improved in regards to overall usability.

## 5. Further reading

- AAMI TIR 50 Post-Market Surveillance
- AAMI/IEC TIR 62366-2:2016

## Acknowledgments

Special thanks to Audra Wright, Gerard Torenvliet and Robert Garfield for their discussion regarding industry practices in Known Use Error and Post-Market Surveillance. Thanks to Elissa Yancey for editing.

## References

AAMI. (2018). *FDA doubles down on active surveillance, real-world data - association for the advancement of medical instrumentation*. Retrieved from http://www.aami.org/newsviews/newsdetail.aspx?ItemNumber=7252.

AAMI Human Engineering. (2014). *Technical information report AAMI TIR50*.

Advamed (2018). FDA fact sheets: Postmarket surveillance. Retrieved from https://www.lifechanginginnovation.org/sites/www.lifechanginginnovation.org/files/files/Postmarket Surveillance_0.pdf.

FDA. (2016). *Applying human factors and usability engineering to medical devices guidance for industry and food and drug administration staff preface public comment*. Retrieved from http://www.regulations.gov.

Gross, T. (2017). *Human factors (medical devices) - human factors and postmarket surveillance at FDA*. Retrieved from http://www.fda.gov/medicaldevices/deviceregulationandguidance/humanfactors/ucm128189.htm.

Ross, J. S., Blount, K. L., Ritchie, J. D., Hodshon, B., & Krumholz, H. M. (2015). Post-market clinical research conducted by medical device manufacturers: A cross-sectional survey. *Medical Devices: Evidence and Research*. https://doi.org/10.2147/MDER.S82964.

Tariah, I., & Pine, R. (2014). *Effective post-market surveillance Understanding and conducting vigilance and post-market clinical follow-up*. Retrieved from https://www.bsigroup.com/meddev/LocalFiles/en-US/Whitepapers/WP-Post-market-surveillance.pdf.

Zippel, C., & Bohnet-Joschko, S. (2017). Post market surveillance in the German medical device sector — current state and future perspectives. *Health Policy*. https://doi.org/10.1016/j.healthpol.2017.06.005.

# PART VI

# Usability validation & reporting

CHAPTER

# 15

# Human factors validation (summative usability) testing including residual risk analysis

*Mary Beth Privitera*[a], *Merrick Kossack*[b], *Kate Cox*[a]

[a]HS Design, Gladstone, NJ, United States; [b]Emergo by UL - Human Factors Research & Design, Chicago, IL, United States

OUTLINE



*Applied Human Factors in Medical Device Design*
https://doi.org/10.1016/B978-0-12-816163-0.00015-3

*Don't make me think.* ***Steve Krug***

## 1. Introduction

Human factors (HF) validation testing is conducted to demonstrate that the device can be used without serious use errors or problems, for the intended uses and under the expected use conditions (FDA, 2016). The testing should be comprehensive in scope, adequately sensitive to capture use errors caused by the design of the user interface, and should be performed such that the results can be generalized to actual use (FDA, 2016).

It is user testing conducted at the end of the device development process to assess user interactions with a device user interface to identify use errors that would or could result in serious harm to the patient or user (this includes compromised patient care). HF validation testing is also used to assess the effectiveness of risk management measures. HF validation testing represents one portion of design validation (FDA, 2016; IEC 62366-1). It is important to note the FDA Human Factors Guidance generally refers to the activity as "human factors validation" while the IEC 62366-1 standard refers to it as a "summative evaluation." However, these terms are typically interchangeable.

This chapter describes how to prepare and plan for human factors validation (summative study). This includes descriptions of participant recruiting, staffing, protocol development and data collection best practices. It also describes considerations for user/participant training and training decay prior to the initiation of the HF validation study. The chapter concludes with discussions regarding data analysis, reporting, and anecdotal issues that might arise.

## 2. Overview of conducting a HF validation study

As FDA Guidance (FDA, 2016) describes, the HF validation testing should be comprehensive, sensitive enough to capture use errors as a result of the user interface design, and should be performed such that results can be generalized to actual use. Below, Fig. 15.1 illustrates the process of conducting a validation study in four phases: planning, testing post-test activities, and reporting.

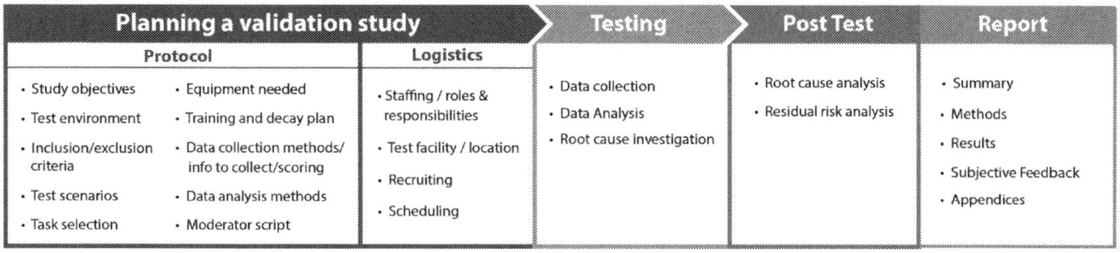


**FIG. 15.1** Process of conducting a HF validation study.

There are several key aspects to consider when designing this type of usability testing. These include the following (FDA, 2016):

- Test participants should represent the intended (actual) users
- All critical tasks should be performed
- The test articles are considered production equivalent including the device user interface which represents the final design
- Test conditions are sufficiently representative of actual use conditions.

## 2.1 Comparing HF validation studies and clinical studies

Keeping in mind the realism of the test, HF validation testing is generally conducted as a simulated use study (see Chapter 11). However, the testing may also be collected as part of a clinical study. Which type of study is conducted is typically dependent on the nature of the device and use environment noting that that they might be particularly complicated or poorly understood (FDA, 2016). Examples of when a HF validation study might be combined with a clinical study include when a simulated use study might not provide enough context of use or delivery of actual therapy is necessary to provide important feedback to the user. Table 15.1 below highlights the differences between clinical studies and usability studies.

## 2.2 Combine usability and design validation testing?

As part of a typical quality management system, all medical device user requirements must be validated. In some instances, the question becomes can this be included in the final usability test (i.e., human factors validation test) or are there alternate methods? See box "The Relationship of Design Validation and Human Factors Validation" in Chapter 3.

In the event that items for design validation are to be "tested" with the participants of the summative test, the following tips should be considered:

1. Create separate protocols, one for the HF validation activities and the other for the design validation activities.
2. Clearly separate summative test tasks from design validation tasks. Separate the data collection (i.e., documentation and collectors) for each protocol.
3. Generate separate reports

**TABLE 15.1** Differences between clinical and HF validation studies.

| Attribute | Clinical study | HF validation study |
|---|---|---|
| Goal | Evaluate efficacy and safety | Evaluate safe and effective use; Identify use errors, close calls and difficulties for intended device uses under expected use conditions |
| Sample size | Enough for statistical significance; dependent upon clinical goal | Following Annex K (AMMI/IEC TIR 62366-2 p. 44) and FDA HF Guidance (2016) a minimum of 15 participants with the caveat of a homogenous user group. Heterogenous user groups result in additional participants e.g. 15 of each user type. |
| Users | Might be trained differently, performance closely monitored or supervised | Training replicates training upon device release; no supervised performance |
| IRB/IDE | IRB will be required; IDE might be required | IRB approval is best practice |
| Environment | Actual use environment | Typically simulated use environment (e.g., in surgical lab, home use usability lab) |
| Use observations | Limited opportunities or no access depending on nature of use | Ability to closely watch and monitor user device interactions |
| Use error | Might not be reported or observable; Might be prevented by study protocol | Observable, documented, analyzed for root cause, and assessed for residual risk |
| Reporting | Self-reporting is allowed | Reporting is typically conducted by an HFE professional (a 3rd party) to assure accuracy and completeness |

## 3. Developing a test plan

The test plan for a device's HF validation study is very similar to previous human factors studies for that device. It also relies on many of the analysis activities performed during the course of the design and development of the device. This includes defining the intended user population, which can be documented as user profiles and use environments (See INTENDED USERS and USE ENVIRONMENTS AAMI/IEC TIR 62366-2). Additionally it includes the use scenarios and knowledge tasks which are often derived from task analyses, hazard analyses, and formative evaluations that identify potential use errors that could case potential harms (Wiklund, Birmingham, & Larsen, 2018) (See "Knowledge Task Data" in FDA's HFE Guidance and "Select the HAZARD-RELATED USE SCENARIOS for SUMMATIVE EVALUATION" in AAMI/IEC TIR 62366-2). Fig. 15.2 below highlights the information that should be gathered prior to planning the HF validation study.

The purpose of a summative evaluation is to evaluate the usability of the user interface as it relates to the successful (e.g., safe and effective) completion of tasks associated with hazard-related use scenarios. As such, the test has no testable requirements as with a laboratory or benchtop test. Rather, the data from the summative evaluation is analyzed to determine the

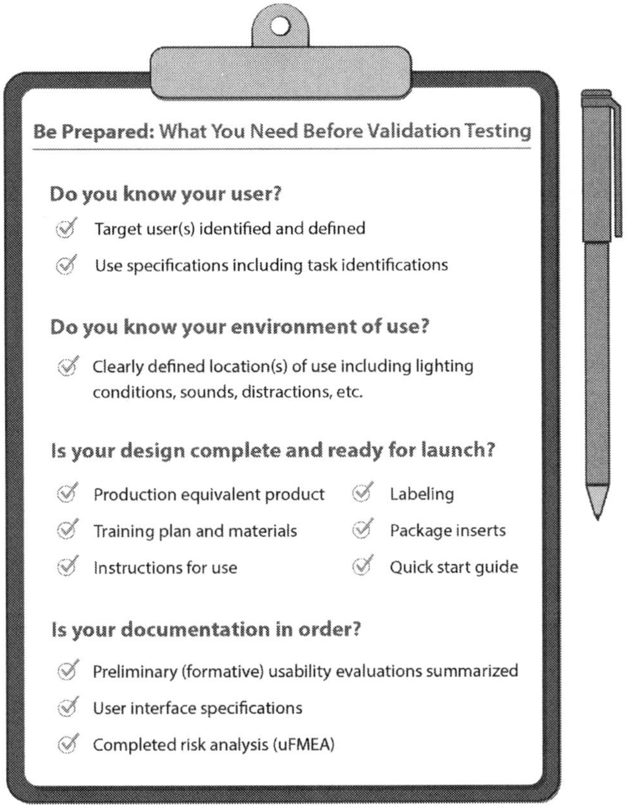


FIG. 15.2 Checklist of what needs to be completed prior to planning the human factors validation study.

root causes of any observed use-related issues (e.g., use errors, difficulties). These results are then further assessed to determine whether the residual risk is acceptable as part of the risk management process (AAMI/IEC TIR 62366-2). In other words, the goal of the HF validation test is to conclude that no further improvement of the user interface is necessary or practicable to reduce use-related risk.

When developing the test plan, a key component is defining what constitutes correct performance of a task, For an HF validation test, the acceptance criterion is the usability testing will prove critical tasks can be performed as intended by representative users without patterns of use errors that could potentially result in harm. Simply stated, the device manufacturer must define what is considered correct performance. Notably, in some circumstance, a successful outcome can result from correctly performing different sets of actions.

An HF validation test is intended to be as bias-free as possible with little or no interruption from test personnel. During task performance, test personnel must only observe the patient's actions or inactions. And only after the test has been completed should the test moderator probe the participant about their performance to better understand the root cause of any demonstrated use-related issues. It is important the test moderator ask questions of the participant in a bias-free, non-leading manner. A well-developed test plan will typically include a moderator script and pre-worded instructions and questions intended to help minimize the chance of bias the participant in any way.

## 4. Developing the protocol

A well written study protocol is direct and straightforward, it communicates clearly with sound organization. The structure should be simple and self-evident. Complex mega-sentences which illustrate complexity with esoteric vocabulary should be avoided. If developing the protocol intended for a pre-submission meeting with FDA, the author should consider that the reader is likely to be a regulatory agent trained in the field of human factors and therefore an expert in the research process. However, they may not be an expert in the specific clinical expertize or technology being evaluated. In other instances, the reviewer may not be familiar with the human factors validation process therefore it may be important to provide further explanation.

The contents of the protocol should directly respond to the requirements found in FDA human factors guidance, AAMI/IEC 62366-1 and AAMI/IEC TIR 62366-2. It is the manufacturer's opportunity to assure regulatory agency reviewers that the validation study is comprehensively thought out and all study issues are resolved. Constructing a solid validation protocol takes time and many revisions. It should be a collaborative process including participants across the cross-functional development team with reviews of study approach, language style and form in order to affirm the plan is as effective as possible.

Important details include discussion of the critical tasks and the methods used to collect data. A full description of the training content, delivery modes, and length of time elapsed prior to testing should be included along with the rationale for number/extent of device uses (i.e., the number of times that participants will use the device). The test protocol should also specify the types of data collected during the test, how use errors and other use problems will be defined, identified, recorded, and reported. In addition, the test protocol should indicate how and when participant interviews will occur. Finally, the study design should indicate that unanticipated errors will be observed, recorded and included during the participant interview.

The sections below describe common contents of a test protocol. Wiklund et al. (2018) provide a recommended structure for the test protocol and format. Below are further details regarding the contents.

### 4.1 Introduction & test purpose description

This section of a protocol provides an overview of the protocol document, the purpose of the validation test, a brief overview of the device user interface, intended use, the test description and brief explanation on all aspects of the test. This includes:

- what standards/guidance the test will be in accordance with,
- an overview of each participant session (e.g., each session will include one participant and last up to 3 h),

- details regarding protection of human subjects,
- data which will be documented (e.g., each use error, close call, difficulty, any test administrator assistance)
- analysis process of each use scenarios and knowledge task performance data with emphasis on critical tasks

The test purpose should describe the goal of the test (e.g., device can be used safely without committing potentially harmful use errors that could lead to harm or disruptions in therapy) and to demonstrate effectiveness. It should also include a list of standards and guidance that the test method is based upon and a rationale regarding the test methodology. The rationale should clearly communicate why the selected research approach is adequate in reflecting actual conditions upon market release and assurance that all identified critical tasks will be assessed.

## 4.2 Primary test data (data collected during the test)

This section should describe how the test is designed in order to uncover use -related issues as well as capture subjective feedback from test participants. This includes a focus on critical tasks within the selected use scenario(s) or knowledge-based task(s).

This section also highlights the means of detection and documentation of use-related issues, the capturing of unanticipated use errors and the need to debrief participants concerning observed use errors.

Additionally, this section can include the definition of use problems (use error, close call, difficulties, administrator assistance), detection methods (e.g., observations made by test personnel, conclusions from participant responses) and participants subjective assessment (e.g., root causes as described by participant, anecdotal feedback and answers to interview questions).

## 4.3 System and user interface overview

This section includes a detailed description of the device, its intended device use, users, training and use environment. It should include a reference to all aspects of a device user interface such as training videos, quick reference guides, etc. and/or ancillary equipment required for correct device use. This section should be comprehensive in scope and clearly communicates all aspects.

## 4.4 Required testing materials

Test articles include a list of items that will be evaluated during the test. In a complex system, this may include a controlling capital good along with the disposables required to operate the device.

Test equipment and materials include a list and description of items required in order to simulate use such as a patient bed, mannequin, cart, bed linens, etc.

This section in the protocol also includes a description of what constitutes a production equivalent device, test articles, test equipment and materials. It might also include a description on any limitations of device use as a result of study design (e.g., using saline instead of a drug) and any impact these might have on the validity of the results.

A production equivalent device accurately represents a final system in all aspects related to user interaction. This includes all production labeling and training materials. The use of prototype or non-finalized designs might be acceptable as long as they do not impact user interactions.

## 4.5 Human subject considerations

This section includes a description of all risks to human subjects (i.e., test participants), protections for human subjects, identity protection, and IRB approval. Since all human factor validation tests involve human subjects, it is best practice to seek IRB approval. In most instances, this can be expedited provided there are minimal risks to participants.

At minimum, best practices for the protection of human subjects should always be included in conducting an HF validation study. These include informed consent, the ability for participants to withdrawal at any time without cause or explanation and without the loss of compensation. It should also be communicated that the focus of the test is on the device and not the performance of the individual, and that the test administer will intervene should a participant appear at risk of injury.

Participant names should be recorded. However, none of the test data collected should be linked to a participant. Care should always be taken to maintain anonymity for all human factor tests. Additionally, adherance to GDPR (General Data Protection Regulation) may be required for EU based companies. GDPR best practices for data privacy includes awareness of the rules, the identification and training of a data protection officer, data tracking to report data breaches, the identification of where information is shared with other organizations, the communication of privacy information (including the processing of EU citizen's data, data retention period and that an individual can complain to the Information Commisioner's Office (ICO) if there is a problem), ensuring data procedures cover rights of the individual, response to subject access requests quickly, consent is provided. Note, there is additional data protection for children.

## 4.6 Participant recruiting

This section should include a description regarding the total number and type of participants, recruitment and compensation. The number of participants is dependent upon the defined user group(s). A minimum of 15 homogenous users i.e. users with similar demographics, is required (ANSI/AAMI/IEC, 2015; FDA, 2016). For example, if a device requires interactions by three distinct user groups, the HF validation test should include at a minimum 45 test participants (15 participants for each of the three groups). It is generally best to over recruit to accommodate for cancellations, no-shows or any disqualified participants.

The intended user should be defined according to specific criteria (e.g., job responsibilities, experience level, confirmed degree or training, in good health, age, etc.). The method of recruitment and confirmation of accordance to screening criteria should be described. Additionally, the amount of compensation for completion of the training and test sessions should be listed by user group given that compensation for time can vary depending upon type of participant (e.g., physician or technologist).

Notably, it is important to start recruiting early for specialized user groups (e.g., those with specific expertize or disease state). This can start with a recruiter first soliciting potential participants' interest level and then replying back later to determine their availability if interested. It might be necessary to include multiple test locations, increase honoraria, or provide travel incentives in order to achieve the required number of test participants.

## 4.7 Staffing or test personnel

This section of the protocol should describe all members of the study team. Best practices are to include at a minimum a test moderator and data collector/analyst. A third team member, a test administrator, might be necessary to greet study participants. Additional study personnel might be required should a test include multiple participants at once as in the case of a team study (e.g., an OR team interacting with a robotic assisted surgical system).

The moderator should be experienced and trained in conducting HF validation tests and be responsible for guiding the test session per the study script. The moderator should use a checklist to assure activities are fully completed. In addition, they are responsible for conducting the final interviews.

The data collector/analyst should be responsible for data collection and note-taking in real time and ensuring the study is carried out as intended and all objectives of the study are met.

The test administrator is responsible for greeting participants upon arrival and in charge of all participant paperwork and honoraria. In some cases, the test administrator might moderate the test or perform data entry.

Often included in this section is a description of the training test personnel receive to prevent undue influence on the test participants or test results during execution of the test protocol.

Additional observers which may be visible to participants should be limited in order to minimize the potential for test artifact due to participants being overly nervous about being observed.

## 4.8 Test environment: simulated use versus actual use environments

The test location, a description of the test environment (e.g., sound level, lighting level), and any intended distractions should be described in this section.

As stated in regulatory agency guidance, the test should be performed under conditions of actual use when simulated-use test methods are inadequate to evaluate users' interactions with the device. This determination should be based on the results of preliminary analyses (FDA, 2016). Furthermore, HF testing performed under actual use conditions should be preceded by appropriate simulated-use testing to ensure that the device is sufficiently well designed to be safe during actual use (to the degree that simulated-use testing can provide such assurance) (FDA, 2016).

In this section of the protocol, it is important to describe any aspects of device use that are critical, but difficult to capture/evaluate through simulation. There is often a question of "how realistic" when planning the test environment. For example, what if the lighting levels for an at-home device are varied. Does the test then need to have a number of different light levels? This requires thoughtful reflection regarding the actual use environment in order to determine when those elements can impact device use. In the instance that light levels do not impact device use, then there would not need to match any light level per se. The same reflection would be required regarding sound, temperature, etc. The key to the environment is to test under those conditions that impact device use.

## 4.9 Test agenda

This section includes a brief overview of the order of test events. As an example, this section may include the following:

- Welcome
- Complete relevant study documentation (e.g., non-disclosure, informed consent)
- Participant Training
- Training Decay

Begin test session.

- Brief study overview (introduction of device and test personnel)
- Confirm participant background and demographics
- Explain purpose of test
- Participant completes all use-scenarios
- Participant completes all knowledge task assessments as applicable
- Conduct post-test interview
- Thank and dismiss participant

## 4.10 Critical task identification

This section describes the comprehensive use-related risk assessments that were used to identify and analyze potential use errors that may occur with device use. The section should include a table of the severity levels and those tasks which were deemed critical tasks (severity of harm) and are included in the test. Knowledge based tasks e.g., troubleshooting instances and the need to reference labeling, are also included and often presented in a separate table.

Critical task identification is reliant upon use-related information determined in previous human factors activities including the use-related risk analysis (see Chapter 12). A uFMEA identifies potential use errors that could occur during the use of a device. Use steps in the uFMEA should be derived from completed task and workflow analyses and use case specifications, task analysis as well as the device's IFU.

## 4.11 Use scenarios

This section of the protocol describes the hazard use scenarios to be provided to participants in the study. The use scenarios must include all critical tasks identified by the use-related risk analysis. For more information on selecting scenarios see IEC 62366-2.

## 4.12 Data collection and analysis plan

This section should detail practices to assure information gathered during the test will remain confidential, be complete and stored securely in accordance with the manufacturer's internal procedures and may include GDPR.

Typical data collected during an HF validation study include the following:

- Participant demographic information
- Observed and participant reported use-related issues including:
  - Use errors
  - Test moderator assistance
  - Close calls
  - Difficulties
- Participant subjective assessment of root causes for all use-related issues
- Any observations by test personnel, including any device malfunctions
- Participant answers to interview questions

# 5. Conducting the evaluation

Conducting an efficient validation study requires proper planning. This may include conducting a pilot test to fine-tune the study methodology including the moderator's script, study flow, and logistics.

Below are recommendations and best practices for the execution of a validation study.

## 5.1 Moderating test sessions

The moderator of an HF validation study is responsible for facilitating the test sessions such that test participant performances yield valid, useful data. Executing to a well-designed study protocol is just the first step. The moderator must keep the participant within the scope of the study while also allowing the participant to interact with the device as naturally as possible. This helps to ensure the participant's experience with the device is as realistic as possible. (i.e., they are allowed to interact with the device based on their expectations). A strong moderator will set expectations with the test participants by reviewing the purpose of the test, introducing the test personnel, and describing their role.

The test moderator is responsible for observing participant performance during the study, noting any use related issues, and recording any participant comments or feedback. The moderator is also tasked with asking follow-up questions and conducting a post-test interview to help determine potential root causes, solicit subjective opinions, or to conduct knowledge-based tasks. It is during this time that the moderator must be most careful not to inadvertently introduce bias into the study. The wording, tone, and timing of questions can lead participants to answer them in a way they believe is proper (e.g., provide the answer they think the moderator wants to hear) as opposed to being fully truthful. A moderator that can establish a good rapport with the participants is more likely to earn their trust, a necessity to ensure meaningful discussions and honest feedback.

A good moderator will use consistent and neutral expressions and reactions while observing or talking with the participants. The moderator will be positioned such that they are afforded a good view of the participants as they interact with the device yet not be too close as to interfere. Unless requested by the participant, and only after sufficient time has passed to perform a task, the moderator should not assist the participant. This is consider a "test administrator assist" and should be handled in a similar fashion as a task failure. If the moderator had not assisted, the participant is likely to have not successfully completed the task.

The participants might ask the moderator questions whenever they experience an issue. Although it might be counter to their nature, the moderator should refrain from answering any questions to avoid interfering in the test and potentially inserting bias into the process. The moderator can avoid these situations by letting participants know, that should they ask a question, it will not be answered. Should a question be asked, the moderator can redirect the participant by asking a question in return (e.g., "what would you do if this were to happen in real life?").

The test moderator should be careful to limit their amount of talking during a test session. The purpose of the HF validation test is to observe individuals as they conduct use scenarios in the most realistic manner possible. Constant interruptions or distractions from the test moderator is likely to detract from the realistic use environment. There are exceptions to this when distractions are common place in the use environment in which case distractions might be intentionally introduced into the use scenarios.

Additionally, a participant may use the device in an incorrect manner or perform tasks out of order. The moderator should refrain from explaining the intended task order or defending the design at any time. Rather, the moderator should wait until the conclusion of the test session and debrief the participant seeking to understand their choice of actions.

When debriefing with a test participant at the end of the session, the moderator can ask some of the following example questions to limit the risk of introducing bias into the interview:

- "What did you think of the device overall?"
- "Did you have any trouble using it? If so, what kind of trouble did you have?"
- "Did you encounter anything confusing during the test session? If so, what was confusing?"
- "Please tell me about [observed use-related issue]. What happened? How did that happen?" (Note: The interview should include this question for each use error or other issue observed for that test participant. (FDA, 2016))

## 5.2 Conducting a pilot test

If circumstances where a robust simulated use study, or any type of formative usability test, was not performed prior to the HF validation test, it might be advantageous to perform a pilot test (e.g., pre-summative usability test) with a limited number of participants. Conducting such a test provides a variety of benefits including:

1. New or untrained colleagues can gain experience with the device, test protocol, and test script
2. Test personnel can practice executing the test protocol to ensure that it is structured well, contains all necessary tasks, can be completed in the allotted time, and to determine the best choreography among test team members

3. Test moderators can practice executing the test sessions and receive feedback regarding their performance
4. The project team can determine if the device is ready for final HF validation or if additional use-related risk mitigations should be implemented before validation occurs

At the conclusion of a pilot test, the team should determine if the device design is ready for validation and if all test materials are ready and acceptable. If not, the team can then decide on a proper course of action to rectify any observed issues (e.g., refine design, modify script, adjust tasks to shorten test time) with the test.

## 5.3 Training

Training conducted prior to the summative test must be realistic of the real-world training that will be provided to users once the device is on market. This includes the training duration, topics covered, presentation (e.g., video, web-based, hands-on) and teaching style. It is critical that the participants not be trained solely based on what will be covered in the test session. This can easily result in misleading performance since the participants will know beforehand what to focus on. As a result, true use-related issues are likely to go unseen or undiscovered.

### Importance of training by Paula Labat-Rochecouste

Training is an important part of a Human Factors assessment. Manufacturers should take special consideration on how their products are intended to be distributed to their end users. If a device is intended to be given to trained users, then manufacturers should consider and plan how training will be delivered to users.

In some cases, manufacturers may intend to provide training to healthcare providers on the device who will then subsequently coach and advise patients on the use of the device. In these cases, following manufacturer training, healthcare providers will then train their patients on the use of the system and be required to make a professional assessment on whether the system is fit for use by the patient.

In these cases, it is important to consider how training and knowledge is transferred from Manufacturers to Healthcare Providers and then to Patients. This sequence of knowledge transfer has significant opportunities for use error to occur if training programs are not comprehensive by design and verified through testing.

As noted in agency guidance (IEC 62366, FDA HF Guidance 2016), regulators expect manufacturers to document all root-causes for any Use Errors, Use Difficulties or Close Calls encountered as part of a HF validation study. In some instances within a usability study, the root-cause may be attributed to knowledge transfer as a result of poor training or inadequate training materials such as Instructions for Use.

If training is intended to be provided to users; and manufacturers wish to make the claim that their device will only be used by trained users, then it is vital to ensure that the sequence of knowledge transfer and the exact content of the training replicates the intended training that will be delivered to Patients (the users) post market release.

During the HF validation study, it is also important to replicate real-world use and therefore training should be delivered by the intended trainer or training organization that

*Continued*

**Importance of training by Paula Labat-Rochecouste** *(cont'd)*

will be used post market release (e.g., manufacturers train healthcare providers and in turn healthcare providers train patients on proper device use).

As an example, A manufacturer for an insulin delivery system for type 1 diabetes intends their device to be prescribed to patients by a diabetes educator or nurse. To approximate real-world use, during the HF validation study, the manufacturer should provide training to the healthcare provider participants who then train the patient participants taking part in the HF validation study.

As part of the study design, the healthcare providers will be assigned a patient to train. During this process, the healthcare provider may use their expertize and clinical judgment to determine whether or not the system would be prescribed to the individual in a real-world use setting or whether any further training would be required. During this competence assessment, if a patient is deemed to require more training then this should also be simulated as part of the HF validation study. Likewise, if the healthcare provider deems the system unsuitable for use by the patient participant, then this constitute grounds for disqualification from the study.

### *5.3.1 Training decay period*

Determining an appropriate training decay period can be challenging. This is the time test participants must wait between training and the start of performing tasks during the test session. The intent is to ensure that participants are relying on retained knowledge, and not just the recency of the training, to accomplish the tasks. Test teams often face pressures to shorten the overall length of the study or the amount of time users are participating in the test sessions especially when project timelines and participant availability are short. Understanding how training is realistically delivered and utilized, along with literature references, can be the best justification for establishing a learning decay period.

When determining training decay periods, there are a few key points to consider, including:

- How long will real users wait after training before using a device for the first time? If it is no time at all, then one might be able to justify no training decay during a validation study.
- If there is a real-world decay period, the amount of time for the validation test will depend on what is conducive during the test session (e.g., not too long), acknowledges required participant total time over the course of the study (e.g., does not require a participant to come back in an unreasonable amount of time later), and what is feasible from a scheduling perspective.

- In some instances, it may be prudent to train study participants and then wait days or weeks prior to conducting the HF validation test sessions. In the instances of a shorter decay period, participants typically wait on-site between the training and test session.
- Depending on the nature of the device, consider whether training decay needs to be exactly the same among test participants or if some variation is acceptable.
- Having participants sit in an isolated room during a decay period without distraction is not indicative of real life and normal learning decay. Consider having participants engage in other activities or discussions not related to the device being tested. Forgetting is produced by characteristics of the events of filled time, not by time itself (Reitman, 2017).

## 5.4 Data collection

It is vitally important to thoroughly document all aspects of an HF validation study. Data collection sheets are used to provide evidence of task performance including the amount, circumstances, and root causes of any use-related issues. Recording participant pre-screening information and demographics demonstrates that the appropriate user group(s) and number of participants were included in the test. Photographs taken during the test provide proof that the test environment was representative of the real use environment. Photographs of participants' performances are also useful for illustrating use-related issues in the final test report.

### *5.4.1 Data collection sheets*

Unlike the quantitative data typically collected during benchtop validation testing, HF validation data is quite qualitative. As a result, data collection tools tend to be more complex than simple checkboxes for recording pass/fail results or blank spaces for recording values. Efficient HF validation data collection requires in the least, a means to indicate whether a task was performed successfully or not. The tool should also allow for easy notation of any use-related issues and any notes describing what happened. Sometimes a sheet dedicated to each task with a checkbox and an area for notes is enough. Complex tests, with many tasks, might require a more detailed, integrated means of collecting data. In these cases, a spreadsheet might be advantageous. Both paper- and computer-based sheets have their advantages. Paper-based sheets are useful when the data collector must be able to easily move around the test environment. This might be because the test participant is also moving around or it might be to achieve a proper view of the device, its user interface, and the participant's interactions with it. A computer-based sheet though typically allows the data collector to record and analyze data in a quicker and more efficient manner.

Regardless of the medium, it is important for the data collection sheets to accommodate the following:

- Observational data including task performance success and failure
- All use-related issues including use errors, close calls, difficulties, and any instances of test moderator assistance

- Participants' self-reported root causes for use-related issues
- Participants' subjective feedback regarding the device.

### *5.4.2 Video/audio recordings*

Video and audio recordings might also be used for data collection. Such recordings offer the ability to replay and review specific instances during and after the test. In some instances, a web enabled camera can provide a live feed for observers outside the testing room or off-site. This facilitates a broader audience seeing first-hand how the validation study is progressing. For example, important stakeholders can witness critical use errors as they occur, allowing them to prepare for necessary discussions (e.g., to further mitigate or not) without waiting for the final test report.

It should be noted that while video data collection has its benefits, it can be problematic. Many manufacturers, as part of their quality management system and to follow proper design control procedures, are required to archive all data from any validation study. As such, maintaining and managing large amounts of video data can quickly become unwieldy and burdensome, especially when considering such video data is rarely referenced.

### *5.4.3 Recording use-related issues*

It is common for HF validation test results to include instances of use errors, close calls and use difficulties suggesting that no medical devices are perfectly designed (TIR 62366-2 2016). When these occur during a test, it is important for the study team not to panic. Rather, the team should continue with the study to collect and analyze the data to determine if these issues are commonplace. Furthermore, the team should understand whether such issues pose unacceptable use-related risk. Unless the team is already prepared to develop additional design mitigations for the design, they should refrain from making drastic decisions such as ending the study prematurely. Instead, they should collect as much data as possible to fully understand the scope and extent of any use-related issues.

When debriefing a participant regarding possible root causes for an observed use-related issue, a practiced moderator will do so without introducing any bias. The moderator should first inquire if the participant was aware of any issues in an open manner by asking, for example, "when doing X, can you walk through what you were trying to do and what decisions you were making?" This should be followed with questions that are intended to understand what the user was considering, thinking about or attempting to complete in the instance of an error.

The moderator should also be aware of when it is proper to initiate such questioning. Should the moderator ask questions immediately after completion of a use scenario or task? Or should questioning wait until after the test? Typically, if the participant will be performing the task or a related task again during the test, then the moderator should wait until the end so as not to influence and future test performance or behavior.

## 5.5 Test protocol deviations

It is probable that during the execution of an HF validation study something will not go according to plan. In an HF validation study, it is important to note any deviations from the test protocol that can potentially affect the test results. This is true even if the changes were due to mistakes made by test personnel. Importantly, it is critical to determine whether these affected the test results and if so, how. Examples of deviations include skipping a use scenario or task, either intentionally (e.g., due to exceeding test session length) or unintentionally. Another example is when test personnel trigger an incorrect alarm condition to simulate a specific use scenario.

To help prevent such deviations from occurring, it is important for all test personnel to support the moderator in ensuring the test is executed as planned. Test team members should feel empowered to prompt or remind the moderator if necessary about any tasks or questions remaining prior to dismissing a participant. While such deviations are undesirable, they should be explained in the test report. And if necessary, to ensure a complete set of data is collected, additional test participants might need to be recruited.

## 5.6 Data analysis

Analysis of task performance data collected during an HF validation test can be time consuming. The data, consisting of every instance of a use-related issue (i.e., use errors, close calls, test administrator assists and difficulties) associated with a critical task, must be assessed to determine each's possible root cause(s). Root causes are determined by applying knowledge of human factors engineering to understand the fundamental reason an issue occurred. These might be related to design aspects of the user interface, instructional materials, or training. Or they might be due to test artifact.

Noting patterns of use-related issues and participant behaviors (e.g., confusion, questions) when interacting with the medical device can reveal potential usability issues. Behavior patterns should be analyzed and considered during root cause analyses. Participants' demographic information, training records, observational notes and participant subjective feedback should also be considered.

Performance and behavior patterns are easiest to identify when data can be visually displayed together. This also aids in the process of tallying up the number of use errors, close calls, etc. With enough foresight and planning, the structure of the original data collection sheets can accomplish this. Alternatively, consider creating a separate data analysis document or spreadsheet that can be used to compile participant performance data.

# 6. Reporting results

The HF validation report should present results in a logical manner. All performance data should be reported including details of all use errors, close calls, difficulties and test administrator assists observed on a specific task. The order in which the data is presented should be considered to create a comprehensive report. Some reports are organized according to workflow; some by criticality of tasks; and yet others according to the type of use related issues. The best presentation for a report will depend on a number of factors including the number of critical tasks, the number and nature of use related issues, and the intended audience. For example, for a relatively complex device, if a report reviewer might not fully understand how the device operates, organizing the results according to workflow might be more informative than simply indicating there were issues (with apparently random tasks).

## 6.1 Residual risk analysis

One of the common misconceptions about HF validation results is that the results are quantitative and not qualitative and that pass/fail determinations and acceptance criteria is based solely on numbers of occurrences or percentages. However, acceptance criteria should only be based on associated use-related residual risk. This is an assessment that usually occurs after the completion of the HF validation report. Therefore, the test report itself might not indicate whether the validation effort was considered successful or not. It is common and acceptable to only outline the test results (e.g., performance data, root cause analyses) in the validation report and report on residual risk levels and pass/fail determination in separate documents such as the HFE report (see Chapter 17).

Once the HF validation testing results have been analyzed and root causes determined, one must determine if the observed issues pose any unacceptable risk of harm to users and patients. Therefore, a comprehensive residual risk analysis must be conducted. This is expected by the regulatory agencies. Conducting a risk analysis and determining acceptable risk levels are described in Chapter 12. A residual risk analysis is a collaborative effort involving members of a cross-functional team who can best assess an issue's potential impact and how to best mitigate it further if necessary. The steps of residual risk analysis include:

1. Analyze critical tasks with use-related issues
2. Determine residual use-related risk level
3. Decide whether the residual risk is acceptable or unacceptable

Note that while there might be other use-related issues associated with non-critical tasks, these tasks by definition of being non-critical should already have acceptable risk. In other words, there should be no potential for serious harm to the patient or user if the use-related issues occur during actual use.

According to FDA Human Factors guidance (2016), it is impossible to make any device error-proof or risk-free. A goal of the validation effort is to demonstrate that existing use-related risk mitigations for critical tasks are effective. Observed use-related issues suggest the potential need for additional or different user interface design modifications. If the residual risk level is considered unacceptable, indicating that the mitigations are not effective, then the validation test should be treated as a formative evaluation. In other words, the results and analyses of the test should drive further user interface design changes to help further mitigate use-related risk. As mentioned previously, while conducting a pilot test prior to the actual HF validation test, it is acceptable to stop the test after a few participants, modify the design (i.e., implement additional design mitigations), and then begin again with new participants. Once convinced that the risk mitigations will prove effective, then the validation test can be conducted with the target number of participants.

The residual risk analysis should include commentaries along with the results for each of the key events (e.g., critical tasks) associated with the potential for harm. The commentaries should build upon participants' comments regarding potential root causes for the issues.

They should also leverage clinical opinions, potential test artifacts, or standards of care when discussing whether the residual risk is acceptable.

Ideally, the commentaries conclude with an assertion of acceptable residual risk given the reasoning(s) provided. In the event the conclusion if one of unacceptable risk, the report should indicate such along with an intended course of action to mitigate the risk to acceptable levels. In some situations, a manufacturer might argue that the design is sufficient because:

- There is insufficient time or funding to make any modifications
- Most of the users and clinical advisors like the device as-is
- The device is better than what is currently available
- Any modifications would make the device undesirable (e.g., ugly).

These and other similar arguments are typically not accepted by regulators and should be avoided. Should it be determined that a device's design should remain the same, a risk/benefit comparison should be provided as justification. Factors to consider at this point include the likelihood of a patient deriving benefit from other sources, patient perspectives of the device's benefit (e.g., quality of life impact), the medical necessity, and the magnitude to which the device can have clinical impact (e.g., lifesaving potential).

## 6.2 Post validation modifications

Ideally, there should be no modifications to a device's user interface once HF validation has been completed. However, there may be reasons that the device's design continues to change. For example, when the residual risk levels are found to be unacceptable, post validation modifications to the user interface design will need to be implemented to help further mitigate use-related risk. Any post validation modifications to a device's user interface might require additional usability testing (TIR 62366-2).

A follow-up usability evaluation does not necessarily mean repeating the HF validation study in its entirety. It is the manufacturer's responsibility to determine if the design modifications require revalidation according to the risk analysis. The new design should be assessed to determine how it impacts current use-related risks and if it introduces any new risks. Best practices demand repeating, to some degree, the HF validation testing (e.g., conduct a "delta" usability test). This may encompass testing with less users and/or testing a limited number of tasks as long as all tasks that are impacted by the design change are included.

Solely stating in a premarket submission an intention to mitigate unacceptable use-related risks by providing additional training, revising the instructions for use, or creating or modifying some other element of labeling is not acceptable. Additional usability testing data is required to demonstrate the labeling change would be effective in reducing the use-related risks to acceptable levels (FDA, 2016).

Note that there might be instances where the design change is invisible to the user and would not impact their interaction with the device. Or in some cases, the design removes the opportunity for a use-related issue altogether. In these cases, further usability testing might not be necessary, or possible. In these instances, however, a justification for not conducting testing of the design change should be provided.

## 7. Summary

Human factors validation testing is often required for a device to gain regulatory approval. The FDA provides definitive guidance on conducting such testing. International standards call for it as well. HF validation testing is conducted in order to provide evidence that a device can be used in a safe and effective manner and that use-related risk control measures are adequate.

In conducting such tests, proper planning is key as the logistics can be challenging. The development of a robust test protocol takes time and should build upon previous human factors efforts. Prior to initiating a human factors validation test, it is important that the device's user-interface reflects what it will be upon market release (e.g., production equivalent). A successful test execution relies on the strength and skill of the moderator and the rest of test team. Conducting HF validation tests is as much an art form requiring specialized skills as it is about following a strict procedure or protocol.

Whether an HF validation test is considered successful will be dependent on the nature of observed use-related issues and their residual risks. Should the residual risks be unacceptable, further risk control measures might be necessary. Otherwise, assuming all residual risk is considered acceptable, the device can be considered to have successfully completed the HF validation testing.

## 8. Further reading

- ANSI/AAMI/IEC 62366-1:2015.
- AAMI/IEC TIR 62366-2:2016
- FDA Guidance: Applying Human Factors and Usability Engineering to Medical Devices
- Wiklund, M., Dwyer, A., Davis, E. (2016). *Medical device use error root cause analysis*: CRC Press.
- Wiklund, M., Birmingham, L., Larsen, S. A. (2018). *Writing human factors plans & reports for medical technology development*: Association for the Advancement of Medical Instrumentation.

## Acknowledgments

Thank you to Tim Reeves for providing peer review for content. The graphics in this chapter were provided by Laura Haggerty of HS Design.

## References

ANSI/AAMI/IEC. (2015). *ANSI/AAMI/IEC 62366-1.*

ANSI/AAMI/IEC. ANSI/AAMI/IEC 62366-1. (2015). *Medical devices- Application of usability engineering to medical devices* (Geneva, Switserland).

Association for the Advancement of Medical Instrumentation. (2016). *AAMI/IEC TIR62366-2:2016 medical devices-part 2: Guidance on the application of usability engineering to medical devices.* Arlington, Virginia: AAMI.

FDA. (2016). *Applying human factors and usability engineering to medical devices guidance for industry and food and drug administration staff preface public comment*. Retrieved from http://www.regulations.gov.
Reitman, J. (1971). Mechanisms of forgetting in short-term memory. *Cognitive Psychology*.
Wiklund, M., Birmingham, L., & Larsen, S. (2018). *Writing Human factors plans & reports for medical technology development*. Arlington: Association for the Advancement of Medical Instrumentation.
Wiklund, M., Dwyer, A., & Davis, E. (2016). *Medical device use error cause root analysis*. Boca Raton: CRC Press.

CHAPTER

# 16

# Human factors validation testing of combination products

*Tim Reeves, Christina Mendat, Liz Mauer*

**Human Factors MD, LLC., Charlotte, NC, United States**

OUTLINE



## 1. Introduction

Combination products include therapeutic or diagnostic products that combine a drug or biologic with a drug delivery device.[1] Examples of combination products include prefilled

[1]While most combination products pair a drug or biologic with a delivery device, combination products as a category also include products that combine a drug with a biologic, such as anti-body drug conjugates or progenitor cells combined with a drug to promote homing. This chapter focuses on combination products where the drug or biologics is co-packaged with a device or prefilled within a device.


https://doi.org/10.1016/B978-0-12-816163-0.00016-5

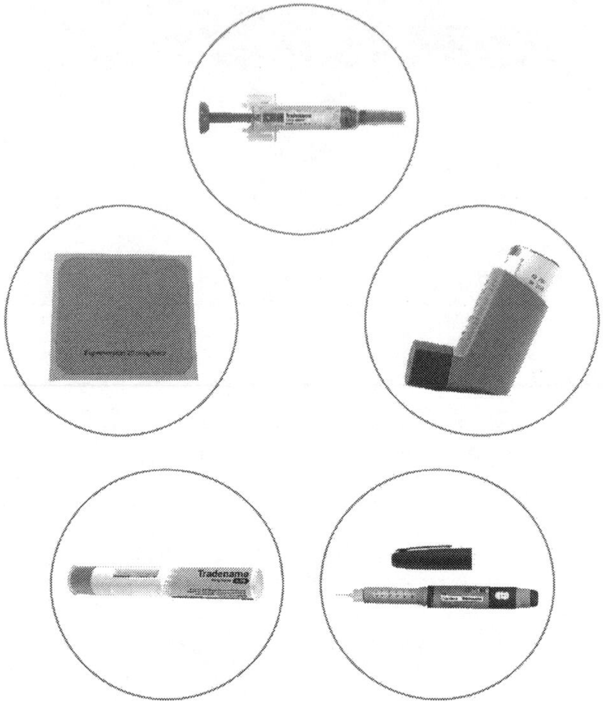


**FIG. 16.1** Examples of combination products.

syringes, pen injectors, auto-injectors, inhalers, transdermal pumps and patches, and kits containing drug administration devices or components (see Fig. 16.1).

Combination products offer a safe, effective, and convenient method of administering drug or biologic therapies. For patients in particular, the ability to safely administer a drug that is only available in an injectable, inhaled, or transdermal form without the assistance of a healthcare provider offers a tremendous convenience and improved quality of life. For a healthcare system with ever-rising costs, combination products offer a more cost-effective means of maintaining patients on drug therapies. The medical device and pharmaceutical industries have responded to this opportunity: the market for combination products will exceed $175 billion by 2024 by some estimates.[2]

As with "pure" medical devices, companies developing combination products should utilize a human factors engineering process during development to maximize the likelihood that their products can be used safely and effectively, including conducting human factors validation testing of their products.

Conceptually, human factors validation tests of combination products are no different than those of other medical devices. The purpose of the test is to gather evidence to support the

[2]https://www.marketwatch.com/press-release/drug-device-combination-product-market-worth-1777-billion-by-2024-grand-view-research-inc-2018-06-13.

claim that all use-related risks associated with use of the product have been mitigated as fully as practicable, and that any residual risks are outweighed by the benefits offered by the product to the patient or user. Human factors validation tests can be conducted in a simulated use (the most common) or actual use manner. Tests should also:

- Engage representative users in a realistic setting;
- Incorporate labeling and training (if provided) in a manner that mirrors what will happen in actual use, when the product is commercialized;
- Utilize a production-equivalent version of the product;
- Have participants complete clinically meaningful scenarios that incorporate all critical tasks.

However, there are noteworthy differences in how to design and execute human factors validation tests for these two classes of product. In this chapter, we discuss the ways in which human factors validation testing of a combination product may differ from that of a medical device. These differences arise in part because combination products include a drug or biologic. The inclusion of a drug introduces use-related risks that are not of concern typically with medical devices.

The inclusion of a drug or biologic in a combination product also means that a different center at the Food and Drug Administration (FDA) may lead the review for approval of the product than the center that leads reviews of medical devices. The Center for Drug Evaluation and Research (CDER) and the Center for Biologics Evaluation and Research (CBER) are responsible for the review of New Drug Applications (NDA) and Biologic License Applications (BLA) involving most combination products, while the Center for Devices and Radiological Health (CDRH) is responsible for reviews of 510(k) and Premarket Approval (PMA) applications for medical devices.[3] These centers have nuanced differences in their approach to human factors that has implications for human factors validation testing. We review these differences in this chapter.

Finally, CDER and CBER have defined another human factors test that may be needed instead of a human factors validation test to support marketing applications for combination products that include a generic drug or biosimilar. Should the company submitting the application for approval of such products request that the product be designated by the FDA as *substitutable*, in the case of a combination product with a generic drug, or *interchangeable* in the case of a combination product with a biosimilar, the FDA *may* recommend that applicants conduct a Comparative Use Human Factors Test instead of a human factors validation test. The purpose of a comparative use human factors test is to establish that the generic version of a combination product is as safe to use as the reference product. We end the chapter with a discussion of Comparative Use Human Factors tests.

[3]Combination products include products for which the "primary mode of action" is the drug, such as prefilled syringes and inhalers, and products were the primary mode of action is the device, such as drug-eluding stents. In this chapter, we focus on combination products for which the primary mode of action is the drug or biologic, and the device itself is a vehicle for delivering the drug. The review of applications for these products are led by CDER and CBER. Reviews of combination products for which the primary mode of action is the device are led by CDRH.

## 2. Combination products and other medical devices

### 2.1 Use-related risks

While the process for identifying use-related risks for combination products is the same as the process for traditional medical devices, the inclusion of a drug constituent introduces risks unique to combination products. Users of combination products who do not understand all the requirements of use may be exposed to drug-related hazards. Examples include:

**Expiration and useful life:** Drugs have limited shelf-life, introducing risks relating to drug expiration. For combination products, for which the drug comes prefilled in the device, the device may be single use or have a limited number of uses. Users may need to understand how many doses are remaining in the device, when to stop using the device, and when to obtain a replacement device.

**Storage requirements:** Some drugs have specific storage requirements, such as requiring refrigeration or storage within a relatively narrow "room temperature" range. Other drugs need to be protected from direct sunlight. Users of some combination products may need to periodically check the appearance of the drug within the device to ensure its integrity.

**Dose differentiation:** Combination products might be stored with other drug products, including other combinations products. Users may need to differentiate between combination products with similar names or that are listed together in computer-based dispensing systems. As with straight drug products, combination products may be marketed as a line of products packaged in different dose strengths or concentrations. Users may need to differentiate between combination products within a family.

**Assembly required to use:** Some combination products, including those packaged as kits, may require assembly in order to use them. Others may require priming to ensure that a full dose is delivered on first-time use. Users need to be able to complete these tasks to ensure accurate dosing.

**Maintaining a dosing regimen:** Users of combination products must deliver complete and accurate doses of the drug, sometimes on a prescribed schedule. This necessitates understanding when and how the drug should be administered including the recommended route of administration. Difficulties using the device during administration may lead to under- or over-dosing, or possibly, patients receiving no dose at all. Users may have difficulties prepping the device, holding the device at the required orientation, placing the device, initiating drug delivery, or holding the device in place during drug delivery.

**Disposal:** Users may need to dispose of the device safely to protect themselves and others from exposure to unused drug or device-related injuries (e.g., from sharps or unintentional exposure to discarded drug product).

Additionally, companies should consider known use problems when identifying use-related risks for their products. To aid in this activity, FDA included a non-exhaustive list of known use issues with injection and inhalation products in the draft guidance for human factors guidance for combination products released in February of 2016 (Draft Guidance for Industry and FDA Staff, 2016). Common use errors with injectables include misunderstanding of dosing instructions, selecting the incorrect product, not priming at all or priming incorrectly, not fully depressing the device plunger, premature removal of the injection device (also known as a "wet injection"), and improper storage and disposal. Known use problems with inhalers include errors assembling the product, not priming or priming incorrectly, improper inhalation technique, failing to wait between doses, failing to clean and maintain the device, and failure to dispose of the spent device properly.

Others have cataloged known use problems with other classes of combination products (Bonds, Asawa, & Ghazi, 2015; Fink & Rubin, 2005; Lee, Pharm, & Ph, 2002; Potera, 2015). For example, Lee, M & Phillips, 2002 reviewed known use problems with transdermal patches. These include users not removing some or all the protective lining, taping patches rather than using the patch adhesive, applying multiple patches at once, placing new patches on top of older ones, cutting patches to adjust the dosing, and failing to safely discard used patches. The latter poses a particular risk to small children if they come in contact with an improperly disposed of patch (see Box).

## Accidental exposure to transdermal patches

In September of 2013, the FDA moved to require changes to the coloring of the labeling used on some fentanyl pain patches in response to reports of accidental exposure to patches causing serious harm and death to children, pets, and others. The choice of colors on some patch labels made the patches difficult to see, including if the patches had inadvertently fallen off the wearer and were accessible by children.

Drug Safety Communications, FDA 9-23-2013

Some of these aforementioned use issues are not unique to combination products. For example, many medical devices have storage and disposal requirements that need to be met to ensure safe and effective use. But combination products present a unique "cluster" of potential use-issues, largely because they include a drug or biologic. Makers of combination products must be sensitive to the degree to which even seemingly negligible use issues in operating their device can have an amplified effect on safety and effectiveness because of the impact of the drug or biologic.

## 2.2 Critical tasks

In February of 2016, CDRH released their final guidance on human factors for medical devices. In that guidance, CDRH defined a critical task as "a user task which, if performed incorrectly or not performed at all, would or could cause *serious* harm to the patient or user, where harm is defined to include compromised medical care" (italics added). In a related guidance on human factors considerations for combination products released in draft that same month, the definition for a critical task was broadened by removing the qualifier "serious". From a combination products perspective, tasks that could lead to ANY harm should be given careful consideration and mitigations for use issues associated with those tasks should be introduced and evaluated.

The broadening of the definition of a critical task to include any harm reflects the seriousness with which the FDA views dosing errors and other use issues that could compromise drug efficacy. For example, while an over-dose of a medication such as fentanyl can be life-threatening, a slight to moderate over- or under-dose of other medications (e.g., the rheumatoid arthritis medication adalimumab) does not pose an immediate threat to a patient. However, chronic under- or over-dosing will be harmful to the patient and will compromise the patient's care. Likewise, depending upon the drug, use errors such as choosing an inappropriate injection site, administering via the wrong route, or improper storage leading to a degraded drug may not expose a patient to physical harm, but could compromise care as a consequence of reduced drug efficacy and consequently, should be taken seriously by the makers of the combination product (see Box).

### Defining critical tasks

In their guidance documents for human factors, the FDA's Center for Devices and Radiological Health (CDRH) and the Centers for Drug Evaluation and Research (CDER) and Biologics Evaluation and Research (CBER) defined "critical tasks" for medical devices and combination products differently. In their guidance, CDRH defined critical tasks for medical devices in relation to *serious* harm, while the draft guidance for human factors considerations for combination products defined critical tasks in relation to *any* harm. The broader definition for combination products reflects the seriousness with which the FDA views dosing errors and other use issues that could compromise drug efficacy.

### Definition of critical tasks for medical devices

*A user task which, if performed incorrectly or not performed at all, would or could cause serious harm to the patient or user, where harm is defined to include compromised medical care.*

Applying Human Factors and Usability Engineering to Medical Devices. Guidance for Industry and Food and Drug Administration Staff. February 2016, page 3.

**Defining critical tasks** *(cont'd)*

### Definition of critical tasks for combination products

*Critical tasks are user tasks that, if performed incorrectly or not performed at all, would or could cause harm to the patient or user, where harm is defined to include compromised medical care.*

Human Factors Studies and Related Clinical Study Considerations in Combination Product Design and Development. Draft Guidance for Industry and FDA Staff. February 2016, page 7.

As is the case when validating other medical devices, test scenarios for human factors validations of combination products should include all critical tasks required to use the product safely.

## 2.3 When to include human factors validation test data in an FDA submission

Per design controls, manufacturers of all medical devices, including those used in combination products are required to validate that the design of their product conforms to defined user needs (21CFR 820.30(g)). Human factors validation testing should be part of any design validation. However, depending upon the device, the FDA may or may not require data from a human factors validation test when reviewing marketing applications for new products.

CDRH and CDER/CBER differ, however, on when companies will need to include data from a human factors validation test in support of a regulatory submission. In their 2016 draft guidance *List of Highest Priority Devices for Human Factors Review*, CDRH listed classes of medical devices for which data from human factors validation testing should be included with a 510(k) or Premarket Approval (PMA) application. Companies developing a medical device which is on this list should submit data from human factors validation testing with their 510(k) or PMA application. However, CDRH may request data from human factors validation tests for other medical devices that are not on this list. To assess if data from human factors validation testing will be required by CDRH, the developer of the medical device should consider the risks associated with use of their device. If the company is developing a new device, and the device has at least one critical task, the company should anticipate that CDRH will require data from human factors validation testing during their review. For companies making a change to an existing medical device, human factors validation test data may be required if the user interface has changed such that a critical task is performed differently with the new version of the device, or the addition of new features and functions has introduced new critical tasks (and new use-related risks).

In their draft guidance for combination products, CDER recommends the same risk-based approach to assessing whether data from human factors validation testing should be included with a New Drug Application (NDA) or Biologic License Application (BLA) for a combination product. However, CDER did offer an 'exemption' for commonly used combination products such as prefilled syringes, if the product is used in a clinical setting, by healthcare providers. In these cases, where the use-related risks are already well understood, CDER is not likely to ask for human factors validation test data when reviewing an NDA, unless that combination product includes novel features that are new to healthcare professionals. By extension, for products used outside of a clinical setting (e.g., in the home) or by lay users (e.g., patients), CDER will almost certainly expect that the results from human factors validation testing will be provided for their review.

CDER is also likely to expect human factors validation test data even if the device constituent of a proposed combination product has its own 510(k) or has been used in another, previously approved combination product. In either case, CDER is likely to require human factors validation test data that establishes that the device can be used safely and effectively for the proposed indication, by the intended users, in intended use environments, given the proposed product's packaging and labeling.

## 2.4 Evaluating training as a risk mitigation

Companies may choose to try to mitigate some use-related risks through training. Indeed, the FDA considers training, when provided with a medical device, to be part of the device's "user interface". If training is incorporated as a risk mitigation (if users must receive training to ensure that they can use the product safely and effectively), then the effectiveness of training in mitigating use-related risks should be evaluated in human factors validation testing.

In our experience, CDRH and CDER differ in their recommendations for how training should be incorporated in validation testing. The difference comes down to how a company will provide training when its product is made available commercially, and the extent to which the company can ensure that all users of the product will receive the training.

CDRH's perspective has been that if most or all users will receive training when the product is commercialized, then it is appropriate to train participants in the validation study provided that the content, format, and method of delivery of the training given to participants is comparable to the training users will receive in actual use. If most or all users will not receive training when the product is commercialized, then participants in a human factors validation test should not be trained. This approach is intended to ensure that the test conditions in the study are representative or "typical" of actual use.

CDER's position has differed from that of CDRH. From CDER's perspective, unless a company can ensure that training is delivered consistently, and to all users, the human factors validation test should include a group of participants who do not receive training. CDER's concern is for the safety of those users, however small in number, who may not receive training in actual use. In essence, CDER adheres to the view that unless the maker of a combination product can ensure that all users will receive the same quality training, training should not be the only mitigation of use-related risks, and users must be able to use the product safely and effectively in the absence of training.

In our experience, unless the company marketing a combination product is in full control of the training, with a system in place to ensure that the content, format, and method of delivery of training in consistent, and that all users will receive the training, CDER will expect an untrained arm in a validation study. Relying on the prescribing physician or the patient's pharmacist to provide training to the patient or their caregiver is not a reliable training program. Survey studies with patients indicate that many do not receive counseling from their physician or pharmacist when prescribed a new drug (Heritage et al., 2006; Svarstad, Bultman, & Mount, 2004). Moreover, when counseling is provided, the information conveyed varies considerably (Stevenson, Cox, Britten, & Dundar, 2004).

## 2.5 Product and dose differentiation with combination products

Combination products may be stored with other drug products, including other combinations products. Like other drug products, combination products may be marketed as a product line with different dose strengths and concentrations but packaged in cartons displaying similar branding and labeling. As a consequence, healthcare professionals may need to differentiate between products with similar names or similar packaging when those products are stored together or are listed together in computer-based dispensing systems. Unfortunately, product selection errors, when users inadvertently choose the wrong product or wrong concentration or dose have been a recurring problem in healthcare, including pharmacy practices (Abood, 1996; Flynn, Barker, & Carnahan, 2003).

In some circumstances, patients themselves need to be able to reliably differentiate between different combination products. Imagine a husband and wife who have different medical conditions. The husband has high blood pressure and high cholesterol and is giving himself a subcutaneous injection every two weeks using an autoinjector per his doctor's prescription. His wife has rheumatoid arthritis and gives herself a weekly injection using an autoinjector per her doctor's prescription. While the medications may be very different, both are stored in the refrigerator, and both product packages have a similar shape and use similar coloring. What could possibly go wrong?

When there is a risk associated with product selection errors, whether in the hospital, retail pharmacy, or patient home, validation testing for combination products should include test scenarios that assess how reliably individuals can discriminate between the combination product being evaluated, and like products that may physically co-exist in the same use environment. Validation test participants should be engaged in realistic selection tasks that evaluate whether users can reliably differentiate the products of interest and other products that may be stored together. These tasks should be evaluated in a test environment that mimics the environment of actual use. For instance, if the combination product being evaluated is to be stored in the refrigerator, other drug products a patient or other family members may be taking should be placed in the refrigerator as well (see Fig. 16.2) and study participants (e.g., patients or caregivers) should be asked to retrieve their product from the refrigerator as they would in actual use.

Similarly, pharmacist participants could be asked to complete ongoing pharmacy tasks that include selecting products off the shelf as they fill prescriptions (see Fig. 16.3). Nurse participants could be asked to retrieve a combination product from a Pixys system to fill a medication order.

FIG. 16.2 Combination products stored in a patient's refrigerator.

FIG. 16.3 Pharmacy environment created for human factors validation testing of dispensing tasks.

Regardless of the participant or environment, test scenarios should include product selection tasks as part of a realistic use scenario. Participants should not be alerted that the purpose of the scenario is to evaluate whether they can differentiate between products. Ideally, participants should be engaged in a realistic scenario where product selection is one of the tasks required to complete the scenario (e.g., fill a prescription, preparing a product for a medication order, or retrieving the medication from wherever it is stored prior to use).

## 3. Comparative use human factors tests for combination products involving a generic drug or biosimilar

### 3.1 Generics and biosimilars

Developers of novel drugs and biologics typically protect their intellectual property through patents. As those patents expire, other companies may make generic versions of the drug, or in the case of biologics, biosimilar versions. This includes generic and biosimilar versions of combination products.

Developers of combination products that include a generic or biosimilar may apply for the product to be designated by the FDA as *substitutable* in the case of a generic, or *interchangeable* in the case of a biologic.[4] If a combination product is substitutable or interchangeable then, depending on the legal jurisdiction, pharmacists may substitute the generic or biosimilar product for the reference product (the branded product) without first obtaining approval from the physician who prescribed the reference product.

In determining whether a combination product is substitutable, the FDA considers whether the generic combination product can be used as safely as the reference product without the intervention of a healthcare provider and/or without additional training.

But while the drug or biologic itself may be "off" patent, patent protections covering the design of the device may prevent companies from simply copying the reference product device user interface. Companies wishing to introduce a generic version of a combination product may need to include a delivery device with a different user interface than the reference product's device. However, this begs the question: Are the products truly substitutable? Can a user of the reference product use the generic version as safely and effectively, or do differences in the user interface of the generic pose use-related risks that differ from those of the reference product? In practical terms: Can the user of the reference product use the generic product without encountering new use-related issues, if the generic product is simply substituted for the reference product by the user's pharmacist, without any training or coaching by a healthcare professional?

To establish whether a generic combination product is substitutable, companies are advised by the FDA to first conduct a *Threshold Analysis* to determine whether there are differences between the user interfaces of their device and the reference product's device that could affect a user's performance of critical tasks. If differences do exist, companies

[4]While similar, the definition of interchangeability is more nuanced, and includes the possibility of patients switching back and forth between the biologic and biosimilar, over the course of treatment, without compromising the patient or their therapy. At the time this chapter was written, the FDA had not yet designated any approved biosimilar as interchangeable.

may need to include the results of a comparative use human factors test. In this type of test, users of the reference product are asked to demonstrate use of both the reference and generic product's devices, in order to demonstrate that the generic product's device can be used as safely as the reference product.

### 3.1.1 *Threshold analysis*

A threshold analysis entails a side-by-side comparison of the proposed generic combination product user interface, and the reference product user interface. The analysis should compare:

- the steps required to administer the drug using the reference product device and the generic device,
- the physical design of the two devices, and
- the labeling (i.e., on-device, carton, and Instructions for Use) that accompanies the two products.

The threshold analysis should focus on attributes of the user interface for the two product devices that could influence a user's ability to perform critical tasks with each product – what the CDER and CBER guidance documents refer to as external critical design attributes. If differences in external critical design attributes are found (i.e., "non-minor" differences in FDA parlance), companies may need to provide additional data from a comparative use human factors test to support a claim of substitutability (or interchangeability, in the case of biosimilars).[5]

### 3.1.2 *Comparative use human factors tests*

The objective of a comparative use human factors test is to confirm that if a pharmacist substitutes the generic version of a combination product, users will be able to use the product as well or better then the reference product. Unlike human factors validation tests which are qualitative, comparative use human factors tests are statistical in nature. These tests should be designed to support a sufficiently powered statistical comparison of the error rates for those critical tasks that may be impacted by differences in the user interface designs of the two products, to determine whether the error rates for the generic product are no worse than for the reference product.

In general, comparative use human factors tests should be simulated-use studies in which representative users are asked to simulate use in representative use scenarios and use environments. Testing should include users of the reference product and *paired* designs, in which the same participant uses both the reference and generic product are appropriate. In addition, should the Threshold Analysis (see Section 2.1.1 above) show that only certain

[5]While the threshold analysis was originally proposed by the FDA as a method for assessing whether combination products involving generics are substitutable, on a limited, case-by-case basis, the FDA has accepted a threshold analysis as evidence of safe and effective use for combination products submitted as new drug applications or NDAs. If the sponsor's device user interface is the same or highly similar to an already approved, on-market comparator device, and the users and use environments are the same or similar, the FDA *may* accept a threshold analysis as evidence that the sponsor's product can be used safely and effectively. The threshold analysis must establish that there are no non-minor differences in external critical design attributes between the two devices.

user subpopulations are likely to experience use difficulties as a consequence of the user interface design differences with the generic product, then participant enrollment can be limited to users of those subpopulations.

## 4. Summary

Prefilled syringes, pen injectors, inhalers, auto-injectors and transdermal patches are all examples of products that combine a drug or biologic with a delivery device. These *combination* products are increasing in popularity and promise to reduce the cost of healthcare delivery while improving patients' quality of life.

As with other medical devices, makers of combination products are required by regulators to validate that their products can be used safely and effectively by intended users in intended use environments. This typically includes conducting human factors validation testing. While conceptually similar, there are differences in the design of human factors validation tests of combination products. These differences stem from the fact that combination products include use-related risks related to the drug itself, as well as to nuanced differences in how the centers of the FDA who lead the review of combination products and medical device submissions view human factors. In this chapter, we reviewed additional considerations for the design of human factors validation tests for combination products relative to tests for other medical devices.

We also discussed a related type of human factors test, that may be used in lieu of a human factors validation test of a combination product involving a generic or biosimilar. For companies that wish to establish that their combination products are substitutable or interchangeable, a comparative use human factors test may be needed.

## 5. Further reading

- Comparative analysis and Related Comparative Use Human Factors Studies for a Drug-Device Combination Product Submitted in an ANDA: Draft Guidance for Industry, January 2017.
- Considerations for Demonstrating Interchangeability with a Reference Product, Draft Guidance for Industry, 2017

## Acknowledgments

Special Thanks to the team at Human Factors MD for their contributions, comments and human factors practice.

## References

Abood, R. R. (1996). *Errors in pharmacy practice*. US Pharmacist.

Bonds, R. S., Asawa, A., & Ghazi, A. I. (2015). Misuse of medical devices: A persistent problem in self-management of asthma and allergic disease. *Annals of Allergy, Asthma, and Immunology, 114*(1), 74–76. https://doi.org/10.1016/j.anai.2014.10.016.

Draft Guidance for Industry and FDA Staff. *Human Factors Studies and Related Clinical Study Considerations in Combination Product Design and Development*, (February 2016).

Fink, J. B., & Rubin, B. K. (2005). Problems with inhaler use: A call for improved clinician and patient education. *Respiratory Care, 50*(10), 1360–1374. Discussion 1374-5 https://doi.org/10.1016/j.chroma.2004.03.011.

Flynn, E. A., Barker, K. N., & Carnahan, B. J. (2003). National observational study of prescription dispensing accuracy and safety in 50 pharmacies. *Journal of the American Pharmacists Association*. https://doi.org/10.1331/108658003321480731.

Lee, M., & Phillips, J. (2002). Transdermal patches: *High risk for error?, FDA: Drug Topics*, 54–55.

Heritage, J., Paterniti, D. A., Wenger, N. S., Tarn, D. M., Hays, R. D., & Kravitz, R. L. (2006). Physician communication when prescribing new medications. *Archives of Internal Medicine, 166*(17), 1855. https://doi.org/10.1001/archinte.166.17.1855.

Lee, M., Pharm, D., & Ph, R. (2002). Safety page transdermal patches: High risk for error? *FDA: Drug Topics*, 54–55.

Potera, C. (2015). Misuse of autoinjectors and inhalers. *American Journal of Nursing, 115*(3), 17. https://doi.org/10.1097/01.NAJ.0000461799.44904.d3.

Stevenson, F. A., Cox, K., Britten, N., & Dundar, Y. (2004). A systematic review of the research on communication between patients and health care professionals about medicines: The consequences for concordance. *Health Expectations*. https://doi.org/10.1111/j.1369-7625.2004.00281.x.

Svarstad, B. L., Bultman, D. C., & Mount, J. K. (2004). Patient counseling provided in community pharmacies: Effects of state regulation, pharmacist age, and busyness. *Journal of the American Pharmacists Association, 44*(1), 22–29. https://doi.org/10.1331/154434504322713192.

CHAPTER

# 17

# Preparing an HFE report for agency submission

*Merrick Kossack*

Emergo by UL - Human Factors Research & Design, Chicago, IL, United States

OUTLINE



*Answering the question: Is the device safe and effective for the intended users, uses, and use environments?*

## 1. Introduction

A human factors engineering (HFE) report, also referred to as a usability engineering (UE) report, is a compilation of the human factors activities performed during the development of a medical device. It is intended to explain to a regulatory agency that not only was a human factors engineering process adhered to during development, but also the efforts' impact on


https://doi.org/10.1016/B978-0-12-816163-0.00017-7

the design of the device. Authoring such a report to be included in an agency submission may seem like a daunting task, especially if the plan is to submit to more than one agency. This chapter provides guidance and tips in order to prepare robust HFE/UE reports to be included with medical device submissions.

## 2. The need to tell a story

An HFE/UE report should convey to the reader that a strong HFE process was followed during the design and development of the medical device resulting in one that is safe and effective to use. Like any good story, the report should have a beginning, a middle, and a satisfying end. In the case of the HFE/UE report, it would be a beginning that explains the purpose of the device and who will be using it; a middle that details how the device design evolved over time; and an end that provides evidence that any use-related risk associated with the device has been mitigated as much as possible. From the report, readers should be able to understand enough about the device such that its usability can be evaluated appropriately. Without this explanation, reviewers are likely to not fully understand the purpose or operation of the device and request additional information during the approval process, in effect, delaying the entire process. If the report is complete, comprehensible, and concise, a reviewer should be convinced that the device is indeed safe and effective.

## 3. HFE/UE report contents

The contents of the HFE/UE report is largely driven by the regulatory agency that will be receiving the report. For a device that will be marketed in the United States, the FDA has outlined the HFE/UE report information it prefers in the agency's guidance *Applying Human Factors and Usability Engineering to Medical Devices* (2016). If the device is to be released into markets outside the United States (OUS), IEC 62366-2:2016 *Medical devices—Part 2: Guidance on the application of usability engineering to medical devices* describes the contents of the Usability Engineering report that can be created to summarize the device's UE process. The United Kingdom's Medicines and Healthcare products Regulatory Agency (MHRA) outlines similar report contents in their 2017 guidance, *Human Factors and Usability Engineering – Guidance for Medical Devices Including Drug-device Combination Products.* Notably, much of the information contained within the reports is the same, albeit perhaps labeled differently. Therefore, when including an HFE/UE report to multiple agencies, it is both convenient and efficient to reuse much of the same information from one report to another.

An HFE/UE report typically contains the following types of information:

- The device's use specification (e.g., user profiles, use environment description)
- A description of the device user interface
- A summary of known use problems
- A summary of the device's use-related hazards and use-related risk analysis

- A summary of preliminary/formative evaluations
- A discussion of the human factors (HF) validation (summative usability) test including any use-related residual risk that remains after completion of HF validation.

## 3.1 Use specification

To design and develop a medical device that can be used safely and effectively, it is important to know what it will be used for, who will use the device, where it will be used, when it will be used, and how it will be used. It is only with this complete understanding that one can confidently say the device has been developed to be safe and effective. Much of this information is contained within the device's **use specification** as defined by IEC 62366-1:2015. A device's use specification might be described in one single document. Or various aspects of the use specification might be found in multiple documents. And while the HFE/UE report might reference one or more of these documents, it should clearly and concisely presents this foundational information so the reader understands how it all served as inputs during the design process.

The HFE/UE report should describe the **intended use** of the device. For example, what is the purpose of the device? Will the device be used to perform certain medical tasks? Or will it be used to diagnosis a disease? Or will it be used to monitor a specific condition? Additionally, the report should indicate who the patients are for this device. Furthermore, what part of the patient will the device interact with? The device's development team, as well as the report reader, requires an awareness of this fundamental purpose to know whether the device's design generated will allow a user to effectively use the device.

The report should also include descriptions of the device's **user population**. Notably, this might be the same group of individuals comprising the patient population or it might be individuals who use the device on patients (e.g., healthcare providers, caregivers). Importantly, the report should explain whether the device users consist of individuals from distinct user groups and what makes them distinct. Do the users have differing capabilities? Different education levels? Different job responsibilities? Or even perform different tasks with the device? The device design must accommodate all possible user groups to ensure that all can use the device safely and effectively. Often times, the HFE/UE report summarizes any user profiles that have been created as design inputs.

The HFE/UE report should address the question whether the **use environment** can impact device use. Such environment considerations include, but are not limited to, temperature, lighting, and noise levels. Other factors include the presence of other people, distractions, stressors, and high levels of activity. Device developers should be aware of such influences and minimize any negative effects on use performance.

The device's **operational context** should be discussed in the HFE/UE report as well. The report should identify the circumstances of device use. For example, will users receive any training prior to use of the device? If so, what kind of training? Additionally, are there any instances when the device might be used differently? Finally, what are the critical aspects of device use? Besides using the device to deliver therapy of some kind, do users also have to set up the device or maintain it or clean and sterilize it? An understanding of these conditions of use is important when ensuring a device design can support the user in all cases.

For a device to be properly designed, its developers must know who will use the device and where the device will be used. Furthermore, they must understand the intended use of the device and its context of use. The reader of the HFE/UE report will be assessing whether the device can be used safely and effectively by the individuals under the described conditions. Presenting this information in a clear manner will help provide the reader with the requisite knowledge necessary to make this determination.

## 3.2 Device user interface description

While individuals who are reviewing a medical device regulatory submission might be familiar with the device's indications for use and its therapy, they might not be familiar with the device design. The reviewers cannot reliably determine whether a device can be used safely and effectively if they do not understand how a user is expected to interact with it. Manufacturers are encouraged to include detailed descriptions of their device's user interface as part of the HFE/UE report so that reviewers can best understand all of its interaction points.

Authors of the HFE/UE report should describe all elements of the user interface necessary for a user to complete essential and critical tasks. After reading such a description, the reviewers should be able to envision exactly how a user provides inputs (e.g., controls, programs) to the device and how the device provides outputs (e.g., results, alarms) to the user. The reviewers should understand everything about the device that the user interacts with including hardware components, software interfaces, and product labeling.

When describing the device's user interface, the author should include the device's **hardware interface components**. Any controls such as knobs, buttons, and switches that facilitate user inputs should be detailed. Accessories such as keyboards, mice, microphones, and touchscreens should also be discussed. In addition to describing how the device facilitates user input, the report should also include any hardware the device uses to provide output to the user. Examples of such hardware are displays, monitors, speakers (e.g., for alarms), and lights. Finally, the report should describe any other physical components essential to device use. Such components might include connectors, retaining clips, doors, latches, and any other item that must be set, primed, connected, or somehow manipulated to ensure proper device operation.

**Software interface components** are most commonly part of a device's graphical user interface. Such elements of the interface generally provide device output and status information to the user, labeling for controls, and warning and caution messages. Users might also provide inputs to the device by selecting options displayed in the graphical interface. These options might be displayed or listed utilizing menus, lists (e.g., drop-down, scrolling), or other types of graphical components (e.g., radio buttons, checkboxes). The HFE/UE report should detail the various graphical interfaces displayed on the device including how the user is expected to interact with them.

When describing the user interface of a device in the HFE/UE report, the author should not limit the discussion to hardware and software components. Recalling that anything the user interacts with is part of the device interface, the author should also describe the device's product labeling. The **product labeling** of a device provides essential information to the user. This labeling typically takes one of the following forms:

- On device labeling – Text and symbols (e.g., icons) are often placed on devices themselves to guide user actions. This might be in addition to or in lieu of similar labeling that might be present in the graphical user interface (assuming the device incorporates such an interface). Examples of such labeling include placing text next to controls to identify their function (e.g., "ON/OFF"), using an icon to indicate a control's function (e.g., ⏻ for power)), or using symbols to help inform users about the device's status (e.g., 🔋 for battery status). Manufacturers will also often place warnings and cautions on devices themselves in an effort to inform users of potentially hazardous conditions or actions attempting to mitigate any associated risk. As these are intended to help drive user behavior, they too should be included in any description of the device's user interface.
- Packaging – Depending on the type of device, users might be required to interact with some form of packaging. A device user might interact with packing in several ways, including opening it to access the device, consulting the packaging for administration or dosing information, and/or referring to the packaging for operation instructions. As a result, the packaging is considered part of the device's user interface and should be described in the HFE/UE report.
- Instructional materials – User manuals, instructions for use (IFUs), and Quick Reference Guides (QRGs) are examples of product labeling that might be included with a device. Users rely on such reference materials for guidance and instruction when using a device. Considering this form of labeling is utilized when users are unfamiliar with a device, are unsure of the proper actions necessary to use the device, or do not understand the output generated by the device, it is essential for this information to be effective so that the device can be used safely and effectively. For this reason, readers of the HFE/UE report will typically want to best know what instructional materials are provided to the user, their content, and how they are presented to the user (e.g., bound manual, placard, electronically).
- Training materials – Device users interact with training materials with the goal of learning how to properly use the device. This material is typically delivered through an instructor. However, training materials might also be presented in a variety of other formats including training videos, interactive models, or simulators. Training can also be self-administered. As with instructional materials, the training material's content and delivery mechanism are critical elements for ensuring proper device usage. Therefore, the HFE/UE report should describe the training plan, content, and method expected for users. When possible, the actual training materials should be included in the report's appendices to provide additional illustration.

Readers of the HFE/UE report are interested in understanding how users will interact with the device. Therefore, the report should describe in detail the device's user interface ensuring all elements critical to safe and effective use are included. Simply describing the hardware and software components and the product labeling, however, might still leave the reader with questions regarding proper device use. The HFE/UE report should explain the user's expected actions as well as the task flow necessary for effective device use. This, along with describing how the device itself responds to the user interactions, provides the reader with the most complete picture of the device's user interface and how it impacts device use.

Noting that critical tasks have the potential for harm or compromised medical care should there be any device use-related issues (e.g., mistakes, errors), the HFE/UE report should clearly detail how use-related risks associated with critical tasks have been mitigated through design. It is recommended that such design mitigations be cross-referenced with other sections of the HFE/UE report that discuss the device's use-related risks and critical tasks. This will help demonstrate to the report's reader that the device's design mitigates use-related risks necessary to best ensure safe and effective use.

## 3.3 Summary of known use problems

As discussed in Chapter 14, an important aspect of the medical device HFE process is to determine whether any predicate and similar devices have known use-related issues. The HFE/UE report should summarize not only the results of this research, but detail how the results informed the design of the device. Importantly, any known use-related problems should be accounted for in the device's use-related risk analysis (See Chapter 12) and mitigated appropriately. These mitigations should then be discussed in the report's User Interface Description (see Section 3.2 above), especially if the mitigations resulted in modifications to a predicate's existing design.

For example, a known use problem that is highlighted in the summary should at a minimum reference its associated risk(s) (e.g., indicating the risk ID as in the use-related risk analysis) and any relevant design mitigations.

## 3.4 Use-related hazards and risk analysis

A goal for applying a HFE process to the development of a medical device is so the device can be used safely and effectively. The purpose of the HFE/UE report is to summarize the HFE process applied during development of the device, thus demonstrating how the device can be used safely and effectively. The use-related risk analysis that is developed during this process serves as a cornerstone of this conclusion. Developing a comprehensive use-related risk analysis is discussed in Chapter 12.

The discussion of the use-related hazards and risk analysis is of vital importance. The reader of the HFE/UE report is likely to continually reference the use-related hazards and risk analysis during their review. The reader will be interested in understanding under what conditions (e.g., use-related errors) hazards are introduced that have the potential to result in risks to user and patient safety. The report should clearly document how specific use-related issues can result in different types of hazards and the hazardous situations that could lead to harm. The report should also describe the methodology the manufacturer followed while conducting the use-related risk analysis (e.g., the manufacturer's risk management procedure). This includes the process used to identify potential use-related risks as well as the basis for classifying those risks – in other words, how were the severity and likelihood of the risks determined. Ideally, this process is similar to one outlined in ISO 14971, the international standard dedicated to the application of risk management to medical devices.

As a device's use-related risk analysis serves as the foundation for many decisions during the development process, the HFE/UE report should identify those risks considered most significant. Generally, these are the **critical tasks**, those that, should a use-error occur, have the potential to result in serious harm to the user or patient or result in compromised medical care. (For discussion on the definition of "harm" between CDRH and CDER see Chapter 16 Section 1).

It is not enough to simply identify the important use-related risks in the HFE/UE report. The reader will want to know how these risks have been mitigated and how these mitigations were proven effective. The report should identify the risk mitigations as well as provide cross-references to other portions of the report where these mitigations might be discussed in more detail. For example, in the case of a design mitigation, the use-related risk should reference the design details in the device user interface description (see Section 3.2). The same is true for any risk mitigated through product labeling. Furthermore, mitigations' effectiveness were demonstrated during validation testing. Therefore, references should also be included that direct the reader to those parts of the validation effort (e.g., test cases, use scenarios) that provide such evidence (see Section 3.6).

It is not necessary to list every use-related risk in the main body of the HFE/UE report as this could be unwieldly and would likely result in simply duplicating the actual use-related risk analysis. However, it is advisable to provide a reference in the report to either an appendix that contains the risk analysis or the risk analysis' document number in the device's design history file. This will facilitate locating the complete risk analysis should the reader desire to do so.

## 3.5 Preliminary/formative evaluations

A proper HFE process is iterative in nature. As device designs are developed, they should be evaluated to determine if they support safe and effective use. Often, these evaluations reveal design shortcomings and opportunities for improvement. Designs are modified and then reassessed. This cycle continues until it is believed with high certainty that the designs properly support proper usage of the device. It is only then that the iterations discontinue and validation is performed on the final design.

It is during this iterative portion of the HFE process that formative evaluations are used to assess the design. "Formative" evaluations serve as design inputs, and, as the name suggests, their results inform the evolving device design. The HFE/UE report should summarize the formative evaluations conducted during development of the device. This includes indicating how the results of the evaluations were used to improve the design.

Chapter 3 describes a number of different types of formative evaluations (for examples of formative evaluations see Chapters 5, 6, 10, and 11). Any and all of these that are conducted during development of the device should be discussed in the HFE/UE report. Such formative evaluations include both analytical approaches, such as heuristic and expert reviews, and empirical approaches, including cognitive walkthroughs, interviews, and, most notably, formative usability testing (Section Four).

The summary of the formative evaluations should indicate the type of evaluations that were conducted, the scope of the evaluations, and a brief overview of the methodologies highlighting such factors as the types of users and the nature of the tasks performed. The summary should describe the results and any conclusions from the evaluations. Most importantly, the HFE/UE report should describe what was then done with these results and conclusions.

It should be clear to the reader how the results informed the evolving design of the device. Any design modifications that result from conducting a formative evaluation, whether the design is present in the final device or not, should be described. If a design modification is part of the final device and was discussed in the device user interface description (see Section 3.2), the evaluation summary should provide a cross-reference to that description. This enables the reader to understand exactly how the results of the formative evaluation were included in the design process.

Conducting formative evaluations might also result in the discovery of previously unknown potential use-related issues. As with other such issues, these should be analyzed as potential use-related risks. Accordingly, these would be documented in the use-related risk analysis along with any applicable design mitigations. The formative evaluation summary should clearly identify the risks by cross-referencing them with their respective entries in the use-related risk analysis.

While it is not necessary to provide exhaustive details of every formative evaluation conducted, the HFE/UE report should summarize each that contributed to the design of the device. It is recommended that each evaluation mentioned in the summary reference relevant evaluation documentation (e.g., formative usability test report). Therefore, should the reader wish to learn more about a particular evaluation, the relevant documentation can be easily identified (and presumably located in the design history file, See Chapter 3). The report should also highlight any use-related risks that were discovered as a result of the evaluations and how the designs were modified to mitigate such risks. Together, the summarized information will facilitate better understanding of the device and how its design changed over time in an effort to ensure the device's safe and effective use.

## 3.6 Human factors validation (summative usability) testing

Development of a medical device in accordance with an HFE/UE process typically culminates with a human factors validation test, also referred to as a summative usability test (See Chapter 15). It is during this effort that developers demonstrate that their device can be used safely and effectively. This is typically done by conducting a usability test intended to determine whether use-related risk mitigations, especially those related to critical tasks, are effective.

Activities conducted during the HFE/UE process help determine the validation test's methodology. Creation of the use specification early in the process is used to determine who the test participants (e.g., users) are and what the representative use environment should be during the test. Workflow and task analyses provide the context when structuring the use scenarios that test participants will perform; and the risk analyses determine what tasks are of critical importance when assessing test participants' performances. A well-structured HF validation protocol should clearly indicate how the test plan accounts for this work.

And the HF validation report should clearly indicate how the test was conducted, ideally according to the test protocol, as well as its results. Chapter 15 provides a thorough review of developing and conducting the HF validation test.

An HF validation test in itself reveals much of the HFE/UE work a team has conducted over the course of the device's development. The HFE/UE report should therefore demonstrate how the HF validation test was planned, how it was conducted, and its results. Since this information is already included in the test protocol and report, it is not necessary to repeat it all in its entirety in the HFE/UE report. Rather, it is recommended that the HFE/UE report contain the following:

- A description of the test methodology;
- a summary of the test results;
- an analysis of all use-related issues that occurred during the test that would have the potential to result in user or patient harm during actual use; and
- descriptions of any design modifications that were made in response to the test results.

Usually, the HFE/UE report will make direct reference to the test protocol and report to facilitate additional reviews if necessary. Additionally, the protocol and report as well as any other test documentation (e.g., scripts, forms) are often included as appendices to the HFE/UE report.

Given the unlikelihood that a device is "perfect," there will be use-related issues observed during the HF validation test. Report reviewers will want to know if any of these issues pose unacceptable risk to users and/or patients. Therefore, it is extremely important that the HFE/UE report includes a complete residual risk analysis of the device accounting for any such issues. As this information is not necessarily found in an HF validation test report, HFE/UE report authors should be sure to include it with any discussion of HF validation results.

## 3.7 Conclusions

The conclusion of the HFE/UE report should include a statement indicating the device is safe and effective. It is highly recommended to use the exact language provided in Appendix A of the FDA guidance *Applying Human Factors and Usability Engineering to Medical Device* (2016):

> The <device> has been found to be safe and effective for the intended users, uses, and use environments.

Everything in the HFE/UE report should support this claim. If this conclusion cannot be made based on the work outlined in the HFE/UE report, then it is likely that additional HFE work must be performed. If this is the case, the device developer should give considerable consideration in determining whether the device is even ready for regulatory submission.

A summary of the followed HFE/UE process, the activities performed, and the results and resulting decisions and actions of those activities should accompany the claim of safety and effectiveness. Given most devices do not possess an error-proof interface, that mistakes can occur, the conclusion should also include a discussion of the residual risk associated with the device. Such a discussion should outline how use-related risk has been mitigated as much as possible, that further mitigation is impossible or impractical, and that any risks associated with the use of the device are outweighed by the benefits the patient is likely to receive.

## 4. HFE/UE report organization

As previously discussed in Section 3, regardless of the regulatory agency receiving the submission, much of the information included in an HFE/UE report is the same, albeit perhaps referred to by a different label. As a reminder, the purpose of such a report, regardless of which agency will be reviewing it, is to document the HFE/UE process followed during the design and development of the device. Considering the straightforward nature of such a process, it is to be expected that similar, if not identical, information would be included in such a document. This is why it typically requires little extra effort to complete HFE/UE reports for other agencies once the first version is authored.

The following table compares the contents of the human factors engineering/usability engineering report submitted to the FDA, the human factors summary report for the MHRA, and the usability engineering report meant to demonstrate conformance to IEC standards. Each row of the table identifies how the different guidance and standards reference similar information (Table 17.1).

**TABLE 17.1** HFE/UE report contents between guidance and standards.

| FDA guidance | MHRA guidance | IEC 62366-2:2016 |
|---|---|---|
| Descriptions of intended users, uses, use environments, and training | Descriptions of intended users, uses, use environments, and training | Summary of the use specification |
| Description of device user interface | Description of device user interface | Description of the user interface |
| Summary of known use problems | Summary of known use problems | Summary of known use problems |
| Analysis of hazards and risks associated with use of the device | User task selection and prioritization based on the risk management file | Description of hazard-related use scenarios |
| Summary of preliminary analyses and evaluations | Summary of formative evaluations | Summary of formative evaluations |
| Description and categorization of critical tasks | User task selection and prioritization | Description of the hazard-related use scenarios evaluated and why they were chosen |
| Details of human factors v alidation testing | Results of summative usability testing | Summary of summative evaluation |
| Conclusions | • The benefit-risk status of the device from the risk management file<br>• Conclusions | • Executive summary<br>• Conclusions |

**TABLE 17.2** Recommended outline for HFE/UE reports per agency.

| Section | FDA | MHRA | IEC |
|---|---|---|---|
| 1 | Conclusion | Intended users, uses, environments and training | Executive summary |
| 2 | Descriptions of intended device users, uses, use environments, and training | Description of the device user interface | Summary of the use specification |
| 3 | Description of device user interface | Summary of known use problems | Description of the user interface |
| 4 | Summary of known use problems | User task selection and prioritization based on the risk management file | Summary of known use problems |
| 5 | Analysis of hazards and risks associated with use of the device | Summary of formative evaluations | Description of the hazard-related use scenarios evaluated and why they were chosen |
| 6 | Summary of preliminary analyses and evaluations | Results of summative usability testing | Summary of the formative evaluations |
| 7 | Description and characterization of critical tasks | The benefit-risk status of the device from the risk management file | Summary of the summative evaluation |
| 8 | Details of human factors validation testing | Conclusions | Conclusion |

In addition to requiring similar content, the FDA, MHRA, and IEC guidance and standards suggest similar methods for organizing and presenting the information. For the most part, they all suggest presenting the information according to the performance order of the activities included in the HFE/UE. The following table lists the recommended outline for HFE/UE reports (Table 17.2).

Of particular note is the organization of the HFE/UE report as recommended by the FDA. The guidance suggests starting the report with the Conclusions. While this might seem counter-intuitive, doing so serves a similar purpose as starting the report with an Executive Summary. Starting with the Conclusions allows the report reader to receive a summary of the of the HFE/UE process, and an overview of the HF validation results including a discussion of the residual-risk of the device, and whether the device is considered safe and effective.

It is recommended that HFE/UE report authors follow the outline prescribed by the guidance or standard most relevant to their regulatory strategy. This might also result in the need to create different versions of the same report that differ only in organization should the plan be to submit the device to multiple agencies. By adhering to the proposed outline, the report is more likely to satisfy a reviewer's expectations (e.g., where to locate specific information). This in turn facilitates their review as they are more likely to perceive the report as efficient, effective, and strong.

## 5. HFE/UE report generation tips

Authoring a comprehensive HFE/UE report can be a daunting task. It is not uncommon for such reports to total hundreds of pages. There are a number of good practices and suggestions though that, if followed, can ease the effort while ensuring an effective report.

- Follow agency guidance and standards regarding the creation of HFE/UE reports. While such guidance is considered non-binding, the suggested report contents represent the agencies' most current thinking regarding what information is necessary to determine whether a device can be used safely and effectively.
- Organize the HFE/UE report following the recommended report structure of the agency receiving the submission (see Table 17.2). The provided outlines reflect reviewers' expectations for the report. To facilitate their review, it is advisable to match those expectations.
- Begin drafting the report during development. Much of the information included in the HFE/UE report is created during early conceptualization, design, and requirement generation efforts. The use-related risk analysis and formative evaluations are on-going efforts that start early in the process. Each time any design input (e.g., use specification, user profiles, risk analysis, formative evaluation) is created, that information should immediately be added to the HFE/UE report. This will help reduce the report writing effort necessary once HF validation is complete. Notably, should the HF validation protocol be submitted to the FDA for review, the preliminary information (i.e., Sections 2–7 of the FDA HFE report) is often requested to aid the reviewer in understanding the device.
- Be concise. While the goal should be to provide enough detail in the HFE/UE report, care should be taken not to add redundant or unrelated information. Redundancy and non-relevant information adds bloat to the report without providing any additional value. A concise report will make it easier to read and understand.
- Use images. It is recommended that the HFE/UE report contains as many images (e.g., illustrations, photographs) as needed to effectively convey the device's design. Sometimes an image, combined with an explanation, can be more effective compared to a narrative description alone. Using images will help ensure the agency reviewer best understands the nature of the device and how users are intended to interact with it. Failure to accomplish this can result in reviewer confusion and unnecessary requests for additional information and/or clarification.
- Provide traceability. As previously stated, the goal of the HFE/UE report is to demonstrate how the device design evolved over time as a result of following an HFE process. As different HFE activities are conducted, the results likely impact other HFE aspects. These linkages should be clear by using effective cross-references between sections.

One example of this linkage is how a known use-related issue (e.g., Section 4 of the FDA HFE/UE report) is captured in the use-related risk analysis (e.g., Section 5 of the FDA HFE/UE report). The fact that the known issue was captured in the risk analysis should be indicated in Section 4. That risk in turn might be mitigated through design (e.g., described in Section 3 of the FDA HFE/UE report). The design description should reference that it is a mitigation for a risk indicated in the use-related risk analysis. Any formative evaluations of that design should be summarized (e.g., Section 6 of the FDA HFE/UE report) and cross-referenced with the design discussion in Section 3 and risk analysis in Section 5. This example can be further extended if the risk is associated with a critical task (e.g., Section 7) which is then included in the HF validation study (e.g., Section 8). As much cross-referencing should be utilized when appropriate to help demonstrate this traceability across multiple HFE activities.

Good traceability will highlight to the report reviewer that not only were the proper HFE activities performed, but their results were utilized to make informed design decisions.

## 6. Summary

Writing and HFE/UE report can be a long, arduous task. When this happens, it is usually because the report is just being started after completion of the HF validation study. Even if the report is begun during the validation effort, it can still be a significant effort. However, this does not need to be the case. Much of the information contained in the report is created early in the design process and constitutes design inputs. By utilizing these inputs and starting the report early in the design and development process, project teams can minimize the stress, and even angst, that comes with writing such a large report soon before the entire project is to be submitted to an agency.

The HFE/UE guidance and standards provide direction and recommendations for constructing a comprehensive HFE/UE report. Following this guidance is highly recommended to ensure the proper content is included in the report and that the report meets reviewers' expectations.

Finally, the HFE/UE report is much easier to write if a proper HFE process was followed during device design and development. A proper process will result in the right information being generated that can be easily incorporated into a report. This will more than likely result in a strong HFE/UE report. From the agency reviewers' perspective, a strong HFE/UE report will convey that the developer has indeed followed a good HFE/UE process. This can only undoubtedly increase their confidence that the stated conclusion that the device "has been found to be safe and effective for the intended users, uses, and use environments" is indeed accurate.

## 7. Further reading

**Food and Drug Administration.** Applying Human Factors and Usability Engineering to Medical Devices, Guidance for Industry and Food and Drug Administration Staff. Rockville, MD: FDA, February 3, 2016.

**IEC TR 62366-2:2016.** Medical devices—Part 2: Guidance on the application of usability engineering to medical devices. Geneva: International Electrotechnical Commission. (Adopted in the United States as AAMI/IEC TIR62366-2:2016.) Section 18: Document the Usability Engineering Project.

**Wiklund M, Birmingham L, Alpert Larsen S.** Writing Human Factors Plans & Reports for Medical Technology Development. Arlington, VA: AAMI, 2018.

## References

Food and Drug Administration. (February 3, 2016). *Applying human factors and usability engineering to medical devices, guidance for industry and food and drug administration staff.* Rockville, MD: FDA. Appendix A: HFE/UE Report).

IEC 62366-1. (2015). *Medical devices—application of usability engineering to medical devices.* Geneva: International Electrotechnical Commission (Adopted in the United States as ANSI/AAMI/IEC 62366:2015.).

IEC TR 62366-2. (2016). *Medical devices—part 2: Guidance on the application of usability engineering to medical devices.* Geneva: International Electrotechnical Commission (Adopted in the United States as AAMI/IEC TIR62366-2:2016.) Section 18: Document the Usability Engineering Project.

ISO 14971. (2007). *Medical devices—application of risk management to medical devices.* Geneva: International Organization for Standardization (Adopted in the United States as ANSI/AAMI/ISO 14971:2007.).

Medicines & Healthcare Products Regulatory Agency. (September 2017). *Human factors and usability engineering-guidance for medical devices and including drug-device combination products.* London: MHRA. Section 5.4.6: Human Factors Summary Report.

# PART VII

# Special cases

CHAPTER

# 18

# Special cases: introduction

*Mary Beth Privitera*

HS Design, Gladstone, NJ, United States

OUTLINE



## 1. Introduction

The next section of this book provides insights on situations or advanced technology development that are unique. They represent situations in which human factors may not have been considered (at the onset) for the entire lifecycle of device use (Chapter 19); they are best practices for maximum inclusion of users (Chapter 20), especially for home use devices; and there is little or no human factors data published to support an advanced technology (Chapter 21).

The section below highlights their importance when considering applying human factors in medical device design.

## 2. Human factors of reusable medical equipment

There is little guidance in regards to applying human factors to the unique situation of reprocessing medical devices. As such, Chapter 19 describes the challenges found in

*Applied Human Factors in Medical Device Design*
https://doi.org/10.1016/B978-0-12-816163-0.00018-9

the execution of reprocessing. This topic, is highly debated in some circles and is no doubt a matter of patient safety. Key takeaways include:

- Consider applying human factors beyond the product itself and into the requirements of reprocessing.
- Support reprocessing requirements with design features that readily enable disassembly-reassembly without compromise to functionality and incorporate principles of use obviousness.

## 3. Considerations for users with limitations

Knowing the user is the first principal in the application of human factors, however often it is forgotten that users may not look typical or fit into charts that represent healthy adults. Chapter 20 highlights the importance of defining user groups that may warrant further delineation in regards to sensory, cognitive, movement or environmental conditions. Key takeaways include:

- Importance of engaging users with limitations throughout the product design process
- Leveraging community resources, discussions with clinicians and caretakers might be required and ultimately their feedback can improve design

## 4. Augmented reality in medical devices

Augmented reality is increasingly used through smartphones, tablets and head-mounted displays with the goal of enhancing the real world with virtual elements in an interactive manner. This technology is quickly being adopted in the medical field through telemedicine and remote collaboration as well as through visual aids in procedures. Chapter 21 provides a detailed definition on augmented reality and requirements for designing for augmented reality medical devices. Key takeaways include:

- Considering three new modes of interaction afforded by the technology: gaze, gesture and voice.
- Influences of existing technology, the maintenance of sterility and environmental conditions which can be maximized by this hands-free technology
- Human factors of taking a typically complex cognitive task and revising it with interactivity and exploration has the potential to enhance clinical experiences.

CHAPTER

# 19

# The Human Factors of Reprocessing Reusable Medical Equipment

*Emily A. Hildebrand*[a], *L. Bryant Foster*[a], *Russell J. Branaghan*[b]

[a]Research Collective, Tempe, AZ, United States; [b]Arizona State University, Mesa, AZ, United States

OUTLINE



## 1. Introduction

Infection control is critical to patient safety. In the United States each year brings approximately two million healthcare-associated infections (HAIs) resulting in approximately 90,000 deaths (Stone, 2009). Single-use medical devices, which are used on one patient and then discarded, have helped reduce the number of HAIs. Unfortunately, not all devices can be made disposable. For example, endoscopes are complex devices which cost thousands of

*Applied Human Factors in Medical Device Design*
https://doi.org/10.1016/B978-0-12-816163-0.00019-0

dollars, so making them disposable is prohibitively expensive. As a result, devices like endoscopes must be reprocessed, or cleaned and sterilized after each use. The problem is that many HAIs have been traced to errors in reprocessing procedures resulting in contaminated reusable devices. Failure to properly reprocess a medical device carries the risk of person-to-person transmission of viruses such as HIV, Hepatitis C, Hepatitis B, or other bacteria, as well as the transmission of environmental pathogens such as pseudomonas (Mehta et al., 2006; Nelson, 2003; Weber & Rutala, 2001).

Advances in medical science, minimally invasive surgical techniques, and diagnostic procedures have led to improved complex reusable medical devices, such as flexible video duodenoscopes and robotic surgical instruments, which require equally complex reprocessing procedures. Improving reprocessing is not only critical for reducing transmission of HAIs, but is imperative for maintaining trust in healthcare (i.e., if patients are afraid of acquiring infections they may be less likely to pursue preventative or needed medical treatments) and reducing healthcare costs. Healthcare facilities in the United States spend 9.8 billion dollars annually to treat the top five HAIs, and spend countless dollars more to address malpractice and wrongful death lawsuits associated with HAIs (AMA, 2013).

Hazard analyses and incident reports following reprocessing errors often cite use error in complying with manufacturer recommendations. However, design flaws are often disguised as human error or negligence in reprocessing of reusable devices. This chapter presents and discusses the human factors issues, best practices and guidelines for reprocessing reusable medical devices. Although endoscopes are used as an example, the information presented is applicable to all reusable medical devices.

## 2. Why is endoscope reprocessing a problem?

Endoscopes are not the only devices that are reprocessed, however they have become the poster child for the "issues in reprocessing" movement. Endoscope reprocessing has been listed as a top health technology hazard by the ECRI Institute over the last 10 years (ECRI, 2018). Recurring infectious outbreaks due to errors in endoscope reprocessing have been well publicized resulting in the scientific literature and the popular media asking "how clean are those reusable medical scopes" (Daly, 2017); and research shows that they are not clean enough. A 2018 study by Ofstead et al. studied endoscopes deemed "ready for use" in three separate hospitals. They found 71% tested positive for bacteria, indicating that the reprocessing procedure was not effective. The authors concluded that "unfortunately, there has been little to no improvement in this area over the past 10 years" (Ofstead et al., 2018).

### 2.1 What is reprocessing?

Reprocessing is the process which renders a clinically used device safe and ready for its intended reuse (AAMI ST91). The Spaulding classification system, developed in the 1960s, has been used to classify medical devices into categories based on their use and defines the level of cleaning, disinfection, and/or sterilization needed (Ramakrishna, 2002; AAMI ST91) based on the risk of infection involved with their reuse (Petersen et al., 2017). Using the Spaulding classification system, devices are identified as critical, semi-critical, or non-critical as described below (Ramakrishna, 2002; AAMI ST91; FDA Reprocessing; Petersen et al., 2017):

- Critical devices penetrate skin or mucosa and carry a high risk of infection. Critical devices should be sterilized between uses. Examples include biopsy forceps.
- Semi-critical devices, such as endoscopes, do not penetrate intact mucous membranes and should be sterilized or high-level disinfected.
- Non-critical devices, such as blood pressure cuffs or patient carts, only contact intact skin and have very little risk of transmitting infection. Non-critical devices do not need to be sterilized and may be cleaned using low-level disinfection.

## 2.2 Endoscope reprocessing

***Endoscopes***, which fall into the semi-critical category, can be sensitive to heat, thus high-level disinfection is a commonly practiced method of reprocessing.

Endoscope reprocessing typically includes six tasks (Fig. 19.1).

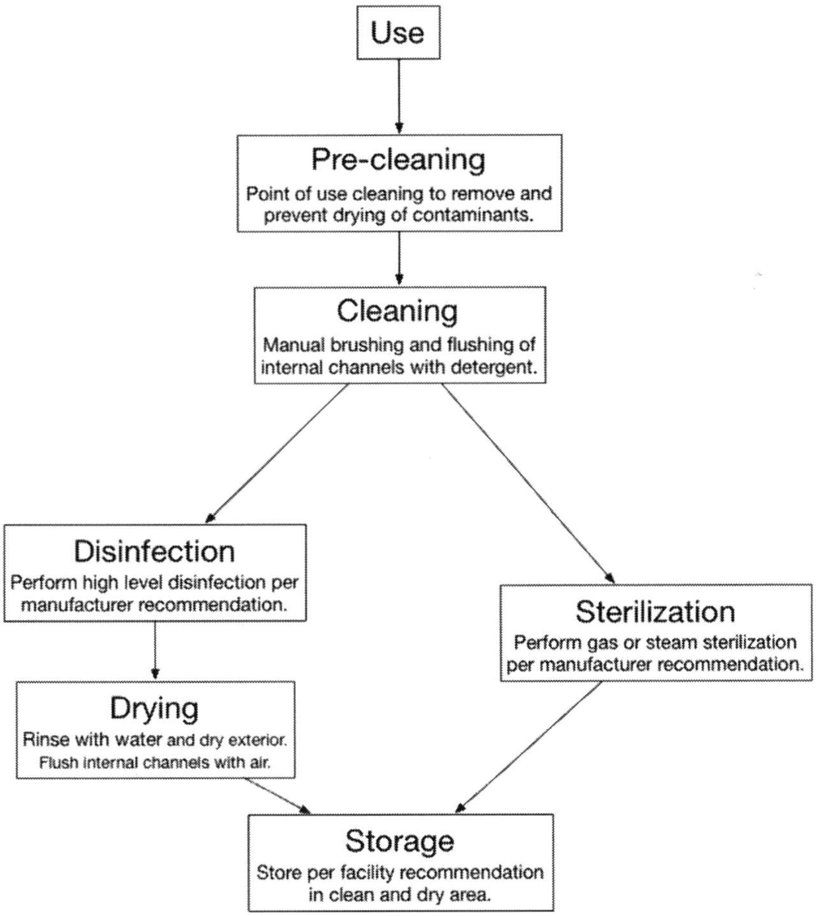


FIG. 19.1 Overview of the endoscope reprocessing procedure.

1. Precleaning or point of use cleaning: This happens immediately after a patient procedure, while the device is still in the procedure room. Its purpose is to remove gross contaminants from the endoscope before it dries by wiping the exterior and flushing the internal channels with water or detergent solution.
2. Leak Testing: This process detects damage in the endoscope prior to manual cleaning. Ensuring an endoscope is mechanically sound is necessary for the cleaning process to be effective.
3. Manual Cleaning: The endoscope is soaked in detergent solution and remaining gross contaminants are removed by manually and thoroughly wiping, brushing, flushing, and rinsing the exterior and internal channels of the device.
4. Disinfection or Sterilization: The choice to perform high-level disinfection or sterilization is made based on manufacturer recommendations. Whether or not an endoscope can be sterilized depends its materials and its ability to withstand different sterilization methods. High-level disinfection can be conducted manually, which involves soaking and flushing the endoscope with a germicidal agent. Alternatively, an automated endoscope reprocessor (AER), which performs that same steps in an automated machine, can be used.
5. Drying: After high-level disinfection, the endoscope is rinsed to remove residual cleaning agents. Then, endoscopes are flushed with forced air and typically flushed with alcohol to promote drying.
6. Storage: Proper handling during storage is important to ensure that contaminants are not reintroduced to the clean endoscope. Endoscopes should be hung vertically in clean, dry, well-ventilated, and humidity-free areas.

Depending on the type of endoscope and the manufacturer, the steps above can vary widely. Further, although not addressed in the steps above, endoscopes come equipped with disposable and reusable accessories. Technicians must identify and discard disposable accessories and reprocess the reusable accessories at the appropriate points in the reprocessing procedure.

## 3. Human factors issues in endoscope reprocessing

Reviews of existing literature on the causes of reprocessing errors usually places the blame on the human operator or "user" error. These analyses rarely take into account the interaction between the user and the device design, nor do they address the instructions-for-use (IFU), training, or environment (Hildebrand et al., 2010). Instead, most reports cite noncompliance with guidelines or reprocessing instructions as the primary culprit of errors (Nelson, 2003; Ofstead et al., 2010; Ramakrishna, 2002). Human factors practitioners, on the other hand, know that assigning blame to the user does nothing to mitigate problems and many problems classified as human errors are actually the result of poor device design.

The following sections discuss in detail three human factors issues that create difficulties for reprocessing technicians when interacting with reusable medical devices. Design principles and guidelines for these issues are also addressed. Understanding the challenges and difficulties users currently encounter help identify and develop solutions to minimize errors and improve reprocessing safety.

## 3.1 Device design

Put simply, ensuring safe reprocessing begins with the design of the device (AAMI TIR12). At present, unfortunately, medical device manufacturers focus foremost on preventing, diagnosing, monitoring and treating diseases and injuries (Malchesky, 1995). Reprocessing is at best an afterthought. Thus, it is no surprise that devices are difficult to clean. For example, some endoscopes have two internal channels accessed through the same port. During cleaning, it is difficult for users to distinguish the two channels and determine which one they are cleaning. Indeed, if distracted, users may skip cleaning one of these channels or accidently clean the same channel twice. One simple way to facilitate cleaning, is to add device labeling (e.g., text labels, color coding, etc.) related to reprocessing. Endoscopes already have multiple labels which serve as cues to clinicians during procedures, but there are rarely labels designed to cue reprocessing technicians.

A human centered design (HCD) process, including reprocessing technicians as a user group, could greatly reduce the likelihood of patient harm. HCD considers each device user early in the design process. In the case of reusable devices, reprocessing technicians should be considered as a primary user group. Manufacturers should learn about the needs, capabilities, limitations, and working environments of these technicians. General principles and guidelines for conducting and applying such user research are detailed in medical device guidance documents (AAMI HE75 2009; FDA, 2016; IEC/ISO 2015). Device design teams should contain at least one member who is responsible for ensuring the human factors of reprocessing. Further, reprocessing procedures should be evaluated through simulated-use usability studies with representative users as early as possible (AAMI TIR12).

Manufacturers should take care to design characteristics that have been identified as problematic for cleaning. The FDA (2016) has published lists of features that pose challenges for reprocessing, including hinges, shafts with lumens, elevator channels, porous materials, and ridges or grooves to name a few. Defining reprocessing design requirements during product development will help to avoid these design pitfalls.

In recent years, endoscope manufacturers have been under intense scrutiny for issues related to cleaning the forceps elevators on duodenoscopes. Despite following cleaning instructions appropriately, debris is often retained in these components on the devices. As a result, patients were exposed to infectious agents (FDA, 2018). Some manufacturers answered the call by designing single-use covers for the distal-end of the endoscope, to promote better cleaning (Fig. 19.2). Another manufacturer went a step further and designed a duodenoscope with a completely detachable and removable, single-use elevator

**FIG. 19.2** Pentax recently released a duodenoscope with Cap H.D., a single-use, removable, cover for the distal end. This cover affords users the ability to more easily clean the forceps elevator, which is one of the key areas for potential recontamination. *Image source: Pentax Medical (2018).*

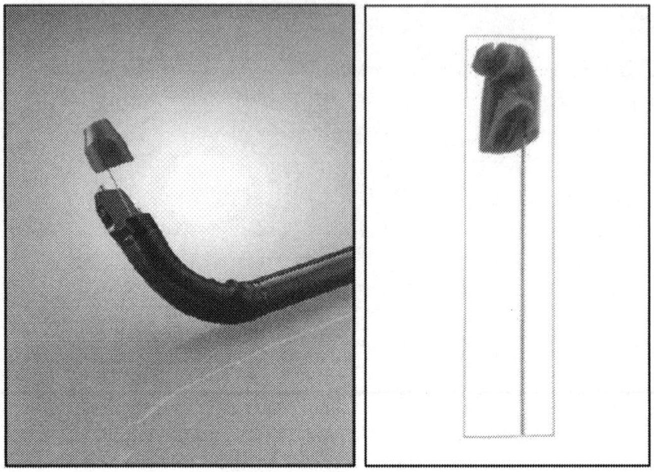

FIG. 19.3 KARL STORZ has designed a duodenoscope with an Albarran Module, a single-use (removable and disposable) elevator wire and forceps. This module importantly facilitates reprocessing by removing a component that is commonly difficult to clean. *Image courtesy of KARL STORZ.*

forceps module (Fig. 19.3). These types of innovations better support user performance in cleaning tasks and improve reprocessing safety.

Failing to design for reprocessing not only leads to complex devices, but also complex reprocessing procedures. For example, reprocessing a flexible endoscope can require over 200 individual steps and the use of over 20 different cleaning accessories (Hildebrand et al., 2010; Jolly et al., 2012). Moreover, facilities may reprocess as many as 250 endoscopes a week (Hildebrand et al. 2010, 2011; Ofstead et al., 2010). Taking an average, this could mean technicians are performing 10,000 steps per day. As human attention and working memory have limited capacities, each additional step in a cleaning process increases the likelihood of the scope being cleaned improperly due to steps being missed or completed incorrectly. Devising methods that provide immediate feedback and alert reprocessing technicians to failures could be critical for preventing potential errors. Further, methods for quickly evaluating the effectiveness of a reprocessing procedure, such as rapid cleaning monitors, could facilitate the recognition of improperly cleaned medical devices (Alfa, 2016).

## 3.2 Providing adequate instructions for use

In 2015, the FDA published guidance placing the responsibility for providing reprocessing instructions on the manufacturers of the reusable devices. Not only are manufacturers expected to provide instructions for reprocessing, but those instructions must be validated through usability testing to ensure that they can be understood by end users. (Alfa, 2016; FDA, 2015). Requiring manufacturers to provide useable instructions is important, however instructions should not be considered a primary means of mitigating potential harm as even the best designed instructions are unlikely to be used at the point of reprocessing. For example, it is difficult for users to reference a paper booklet in the wet environment of a reprocessing suite. Innovation beyond paper IFUs is needed to continue evolving and improving reprocessing practices. Design considerations for alternative methods of supporting and guiding users through reprocessing procedures could have substantial effects on user performance.

## 3.3 Training

Manufacturers commonly provide training via in-service sessions for the reprocessing staff. Regrettably, it is often challenging to assemble all of the reprocessing staff at once for training. Consequently, those technicians who attended training are left to train those who could not attend which may result in inconsistencies. Ultimately, healthcare facilities are responsible for ensuring that staff is trained to perform reprocessing as intended by the instruction and the manufacturer. Ideally, facilities have established practices for training technicians in reprocessing. Many incorporate an apprenticeship style approach where technicians learn from their predecessor. Unless training can be guaranteed by manufacturers across all facilities, it should not be relied upon as a method of mitigating harm when considering reprocessing in the device design process.

ANSI/AAMI ST91 provides guidance on implementing training and educational programs for flexible and semi-rigid endoscope processing in health care facilities. When developing training programs, whether it be an in-service, educational video, or visual aids, manufacturers should apply the same human factors approach to the design of instructional materials. Designers should ensure that information across all training mediums and the IFU are consistent and do not promote contradictory ideas. Finally, as with IFUs, manufacturers should seek to validate their training materials through use testing with representative users.

# 4. Case study

In 2009, the Department of Veterans Affairs (VA) mailed 10,000-plus letters to veterans, notifying them that they may have been exposed to or infected with bloodborne pathogens (including hepatitis B, hepatitis C, or HIV) due to improperly reprocessed endoscopes from procedures performed between 2003 and 2008 (ECRI, 2009, 2010). Most incident reports cited "reprocessing technicians inability to adhere to manufacturer cleaning instructions" (Hildebrand et al., 2010; Jolly, Hildebrand, & Branaghan, 2013; VA, 2009) (Fig. 19.4). In response, the VA National Center for Patient Safety funded a Patient Safety Center of Inquiry entitled the Center for Evaluation of Human Factors in Reprocessing Safety (CEHFRS). The goal of CEHFRS was to understand the issues involved in endoscope reprocessing from a human factors perspective. Not surprisingly, many, if not all, of the contributors to the problem listed above were identified quickly (Hildebrand et al., 2010).

One research effort identified and characterized challenges associated with IFUs for reprocessing. An initial usability study asked novice users to simulate endoscope reprocessing relying only on the reprocessing IFU and a supporting visual aid (i.e., manufacturer-provided poster). The results showed that none of the participants were able to reprocess the endoscope without error. Qualitative data analysis confirmed earlier research efforts (Hildebrand et al., 2010, 2011) and found that there were three major root causes: high memory demands; lack of visibility; and insufficient feedback (Jolly et al., 2012).

Norman (1993) stated, "The power of the unaided mind is highly overrated" (p. 43). Without external aids, our memory, thought, and reasoning are constrained. When well designed, external aids can complement our abilities, strengthen our mental powers, and help us overcome our own limits. However, in the case of reprocessing, poorly designed

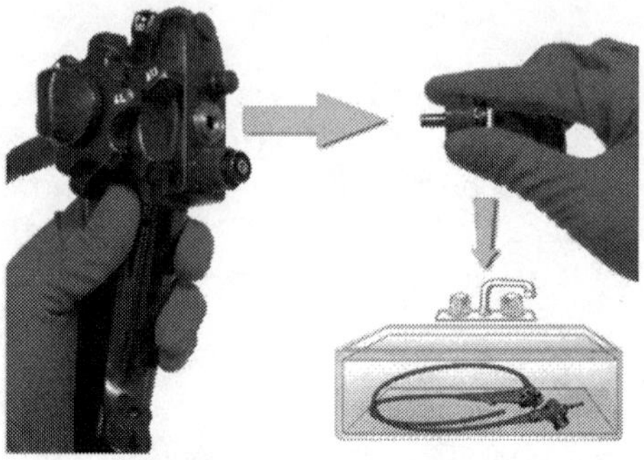

FIG. 19.4 Vision was supported by using consistent design and formatting conventions. The image above depicts green "action arrows" which were used to consistently inform the user of when and where to move a component of the endoscope. (For interpretation of the references to colour in this figure legend, the reader is referred to the web version of this article.) *Image source: Jolly, Hildebrand, & Branaghan, 2013.*

support materials contributed to improperly cleaned endoscopes (Jolly et al., 2012). Accordingly, CEHFRS set out to redesign a cognitive aid, utilizing human factors design principles, that could effectively guide a user through the entire reprocessing procedure. During the design efforts, specific focus was given to address known usability issues by:

- Increasing visibility
- Reducing memory demands
- Providing adequate feedback

*Visibility* was supported by ensuring that visual aspects of the IFU were consistent throughout (Fig. 19.5). The same design elements (e.g. green arrows) were used repeatedly to indicate actions. Additionally, showing part tools with similar names up close, with clear pictures, made it easy to discriminate them.

*Memory demands* were addressed by reducing complexity in the instructions, while at the same time avoiding oversimplification. These improvements replaced the need to remember details by placing necessary knowledge in front of the user instead of requiring knowledge to be maintained in their memory. This was achieved by focusing on the following concepts:

- Orientation: The endoscope and tools were shown in a first person point of view to limit the need for mental rotation.

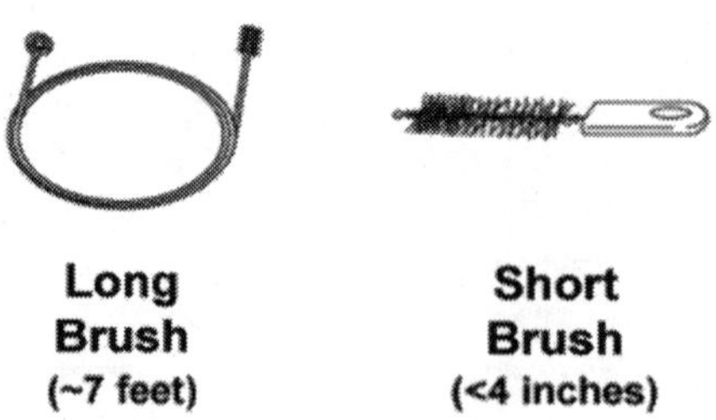


FIG. 19.5 Brushes with similar names are illustrated next to each other to assist the reader in discriminating between them. *Image source: Jolly et al. (2013).*

FIG. 19.6 This image depicts the attachment of a cleaning adapter to the endoscope. Visual and textual cues about how to physically manipulate the adapter (e.g., "push" and "slide") afford users the necessary steps to complete this task. *Image source: Jolly et al. (2013).*

- Visuals & Words: Text referring to a visual was placed in closeness, proximity to support understanding.
- Vocabulary: Text facilitated memory by removing similar sounding or confusing terms and limiting technical jargon (Fig. 19.6).

*Feedback* was incorporated by leveraging visual and auditory clues that users could rely on to ensure they completed a step correctly and completely (Fig. 19.7).

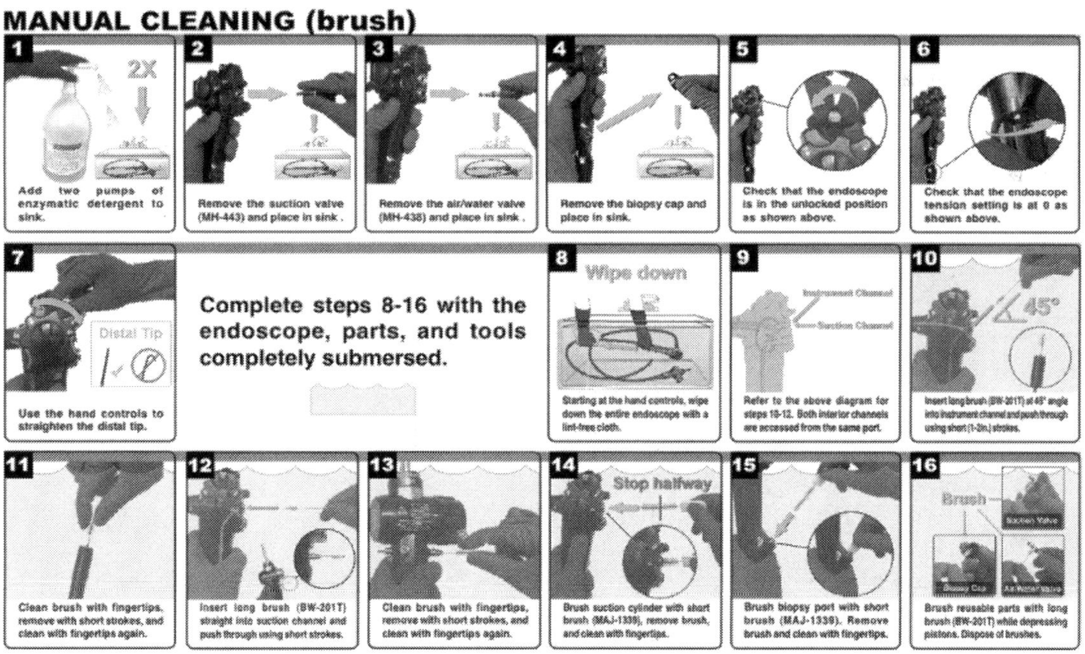


FIG. 19.7 Cognitive aid prototype for the "Manual Cleaning – Brushing" instructions provided to participants during usability testing of an endoscope reprocessing procedure. *Image source: Jollyet al. (2013).*

Below is an example of one cognitive aid prototype that was created utilizing the design goals and considerations above. In total, three cognitive aids were created to represent larger complex steps in the endoscope reprocessing procedure: leak testing, manual cleaning (brushing) and manual cleaning (flushing).

To assess the effectiveness of the cognitive aids, a comparative (between subjects) usability test was conducted in which novice participants simulated reprocessing relying only on the cognitive aids as a guide. The human factors-based cognitive aids were found to provide participants with significantly better support to perform reprocessing. Participants in the cognitive aid usability study completed more than 87% of the reprocessing tasks correctly, compared to less than 45% in an initial usability study. Participants using the cognitive aids were also significantly faster. For a detailed write-up of the process, evaluation, and findings see Jolly et al. (2013).

## 5. Summary

As the medical device industry proliferates and the need for complex reusable medical equipment increases, human factors in reprocessing needs to be at the forefront of the design process. If errors are made during reprocessing, devices may do more harm than good. Implementing human factors in the design of reusable medical equipment and reprocessing procedures will result in safer and more effective devices.

Key takeaways:

- Human factors issues for reprocessing should be considered early in the design process and should involve user research, iterative design, and evaluation through usability testing
- Design features that afford cleaning (easy-to-disassemble, disposable) and utilize labeling (color-coding, text, icons) to support reprocessing tasks should be implemented in to the design or reusable medical equipment

## 6. Further reading

Sources of information for applying human factors approach to reprocessing of reusable medical devices includes, but is not limited to, the following:

- AAMI TIR12: Designing, testing, and labeling reusable medical devices for reprocessing
- AAMI TIR55: Human factors engineering for processing medical devices
- AAMI HE75: Human factors engineering - Design of medical devices
- ANSI/AMMI ST91: Flexible and semi-rigid endoscope processing in health care facilities
- FDA guidance: Applying Human Factors and Usability Engineering to Medical Devices
- FDA guidance: Reprocessing Medical Devices in Health Care Settings: Validation Methods and Labeling
- IEC/ISO TR 62366: Application of usability engineering to medical devices

## References


Alfa, M. J. (2016). Current issues result in a paradigm shift in reprocessing medical and surgical instruments. *American Journal of Infection Control, 44*(5), e41–e45.

American Medical Association (AMA). (September 2, 2013). *Costs of healthcare-associated infections: 9.8 billion annually in US*. Science Daily. Retrieved from https://www.sciencedaily.com/releases/2013/09/130902181001.htm.

Association for the Advancement of Medical Instrumentation (AAMI TIR12). (2010). *Designing, testing, and labeling reusable medical devices for reprocessing in health care facilities: A guide for medical device manufacturers*. TIR12.

Association for the Advancement of Medical Instrumentation (AAMI TIR55). (2017). *Human factors engineering for processing medical devices*. TIR55.

Association for the Advancement of Medical Instrumentation (AAMI HE75). (2009). *ANSI/AAMI HE75: 2009. Human factors engineering—Design of medical devices*.

Association for the Advancement of Medical Instrumentation (AAMI ST91). (2015). *ANSI/AAMI ST91: 2015, flexible and semi-rigid endoscope processing in health care facilities* (Arlington, Vol. A).

Daly, J. (July 3, 2017). *Endoscope safety: How clean are those reusable medical scopes?* Pittsburgh Post-Gazette. Retreived from http://www.post-gazette.com/news/health/2017/07/04/Endoscope-safety-Taking-a-good-look-at-a-reusable-medical-scope-key-part-of-effective-cleaning/stories/201706270010.

Department of Veterans Affairs Office of Inspector General (VA). (2009). *Healthcare inspection: Use and reprocessing of flexible fiberoptic endoscopes at VA medical facilities (VAOIG Pub. No. 09-01784-146)*. Washington, DC: Author. Retrieved from http://www.va.gov/oig/54/reports/VAOIG-09-01784-146.pdf.

ECRI Institute (ECRI). (2009). *U.S. Veterans Health Administration announcements highlight need for comprehensive endoscopy-reprocessing protocols* (Special report). Health Devices Alerts 2009 Apr 17. Accession no. S0193.

ECRI Institute (ECRI). (2010). Clear channels: Ensuring effective endoscope reprocessing. *Health Devices, 39*(10), 350–359.

ECRI Institute (ECRI). (2018). *Top 10 health technology hazards for 2018*. Health Devices. Retrieved from https://mtintegraal.nl/media/articles/583/attachments/Haz_top10_ECRI_2018.pdf.

Food and Drug Administration (FDA). (2015). *Reprocessing medical devices in health care Settings: Validation methods and labeling*.

Food and Drug Administration (FDA). (2016). *Guidance for industry and FDA staff: Applying human factors and usability engineering to medical devices*.

Food and Drug Administration (FDA). (2018). *Infections associated with reprocessing duodenoscopes*. Retrieved from https://www.fda.gov/medicaldevices/productsandmedicalprocedures/reprocessingofreusablemedicaldevices/ucm454630.htm.

Hildebrand, E. A., Branaghan, R. J., Neuman, B. L., Jolly, J., Garland, T. B., Taggart, M., et al. (2011). An expert perspective of errors in endoscope reprocessing. In *Proceedings of the human factors and ergonomics society annual meeting* (pp. 748–752). Sage CA: Los Angeles, CA: SAGE Publications. Vol. 55, No. 1.

Hildebrand, E. A., Branaghan, R. J., Wu, Q., Jolly, J., Garland, T. B., Taggart, M., et al. (2010). Exploring human factors in endoscope reprocessing. In *Proceedings of the human factors and ergonomics society annual meeting* (pp. 894–898). Sage CA: Los Angeles, CA: SAGE Publications. Vol. 54, No. 12.

IEC/ISO. (2015). *62366-1: 2015 medical devices part 1: Application of usability engineering to medical devices*. Geneva: International Organization for Standardization.

Jolly, J. D., Hildebrand, E. A., & Branaghan, R. J. (2013). Better instructions for use to improve reusable medical equipment (RME) sterility. *Human Factors, 55*(2), 397–410.

Jolly, J. D., Hildebrand, E. A., Branaghan, R. J., Garland, T. B., Epstein, D., Babcock-Parziale, J., et al. (2012). Patient safety and reprocessing: A usability test of the endoscope reprocessing procedure. *Human factors and ergonomics in manufacturing & service industries, 22*(1), 39–51.

Malchesky, P. S., Chamberlain, V. C., Scott-Conner, C., Salis, B., & Wallace, C. (1995). Reprocessing of reusable medical devices. *ASAIO Journal (American Society for Artificial Internal Organs: 1992), 41*(2), 146–151.

Mehta, A. C., Prakash, U. B. S., Garland, R., Haponik, E., Moses, L., Schaffner, W., et al. (2006). Prevention of flexible bronchoscopy-associated infection. *Chest, 128*, 1742.

Nelson, D. B. (2003). Infectious disease complications of GI endoscopy: Part II, exogenous infections. *Gastrointestinal Endoscopy, 57*, 695–711.

Norman, D. (1993). *Things that make us smart: Defending human attributes in the age of the machine*. New York, NY: Basic Books.

Ofstead, C. L., Heymann, O. L., Quick, M. R., Eiland, J. E., & Wetzler, H. P. (2018). Residual moisture and waterborne pathogens inside flexible endoscopes: Evidence from a multisite study of endoscope drying effectiveness. *American Journal of Infection Control, 46*(6), 689–696.

Ofstead, C. L., Wetzler, H. P., Snyder, A. K., & Horton, R. A. (2010). Endoscope reprocessing methods: A prospective study on the impact of human factors and automation. *Gastroenterology Nursing, 33*(4), 304–311.

Pentax Medical. (October 2, 2018). *Pentax medical receives breakthrough technology award from Premier Inc. For C.A.P. HD Duodenoscope*. Retrieved from: https://www.pentaxmedical.com/pentax/en/99/1/PENTAX-MEDICAL-RECEIVES-BREAKTHROUGH-TECHNOLOGY-AWARD-FROM-PREMIER-INC-FOR-CAP-HD-DUODENOSCOPE.

Petersen, B. T., Cohen, J., Hambrick, R. D., Buttar, N., Greenwald, D. A., Buscaglia, J. M., & Eisen, G. (2017). Multi-society guideline on reprocessing flexible GI endoscopes: 2016 update. *Gastrointestinal Endoscopy, 85*(2), 282–294.

Ramakrishna, B. S. (2002). Safety of technology: Infection control standards in endoscopy. *Journal of Gastroenterology and Hepatology, 17*(4), 361–368.

Rutala, W. A., & Weber, D. J. (2004). Disinfection and sterilization in health care facilities: What clinicians need to know. *Clinical Infectious Diseases, 39*(5), 702–709.

Stone, P. W. (2009). Economic burden of healthcare-associated infections: An American perspective. *Expert Review of Pharmacoeconomics and Outcomes Research, 9*(5), 417–422.

CHAPTER

# 20

# Considerations for users with limitations

*Melissa R. Lemke, Jessie Huisinga*

Agilis Consulting Group, LLC., Cave Creek, AZ, United States

OUTLINE



*When you include the extremes of everybody, that's to say differently abled people of all sorts, then you produce things that are better for all.* **Michael Wolff, Wolff Olins**

## 1. Introduction

Users with limitations are abundant within the medical device field. This is due to our aging society, due to temporary and permanent limiting conditions for which individuals seek healthcare services, and due to environmental conditions that influence user abilities. Healthcare professionals with limitations also exist given the aging workforce, repetitive stress injuries, and environmental conditions which include multitasking, interruptions, noise, etc.

*Applied Human Factors in Medical Device Design*
https://doi.org/10.1016/B978-0-12-816163-0.00020-7

Although topics such as disability, functional limitations, and accessible design are most often considered for "special user populations," users with limitations are typically more the norm than the exception for medical devices.

When considering users with limitations, it is important to acknowledge that health conditions can cause functional or performance related limitations which can be visible or invisible; temporary or long term; static, episodic, or degenerating; painful or inconsequential (WHO, 2011). User limitations also account for diverse user characteristics and capabilities that can impact user interactions with a medical device, which ultimately influences the use related safety and effectiveness of medical devices. In home healthcare environments, for example, a common thinking is that the majority of users are non-disabled (Fig. 20.1: Adapted from Accessibility and Home Healthcare (Lemke & Mendonca, 2013)). However, the more realistic view is that many users do have limitations and also encounter environmental limitations.

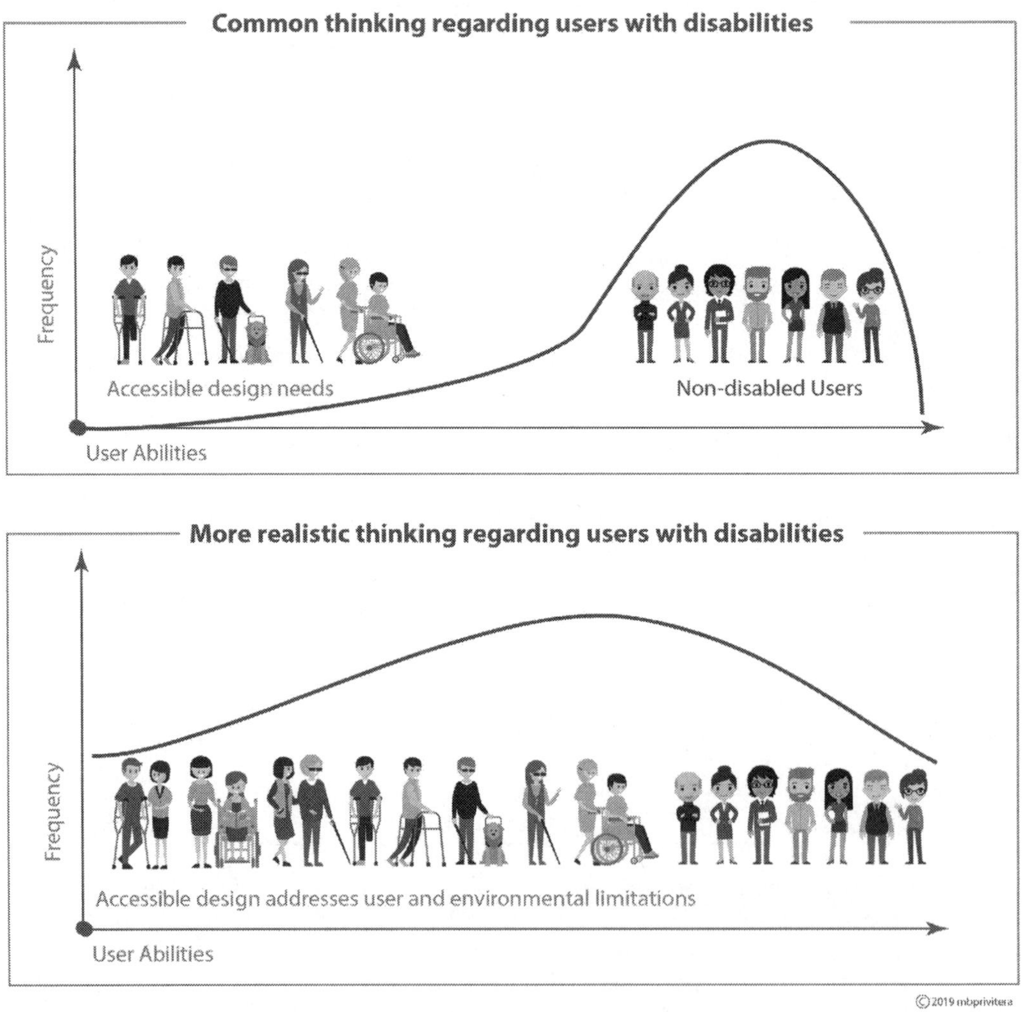


**FIG. 20.1** Representative thinking about users with disabilities in home healthcare. *Adapted from* Biomedical Instrumentation and Technology, *47, 20–25 (2013).*

Importantly, considering users with limitations during device design can expand the market segment for which a product is useful, improve use related safety and effectiveness, better address diverse and dynamic user needs, reduce post market complaints and required updates, etc. Similarly, including representative users with limitations during testing can highlight potential design features which are inconsistent with user abilities during specific tasks and can allow for design mitigations to be introduced before the device is released and post-market issues occur. This chapter focuses on the important considerations during the design and evaluation of most medical devices.

For designers and developers of medical devices, it is necessary to remember that users with limitations, whether due to illness or due to age, need to have access to devices which help to alleviate their health problems. Without this access, individuals are effectively made to be disabled further. Therefore, there is an implicit responsibility for those who are able to optimize medical devices for users with limitations to do so.

## 2. Relevant statistics about users with limitations

The most common interpretation of users with limitations is to equate them to people with disabilities. While people with disabilities fall into the category of users with limitations, it is important to consider those with naturally occurring limitations as well. When considering users with limitations, statistics indicate (WHO, 2018):

- Over one billion people, about 15% of the world's population, have some form of disability.
- Between 110 million and 190 million adults have significant difficulties in functioning.
- Rates of disability are increasing due to population aging and increases in chronic health conditions, among other causes.
- People with disabilities have less access to health care services and therefore experience unmet health care needs.
- The American Community Survey (ACS) estimates the overall rate of people with disabilities in the United States population in 2015 was 12.6% (Institute on Disability, 2016).

Far greater than the number of users who have specific disabilities, the number of elderly individuals highlights a huge population that could have naturally occurring limitations due to aging. By 2050, the global population of people older than 60 is expected to jump to two billion (UN, 2015). In the United States alone, the number of individuals ages 65 and older is projected to more than double from 46 million today to over 98 million by 2060, which is nearly 24% of the total population (PRB, 2015). For medical device developers, the high percentage of aging individuals and the unmet health care needs of persons with disabilities represent a very large market of potential device users.

## 3. Example personas of users with limitations

**Meet Margie.** In her early 50's, Margie had to have hip replacement surgery. This was her first experience in a hospital, and she was nervous about the procedure. For the 30 days following the successful surgery, her surgeon prescribed oxycodone and Percocet. This was another first for Margie as she never 'did drugs' and was unaware how these powerful medications were affecting her. When Margie learned she was ready to move from a walker to a cane, she excitedly opened the instructions to assemble the cane. All she saw were engineering style line drawings with cryptic text that she could not understand. Completely unaware of how her medications were affecting her cognitively. Margie panicked thinking, if I can't assemble this cane, I can't improve and I'll never get be able to walk again! Fortunately, her sister was there to provide the support Margie needed and assembled the very simple cane that was completely intimidating to Margie while she was taking prescription pain medication. Margie experienced a cognitive limitation due to her prescribed medications.

**Meet Mike.** Mike is a husband and father with an MBA from a prestigious business school. His youngest child Matt was born with dwarfism and requires almost constant medical care. Woken in the middle of the night, Mike needed to quickly administer one of his son's many medications. Exhausted and working in low lighting, he accidently picked the injector with the wrong medication—they all looked the same. Fortunately, his wife Susan also awoke and stopped him in time. Mike was relieved that Susan stepped in, or they would have spent another night in the emergency room. Here, Mike experienced both a cognitive limitation as a result of his fatigue and an environmental limitation as a result of the lighting conditions.

**Meet Steve.** Steve is a nurse in his late 20's who works in a large metropolitan hospital with post-operative spinal surgery patients. Because of his age and strength, Steve often assists other nurses in transferring patients into or out of their hospital beds. Steve also assists patients in learning how to independently transfer into and out of bed after their spinal surgery. Having worked in this environment for several years, Steve has intermittent shoulder pain. When Steve's shoulder pain flares up, he cannot lift his arm above 60-degrees and cannot lift objects that weigh more than 15 pounds. During a pain flare up, Steve enters a patient's room to perform the patient's hourly vital checks where he encounters two difficulties caring for his patient. Because of the busy hospital hallway, Steve does not hear the alert of the vital signs monitor indicating the patient's oxygen saturation is low. Steve also cannot reach his arm up to get the blood pressure cuff that is hung over the elevated monitor in order to take his patient's blood pressure. Here, Steve experienced a movement limitation due to his shoulder injury and an environmental limitation due to the noisy hospital.

## 4. Defining user groups with limitations

According to the FDA/CDRH, a user is defined as a "person who interacts (i.e., operates or handles) with the device" (CDRH, 2016). The user's interaction with the device includes all elements of the device's user interface that the user may see, hear, or touch. In addition, the user's interaction includes all sources of information communicated by the device user interface, such as software prompts, instructions for use, packaging and labeling, as well as training and all physical controls and display components of the device, such as alarms and order of operation, for each component and the whole device. Simplified, any human factors evaluation of a device must consider:

- Every part of the device a user might see, hear, or touch
- Every piece of information communicated by the device to the user

- Any user training necessary to use the device
- Any physical controls or displays on the device the user must interact with

The necessary considerations for a successful human factors/usability study for all of these user interface elements is arduous even when the user would be considered completely healthy and without limitations. However, when a user group does have functional limitations, the evaluation of each of these device elements must be considered with even more intention. Importantly, AAMI and the FDA's CDRH guidance recommends that user limitations be considered early in the device design process where users could provide critiques to existing devices and offer insights about what features could optimize user interactions with the device (AAMI, 2012). Prior to evaluating devices with potential users, it is helpful to understand how user limitations and risk are related. CDRH guidelines specifically outline that *"you should evaluate and understand the characteristics of all intended user groups that could affect their interactions with the device and describe them for the purpose of HFE/UE evaluation and design. These characteristics should be taken into account during the medical device development process, so that devices might be more accommodating of the variability and limitations among users."* (CDRH, 2016).

To appropriately consider potential functional limitations of users, the definition of a user must be expanded to include additional considerations for specific functional user groups with limitations and the relevance of this definition to consideration of risk. First, it is necessary to understand how FDA defines use-related hazards which must be considered as part of risk management (FDA, 2016). The use-related hazards are anything that *"might result from aspects of the user interface design that cause the user to fail to adequately or correctly perceive, read, interpret, understand or act on information from the device."* Again, these use-related hazards could be impactful in different ways for users with limitations. As a result, the definition of a user is a person who interacts with the device who may fail to adequately or correctly perceive, read, interpret, understand or act on information due to the design of the device. When considering users, the presence of any user limitations could definitely impact how the user interacts with the device. By extension, the device design itself influences how the user with limitations (and without limitations) perceives, reads, interprets, understands or acts on information from the device (see Chapter 7).

User limitations could be present in a variety of forms, so it is less helpful to consider specific limitations, for example decreased grip strength, and more helpful to consider how general types of limitations could affect how a user interacts with a device. An elderly individual, for example, could have aging related changes to both sensory and motor capabilities that could cause a variety of functional changes including decreased grip strength. To begin to understand the impact of user limitations on device interaction, the Perception, Cognition, Action (PCA) model (discussed in more details in Chapter 6 and Chapter 12) is commonly used to analyze user interactions with a device. Briefly, the PCA model is a model for information processing that is used to outline human-machine interactions (Rauterberg, 1999). The PCA model is used in human factors to evaluate user requirements for individual tasks and subtasks within a task analysis. Here, the PCA model is presented to outline how those perception, cognition, and action steps during device interaction could be affected by different categories of limitations.

## 5. Categories of user limitations

There are various ways that user limitations can be categorized. For simplicity, the categories of user limitations discussed here are Sensory Limitations, Cognitive Limitations, Movement Limitations, and Environmental Limitations caused by environmental factors such as noise and dim lighting. The fourth limitation category, Environmental Limitations, is not user focused like Sensory, Cognitive, and Movement Limitations. Rather, Environmental Limitations are external to the user but could still introduce potential use errors and difficulties during device use, so it is an important user limitation category to consider during product design and testing.

There are specific and general types of pathologies that may cause a user limitation in one, several, or all of these categories. Examples for each category are presented here for consideration.

### 5.1 Sensory limitations

Sensory limitations result from physiological changes that specifically affect how a user's sensory system interacts with a device. Some examples of pathologies that change how a person receives and integrates sensory input are:

- Diabetes results in high blood sugar which leads to nerve fiber damage and results in altered sensation of pain, pressure, temperature, etc. in the hands and feet. This altered sensation could include feeling of numbness or of burning and tingling.
- General sensory neuropathy is the result of a variety of specific disease states which result in a degradation of sensory input. Worsening of vision from glaucoma or worsening of hearing from a viral infection are both examples of general sensory neuropathy.
- Degenerative neurological conditions include conditions which cause progressive worsening of neurological function and by extension, worsening of sensory function. Degenerative neurological condition examples are multiple sclerosis, Parkinson's disease, and dementia.

Table 20.1 outlines how sensory-related impairments would impact the PCA model of device interaction.

### 5.2 Cognitive limitations

Cognitive limitations result when the brain is somehow damaged or impaired and leads to a loss in cognitive function either from physical trauma or neurophysiological changes. It is common for side effects of prescribed medication to impact cognitive function. In addition, general fatigue, emotional stress, and sleep deprivation/insomnia can impact cognitive function. Some examples of conditions which experience cognitive limitations are:

- Aging individuals could experience decline in cognitive function due to brain atrophy that is a part of the aging process. However, not all elderly individual experience a decline in cognitive function so care should be taken to also consider an individual's medication, any other conditions which may affect neurological function, and other conditions that may affect mood, such as depression.

TABLE 20.1 Examples of how limitations experienced by different populations (diabetes, aging, adolescents & children, neurological conditions) impact Perception, Action, and Cognition during device use.

| | Perception | | Cognition | | Action | |
|---|---|---|---|---|---|---|
| | *Limitation* | *How performance is affected* | *Limitation* | *How performance is affected* | *Limitation* | *How performance is affected* |
| **Diabetes** | Loss of tactile discrimination; Loss of visual acuity | User cannot distinguish the edges of buttons so pushes multiple button concurrently | Loss of executive functioning, e.g., ability to focus | User has difficulty understanding training with nurse and does not fill syringe with correct dose | Loss of strength | User cannot grip an autoinjector cap with enough force to remove it |
| **Aging Individual** | Loss of visual acuity; Loss of color discrimination | User cannot distinguish different colors of text or device labels | Loss of executive functioning, i.e., ability to multitask | User cannot perform multiple tasks concurrently like observing an alarm and pressing a button to silence it | Loss of strength; Increased fatigability | User cannot open product packaging |
| **Adolescent/ Children** | Developing visual acuity and color discrimination | User cannot distinguish visual cues in instructions | Developing cognitive abilities; Range of reading levels | User cannot read and understand instructions for adult device | Developing motor abilities; Decreased strength | User cannot perform fine motor tasks like drawing up medication into a syringe |
| **Neurological conditions** | Loss of visual acuity; Loss of hearing | User does not perceive an audible or visual alarm | Loss of executive functioning; Loss of memory | User does not retain training on how to use a device | Loss of fine motor skills; Loss of strength | User cannot press small buttons like on a keyboard or smartphone |

- Traumatic brain injury results from a specific injury to the brain that impacts how that area of the brain functions. Disturbances to attention, memory, and executive functioning are common for individuals with a traumatic brain injury.
- Neurodegeneration of the brain can be caused by a variety of neurological disorders including Alzheimer's disease, Parkinson's disease, and Huntington's disease which all cause degeneration and/or death of nerve cells. These conditions result in, for example, memory impairment, challenges in solving problems, confusion, difficulty in understanding visual images and spatial relationships.
- Adolescents and children do not have a specific loss in cognitive function, rather their cognitive skills are still being developed. Also, literacy levels in adolescents and children could differ from adults. These are especially important considerations when adult devices are used off-label for children.
- Adults with Low Literacy have lower reading levels and generally have lower overall ability to obtain, process, and understand basic health information and services needed to make appropriate health decisions (i.e. health literacy). These individuals cannot be expected to interpret and utilize instructional materials in the same manner as users with higher literacy.

Table 20.1 outlines how cognitive-related impairments would impact the PCA model of device interaction.

## 5.3 Movement limitations

Movement limitations result from a loss of connectivity between the muscles and the central nervous system or when there is dysfunction of the muscles themselves. Movement limitations also result from orthopedic problems which is anything that affects the muscles, ligaments, and tendons. Some examples of conditions which experience movement limitations are:

- Elderly/geriatric individuals experience aging related loss of muscle and an overall loss of strength. This loss of muscle also results in increased fatigability when performing motor tasks.
- Diabetes causes neuropathy and associated loss of strength. This loss of strength is specifically related to the neuropathy as individuals with diabetes who do not experience neuropathy have normal strength.
- Arthritis causes painful inflammation and stiffness in the joints. The pain and inflammation cause reduced range of motion and stiffness at the affected joints.
- Adolescents and children, as with cognitive limitations, do not have movement impairments, but their motor skills are still being developed.
- Repetitive strain injuries most often experienced in the forearms and elbows, wrists and hands, and neck and shoulders. Health care providers, such as nurses, experience high rates of repetitive strain injuries.
- General orthopedic disorders are anything that affects the body's musculoskeletal system and could include bursitis, fibromyalgia, any bone fractures, and low back pain to name a few. Any of these issues could affect an individual's movement patterns, fine motor skills, or strength.
- General neurological disorders are anything that affects the central or peripheral nervous system and includes stroke, headache disorders, epilepsy, Parkinson's disease. Any of these issues could also affect an individual's movement patterns, fine motor skills, or strength.

Table 20.1 outlines how movement-related impairments would impact the PCA model of device interaction.

## 5.4 Limitations caused by environmental conditions

Environmental limitations are the result of specific environmental conditions but are not specific to one type of pathology. Environmental limitations can be considered "external" to the user while sensory, cognitive, and movement limitations may be considered "internal" to the user. Importantly, anyone can be impacted by environmental conditions which impact their sensory, cognitive. or movement capabilities — not just those individuals with a specific pathology (see Chapter 5 and Chapter 7). A healthy individual who uses a device in low lighting or uses a device while multi-tasking would also likely be affected by these external environmental limitations. In addition, these environmental limitations can be compounded in a user who has a specific pathology.

For example, visual information is impacted when lighting conditions change (external to user), so a person with glaucoma (condition internal to user) who must operate a device in low lighting conditions, would likely have more difficulty with device operation than if there was sufficient ambient light. This type of environmental limitation would specifically affect sensory input. Other examples of environmental limitations that could affect sensory input are excessive noise and lighting glare. Cognitive function can also be impacted by environment limitations. Any person who experiences excessive distraction while operating a device would have increased cognitive load. However, someone with multiple sclerosis (condition internal to the user) has altered cognitive function due to their disease. Therefore, this user would have amplified difficulty when using a device if they were multitasking or experiencing excessive distractions (external to user) in comparison to if the device was used without multi-tasking or interruptions.

## 6. Engage users with limitations during product design and testing

As mentioned earlier, speaking with users who have limitations is an important part of the device design process and both FDA and AAMI guidance recommend that user input is incorporated in early design stages. When considering which users to speak with, it is important to remember that none of the specific conditions or pathologies mentioned previously are comprised of a homogenous symptom presentation or life experience. Therefore, it is necessary to speak with several persons with diabetes, for example, or with several elderly individuals rather than assuming all persons in those groups would give the same feedback. Physician's offices are often a very good resource to recruit individuals from user groups with specific pathologies to give device design feedback. Physicians or other treating clinicians are very aware of the difficulties that device designs can cause their patients and are often very receptive to new devices to assist in treatment. In addition, support groups for specific pathologies are very good places to contact users with functional limitations. These groups are also very aware of how members could benefit from improved medical devices.

During the product testing process, FDA guidelines (CDRH, 2016) require that validation studies "*focus on user characteristics that could have a particular influence on their interactions with elements of the device user interface, such as age, education or literacy level, sensory or physical impairments or occupational specialty.*" All of the categories of user limitations (sensory, cognitive, movement, and environmental) could have an impact on any of these user characteristics. Prior to conducting a validation study, early formative evaluations that involve representative users with limitations can be invaluable to identifying design issues that may affect user-device interactions. Even before a formative evaluation, performing a full task use error analysis including the PCA model with specific consideration of the user's limitations could also highlight potential design issues to optimize prior to testing.

An example of combining this product testing process with user interviews could be for an autoinjector indicated for use by individuals with arthritis. As a first step, talk with a small group of users who have arthritis about their condition, what tasks are challenging, and what they dislike or find problematic about any of their current arthritis treatments or devices used.

As a next step, perform a task use error analysis with PCA model while specifically considering the sensory, cognitive, and movement limitations that someone with arthritis would experience. For example, a user with arthritis would have movement limitation of their hands which could include both range of motion problems and strength deficits. While performing the PCA analysis, consider that the user with arthritis could be over 65 years old and have additional age-related sensory, cognitive, and movement limitations. For example, the user with arthritis who is aging could also have poor eyesight due to glaucoma. At this point, the task use error analysis may have highlighted some potential use errors that are specific to the user with arthritis' limitations. Appropriate design modifications should be implemented to mitigate these potential use errors because the user's limitations, which are the result of a medical condition, cannot be modified. After any necessary design modifications, early formative evaluations can be performed with representative users who have arthritis to assess the effectiveness of the implemented design mitigations.

FDA guidelines also require that any environmental factors which could affect the users' interactions with the device should be incorporated into the simulated use environment for a validation study (CDRH, 2016). Therefore, the product testing process must consider the possible internal limitations of the representative user group in the protocol design and must also consider any environmental limitations in a simulated use scenario. Continuing with the previous example of a user with arthritis, early formative evaluation scenarios should consider the lighting conditions where the device will be used and the body position of the user. For example, should the user with arthritis be sitting or standing when using the autoinjector? Should the instructions indicate options for user positioning for safe and effective use? Because the user has movement limitations, sitting or standing while using the autoinjector could have a large impact on potential use errors and task completion. These types of detailed environmental considerations are important when considering designs for users with internal limitations because the environmental factors could compound the existing internal limitations the user experiences due to their medical condition.

## 7. Engaging users: tips

While engineers and designers are very knowledgeable about the device development process, they are rarely the end-user of the product. Engaging representative users can be very informative during the development process and should result in a much more inclusive device design.

- User engagement should be done at different time points during device development, as early and often as possible
- Involve a broad range of representative users
- Consider all user limitations, including environmental and age-related influences, which may compound limitations from a specific pathology
- Talk with clinicians and caregivers
- Leverage community resources like physician offices and support groups to find users

Feedback from representative users can certainly improve device design. However, incorporating other tools, such as task analysis development with PCA, in concert with user feedback and environmental considerations of use is a more comprehensive approach.

## 8. Summary

Throughout the development process, considerations for users with limitations during both user interface design and user interface testing will improve the usability of any device design. Most importantly and based on risk management, the definition of a user is a person who interacts with the device who may fail to adequately or correctly perceive, read, interpret, understand or act on information due to the design of the device. Users may have sensory limitations, cognitive limitations, movement limitations, and environmental limitations caused by factors such as noise and dim lighting. For designers and developers of medical devices, it is important to remember that users with limitations, whether due to illness, age or environmental conditions, need to have access to devices which help to alleviate their health problems. By considering the different categories of limitations during the design process and during human factors testing, user interface designs will be more safe, effective, and useable for all populations.

## References


Association for the Advancement of Medical Instrumentation (AAMI). (2012). *Human factors engineering – Design of medical devices* (ANSI/AAMI HE75:2012).

FDA. (2016). *Applying human factors and usability engineering to medical devices guidance for industry and food and drug administration staff preface public comment.* Retrieved from http://www.regulations.gov.

Institute on Disability. (2016). *2016 Disability Statistics Annual Report.* Retrieved from https://disabilitycompendium.org/sites/default/files/user-uploads/2016_AnnualReport.pdf.

Lemke, M., & Mendonca, R. (2013). A question of accessibility: understanding lay users of medical devices. *Biomedical Instrumentation and Technology, 47*(s1), 20–25.

Population Reference Bureau (PRB). (2015). *Aging in the United States.* Retrieved from https://www.prb.org/wp-content/uploads/2016/01/aging-us-population-bulletin-1.pdf.

Rauterberg, M. (1999). Perception, cognition, action: an action theoretical approach. *Systematica, 14*(1), 1–8.

United Nations (UN). (2015). *World population aging.* Retrieved from http://www.un.org/en/development/desa/population/publications/pdf/ageing/WPA2015_Report.pdf.

U.S. Department of Health and Human Services Food and Drug Administration Center for Devices and Ratiological Health (CDRH). (2016). Applying human factors and usability engineering to optimize medical device design. *FDA Guidance for Industry and Food and Drug Administration Staff.* Retrieved https://www.fda.gov/downloads/MedicalDevices/DeviceRegulationandGuidance/GuidanceDocuments/UCM259760.pdf.

World Health Organization (WHO). (2011). *World report on disability.* Retrieved from https://www.unicef.org/spanish/protection/World_report_on_disability_eng.pdf.

World Health Organization (WHO). (2018). *Disability and health.* Retrieved from http://www.who.int/news-room/fact-sheets/detail/disability-and-health.

# 21

# Augmented reality in medical devices

*Kate Cox*[a], *Mary Beth Privitera*[a], *Tor Alden*[a], *Jonathan R. Silva*[b], *Jennifer N. Avari Silva*[b]

[a]HS Design, Gladstone, NJ, United States; [b]Washington University in St Louis, SentiAR, Inc., St Louis, MO, United States

OUTLINE



*Augmented reality allows individuals to be present in the world but hopefully allows an improvement on what's happening presently* **Tim Cook**

## 1. Introduction

As computer graphics have progressed to a point where images are indistinguishable from the real world, enhancing mobile computing so that the association with the real world happens automatically is an increasing proposition (Schmalstieg & Höllerer, 2016).

*Applied Human Factors in Medical Device Design*
https://doi.org/10.1016/B978-0-12-816163-0.00021-9

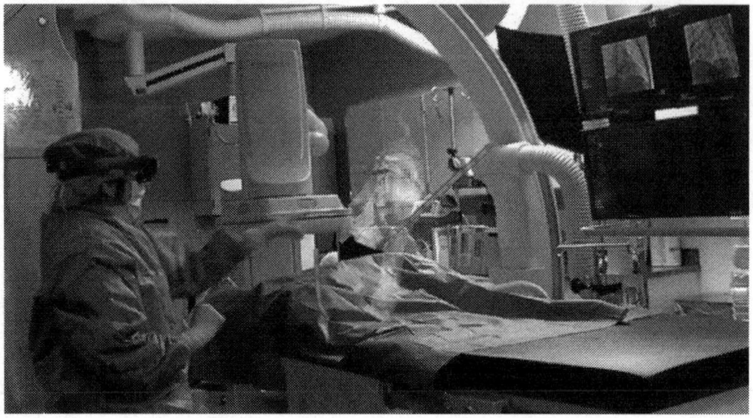

FIG. 21.1 SentiAR SentEP system displaying a geometry of a cardiac chamber with catheters, and the user pointing at a target of interest.

Augmented Reality (AR) is a technology in which computer-generated content is superimposed on a real-world environment. AR is unique in that the users are still fully aware of their surroundings, unlike Virtual reality which is fully immersive and cut off from the real-world. An example AR system is presented below (Fig. 21.1), the SentEP system by SentiAR which displays an interactive holographic visualization of the patient's actual anatomy.

Augmented Reality has been implemented in medicine across the spectrum, ranging from patient and student education, resident and fellow training, pre-procedural planning and intra-procedural use. Clinical benefits of using AR in medical devices starts with aiding in the visualization and control of anatomy allowing for improved understanding of visual-spatial relationships. This chapter describes design considerations for implementing Augmented Reality into medical devices.

## 2. Background of AR

The term "Augmented Reality" was created in 1990 by a Boeing researcher, Tom Caudell (Johnson et al., 2010). In this era, a number of major companies used AR for visualization, training, and other purposes. Early mobile applications began to appear in 2008 (Johnson et al., 2010). Presently, AR technologies are sophisticated enough to be able to deliver AR experiences through smartphones, tablets, and head mounted displays (e.g. Microsoft HoloLens, Google Glass, Meta 2) and have many applications. In order for a system to be considered "augmented reality" it must have the following three characteristics: combine real and virtual, interactive in real time and registration in 3D (Schmalstieg & Höllerer, 2016). It is a blend of interactive computer-generated elements and real-world information in order to amplify human perception and cognition in remarkable new ways (Fig. 21.2).

A complete AR system will require at least three components which include a tracking, registration and a visualization component and in some cases a fourth component which relates to a spatial model or database that stores information about the real world and virtual world (Schmalstieg & Höllerer, 2016). The real-world model is required to serve as a reference for tracking which must also include information regarding the user's location in the real world with the system acting in the same coordinate system (Schmalstieg & Höllerer, 2016).

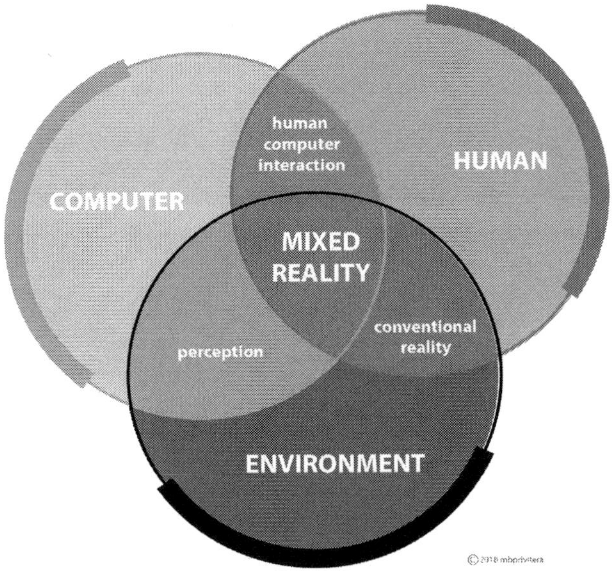


FIG. 21.2 AR overlays digital content over the real world.

## 2.1 AR in the medical field

Augmented reality has made its way into the medical field in a variety of ways: telemedicine and remote collaboration, visualization aids in procedures and education tools. With increasing complexity of neuroimaging data and the need to convey this information to end users, quick and easy AR applications have become available through consumer grade head mounted AR devices, such as Microsoft HoloLens (HoloLens, Microsoft, Inc.) (Karmonik, Boone, & Khavari, 2018) The MS HoloLens is a self-contained holographic computer in the shape of a wearable headset that allows the user to see, hear, and interact with computational objects projected into the environment (Karmonik et al., 2018).

Additional explorations in the medical field have included function learning, wherein interactive mixed reality systems are proposed for facilitating the learning of the muscles of the upper extremity as the structure and function of the muscles are difficult to learn as muscle movement is imperceptible using traditional methods (Meng et al., 2016; Taylor, 2010, pp. 37–41).

# 3. Designing for augmented reality

Based on the lack of predicate data and potential for disruption in the use environment, nontraditional user interfaces provide challenges over established UI's (e.g., visual displays, touchscreens, web, etc.). Nontraditional UI's typically involve a new technology that can pose additional training and time in order for the user to gain familiarity (e.g., Virtual and Augmented Reality).

Augmented reality has not been integrated into medical devices long enough to have recognized standards and guidelines published by regulatory authorities. Until then, designers should look to recommendations from AR hardware developers (e.g., Microsoft, Google), common trends being used across industries, basic usability principles and recognized UI guidelines (e.g., AAMI HE75) for commonalities that can be reciprocated in augmented reality user interface design.

A key to new technology adoption in any field, but especially in medicine, is being able to minimize the amount of new learning for users. Utilizing recognizable icons, intuitive interaction methods and terminology will allow users to adapt to a new technology quicker.

## 3.1 Interaction methods

There are three new modes of interaction for augmented reality technologies. These include gaze, gesture and voice (Fig. 21.3). Each of these are described in detail below.

### *3.1.1 Gaze*

People naturally gaze at the world in conjunction with other activities as eye gaze contains information about the current task and the manipulation of objects (Sibert & Jacob, 2000). The term "gaze" refers to looking steadily and intently at a specific object. Eye gaze interaction technique is faster than selection with a mousse and is convenient in situations where it is important to use hands for other tasks (Sibert & Jacob, 2000). Within AR, users may look at various objects and buttons to manipulate the interface. Gaze may be executed by means of head or eye tracking. The gaze precision required is dependent on the needs and goals of the application and the available technology.

Gaze may be coupled with a gesture or voice command to select an object. For example, (1) a user could gaze at a zoom button and say 'select' or (2) a user could gaze at a zoom button and pinch and drag upwards to zoom in.

#### 3.1.1.1 Gaze-dwell

Gaze-dwell is a variant of selecting objects that does not involve the use of gestures or voice commands. This action consists of a user gazing at an object and remaining fixated on the object. A fixation is a period of time (from less than 100 ms up to several seconds Holmqvist et al. 2011) that the gaze location is static within a specific region.

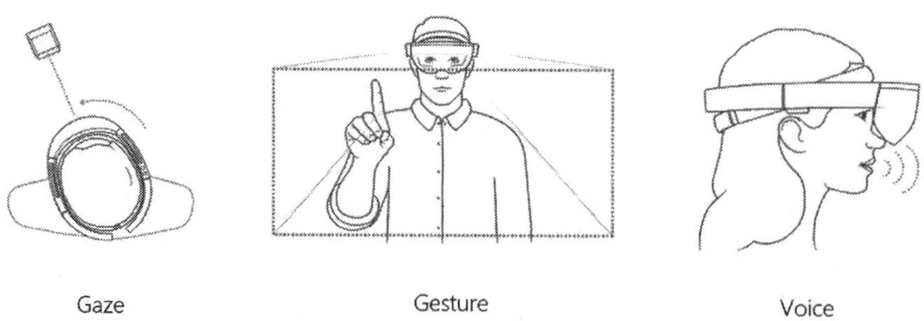


FIG. 21.3 Modes of AR interaction.

Gaze-dwell is utilized to allow the system to register that the object is intended to be selected/activated and has not been actuated inadvertently. Sticking with the same example of zooming in on an object, the user would gaze at the zoom button and continue to gaze at the button until the zoom tool became active.

The period of time that the user will need to 'dwell' is programmable and should be determined based upon the user's physical capabilities to remain fixated on a target, just-noticeable differences between deactivated, selected and activated states and slowing down the workflow unnecessarily causing user annoyance.

### Gaze: tips

The following items should be considered when designing an interface that utilizes gaze selection.

- Be mindful of target sizing
- Highlight object when targeted by gaze cursor
- Be conscious of field of view (FOV) and target placement
- Differentiate hover from selection cues (visual and audible)
- Be mindful of time to confirm selection using gaze-dwell

### 3.1.2 *Gesture*

> A gesture is a form of non-verbal communication or non-vocal communication in which visible bodily actions communicate particular messages, either in place of, or in conjunction with, speech. Gestures include movement of the hands, face, or other parts of the body. Gestures differ from physical non-verbal communication that does not communicate specific messages, such as purely expressive displays, proxemics, or displays of joint attention. ***A. Kendon***

When using gestures as an interaction method for AR, the system needs to be able to detect the movement, identify the gesture and determine when the gesture has been completed. The gestures need to be deliberate and unique from one another and from other natural motions. Within an application the number of gestures should be limited in order to reduce the burden on the system and user's memory.

Gestures should be easy to remember and perform, intuitive, metaphorically and iconically logical towards functionality, and ergonomic and not physically stressing when used often (Signer, 2017).

### Gestures: tips

The following items should be considered when designing an interface that utilizes gestures.

- Be mindful of expected movements
- Utilize motions that are distinguishable from one another
- Be mindful of accidental actuation or closing.
  - Gestures for ending or closing an application should be motions that would not be done accidentally or by another person in the field of view

### 3.1.3 Voice

Voice commands are an intuitive interaction method for users to dictate what they want to happen/how to have the interface respond. However, when utilizing voice commands, the system needs to know when it is being addressed and expected to process and respond to a command. Implementing the use of 'wake' words (e.g., Amazon's Alexa and Apple's Siri) that are uncommon words/phrases for the use environment, will minimize the number of instances that the device responds in an unexpected and/or unwanted manner.

**Voice: tips**

The following items should be considered when designing an interface that uses voice commands.

- Use concise commands
- Use simple vocabulary
- Avoid similar sounding commands
- Test with different accents
- Implement 'wake' words

## 4. SentiAR system development

SentiAR partnered with HS Design to develop a 3D augmented reality platform to visualize patient anatomy for intraprocedural use.

SentiAR's mission is to transform the experience for both patient and clinician in interventional procedures with a 3D augmented reality platform featuring real-time holographic visualization of the patient's actual anatomy, "floating" over the patient. The visualization is fully controllable "hands-free" by the clinician, providing an ergonomic breakthrough for the treatment and analysis of cardiac arrhythmias within an interventional catheter lab environment.

The SentiAR system is an imaging system intended to be used as an adjunct product to assist the clinician in visualization of the heart anatomy during cardiac mapping and ablation procedures. Images are imported from an EAMS (electroanatomic mapping system) and displayed through the SentiAR system. The system includes proprietary software that is loaded on a Data Manager PC and the Head Mounted Display (HMD, Microsoft HoloLens). The HMD enables users to have meaningful interactions with patient specific 3D cardiac models and electrical maps that are presented in a mixed reality environment. In addition, the system displays a real-time representation of catheter locations based on electro-anatomic mapping and uses shared models to allow for collaboration on unified information sharing.

The design process began with interface sketches utilizing transparencies or paper (Fig. 21.4) elements to explore design elements. In some instances, three-dimensionality was added to the paper prototypes and a laser pointer was utilized to serve as the gaze-curser.

Paper prototypes in conjunction with the technology can also be used to e1xplore and ultimately configure the display parameters for menu options (Fig. 21.5). This assures the legibility of menu options and icons that may be planned for successful device use.

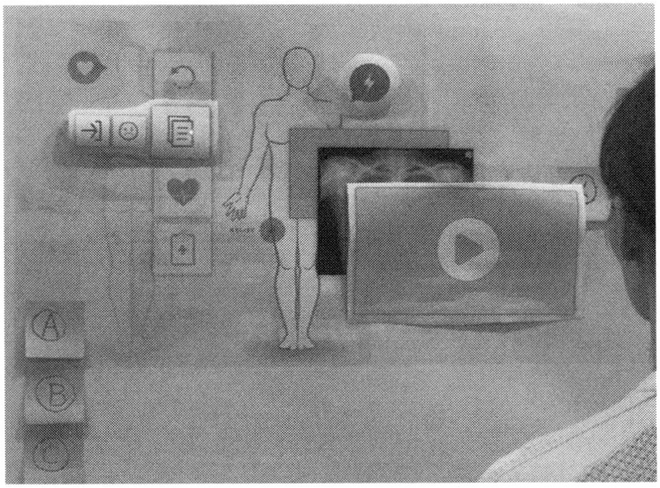


FIG. 21.4 Interface sketch visualization.

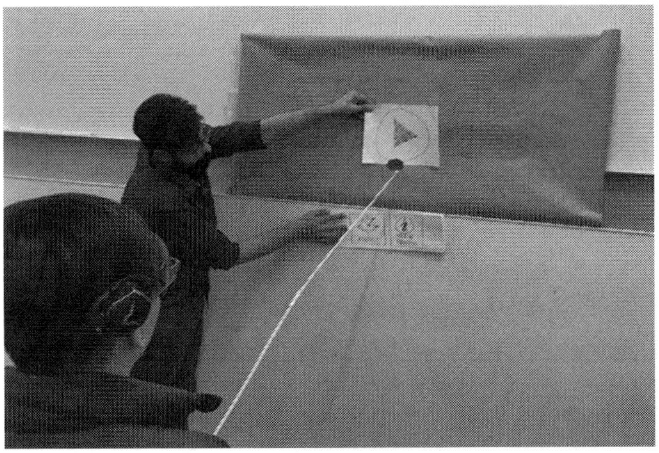

FIG. 21.5 Configuring appropriate display parameters for menu options.

## 4.1 Design considerations

As seen in Fig. 21.1, the Electrophysiology (EP) or interventional catheter lab environments are filled with various types of equipment and monitors that the physicians require to perform the procedures. As a result, the following considerations were made in order to frame the UI requirements, architecture and hierarchy for the SentEP user interface:

- There is a high potential for visual interference/distraction with the various monitors and equipment in the room

This requires that the system allow users to change hologram location.

- Various lighting conditions are utilized within EP/Cath labs

This requires that the system consider color and opacity on multiple, variable backgrounds i.e. the color the holographic image is projected onto will not be solid nor consistent.

- New technology being introduced as an adjunct to existing and familiar system.

The device utilizes existing information found from other devices and takes the cognitive task of piecing together 2-D images and creates an interactive hologram for the physician to interact with. As such, the UI should consider familiar icons within the existing system for easy recognition.

- Physicians will be sterile when using this device and want to maintain sterility.

This requires that the system UI maximize hands-free options.

- Certain manipulation tools are used more frequently than others

This requires attention in the consideration tool hierarchy with the most frequently used icons, the most readily available.

## 4.2 SentEP user interface design

The final UI is seen below in Fig. 21.6 and includes an interactive menu in the upper left, an interactive holographic display of the heart and battery indicator in the lower left. For the purposes of clarity in communication a black background is presented in this image. In reality, the background is the electrophysiology lab with a variable background.

Gaze-dwell is included as an interaction method with button state changes differentiated with color. This is an important interaction requirement as the physician has their hands on the catheters during device use. When the gaze cursor is hovering over a button, the icon turns blue. When the cursor remains fixated on the button for the duration of the dwell period,

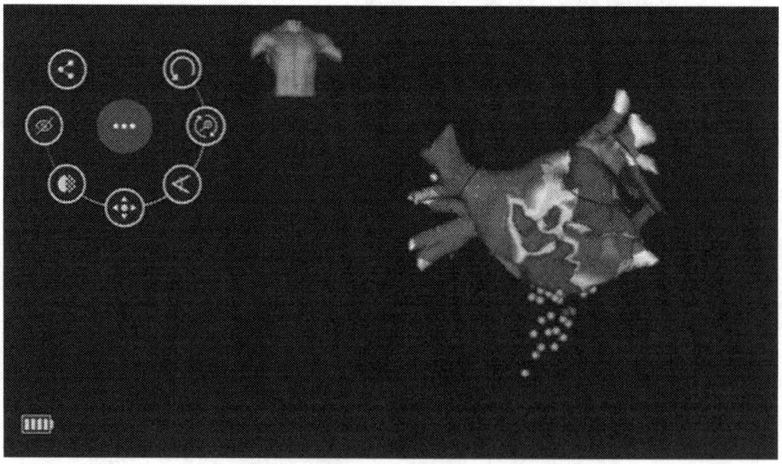

FIG. 21.6 SentiAR SentEP user interface.

the icon turns green to show that the tool is active. For specific items that may take extra time/ effort to correct if activated inadvertently (e.g., profile selection, initiating calibration, training module and share session), a 'long gaze' was employed; the dwell duration for a long gaze is longer than for other tool selections.

An interesting phenomenon occurred in design process in regards to a single mode of interaction so that the physician didn't have to switch back and forth further adding complexity to the interface. This topic of single interaction mode received considerable debate. The situation was that at first, when the initial device was designed it was designed using gestures and once the development and clinical team became proficient, they were reticent to try the gaze interaction even though gaze was found to be a superior interaction modality. The moral here is to be open to try various user interfaces with careful consideration of the clinical context.

## 4.3 Usability testing for SentEP

Throughout the design and development phases, the usability of the SentEP system was evaluated via a task analysis, heuristic evaluation, expert review and usability studies. The menu architecture and hierarchy, visual balance, legibility, iconography and potential use errors were evaluated in each design iteration.

Usability evaluations progressed in fidelity (Fig. 21.7) with the use of animations in order to demonstrate the user interface. This enabled the design team to act out user inputs while the animations portrayed what would happen upon completion of the technical design.

Best practices for early prototyping AR technology include the following:

- Use animations to demonstrate the interface early in conceptual design.
- Act out user inputs while the animations play.
- Pre-record videos of the user inputs using the headsets front facing cameras and animation software to overlay interface concepts. This enables comparison of concepts.
- Know when to add more development time for prototyping.
- If the AR interface is complex, aim to create lo-fidelity prototypes to be used by participants on the headset then move to higher fidelity when appropriate.

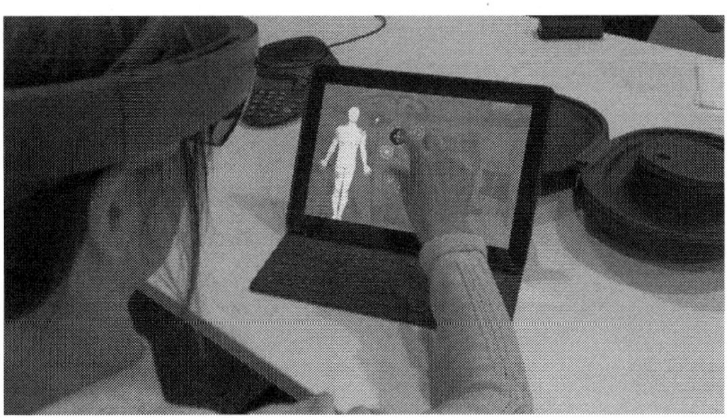

**FIG. 21.7** High fidelity prototypes generated for formative testing.

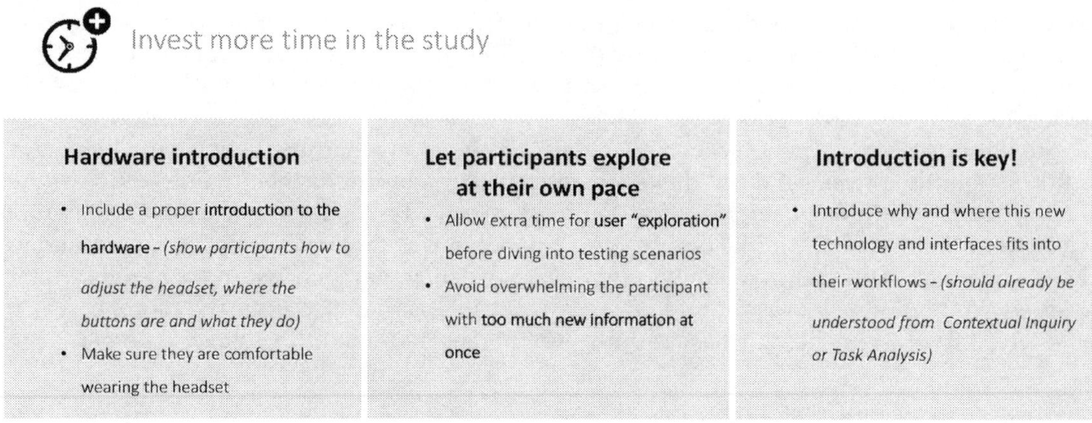


**FIG. 21.8** Best practices for AR usability testing.

Best practices for usability testing include ensuring that the users are familiar with the technology. This may mean that instructional materials might be required at early introduction of prototypes. It might also mean that more time is required for introduction and familiarity of the hardware (Fig. 21.8) prior to evaluation of the UI.

For the HoloLens application, this equates to demonstrating to participants the adjustments and control of the hardware and to make sure they are comfortable prior to assessing the user interface of the AR. The ability to wear a headset comfortably can be the difference between acceptance or reluctance to provide a fair evaluation due to hardware issues.

In order to mitigate this issue, more time is required to enable the participants to become familiar and to integrate the new technology into their workflow.

## 5. Summary

Augmented reality is being utilized increasingly across many industries, with great potential for expansion in medical devices. When implemented successfully, AR has the ability to advance current techniques in surgical and invasive procedures, telemedicine, medical training and education. In developing a device intended for intraprocedural use, such as the SentEP system, data which is typically a complex cognitive task becomes interactive and exploratory, ultimately enhancing the clinical experience.

## Acknowledgments

Special thank you to SentiAR and HS Design UI/UX team for permission to highlight the development of the SentEP system.

# References


Holmqvist, K., Nystrom, M., Andersson, R., Dewhurst, R., Jarodzka, H., & de Weijer, J. (2011). *Eye tracking: A comprehensive guide to methods and measures*. Oxford: OUP.

Johnson, L., Levine, A., Smith, R., & Stone, S. (2010). *The horizon report*. Austin: The New Media Consortium.

Karmonik, C., Boone, T. B., & Khavari, R. (2018). Workflow for visualization of neuroimaging data with an augmented reality device. *Journal of Digital Imaging, 31*(1), 26–31. https://doi.org/10.1007/s10278-017-9991-4.

Kendon, A. (2004). *Gesture: Visible action as utterance*. Cambridge University Press.

Meng, M., Philipp, J., Felix, B., Ina, S., Anna Maria von der, H., Nassir, N., et al. (2016). Interactive mixed reality for muscle structure and function learning. In G. Z. H. Liao, P. J. P. Cattin, & S.-L. Lee (Eds.), *Medical imaging and augmented reality*. Springer.

Schmalstieg, D., & Höllerer, T. (2004). *Augmented reality pinciples and practices* (Chapter 1) (Vol. 25). https://doi.org/10.1080/01639620490266187.

Sibert, L. E., & Jacob, R. J. K. (2000). Evaluation of eye gaze interaction. In *Proceedings of the SIGCHI conference on human factors in computing systems - CHI'00* (pp. 281–288). https://doi.org/10.1145/332040.332445.

Signer, B. (2017). *Gesture-based interaction - next generation user interfaces*.

Taylor, P. (2010). *Co-experience: User experience as interaction* (May 2012).

# Index

*Note*: 'Page numbers followed by "f" indicate figures, "t" indicates tables'.

Made in the USA
Monee, IL
13 September 2026

62977012R00203